IET MATERIALS, CIRCUITS AND DEVICES SERIES 51

Modelling Methodologies in Analogue Integrated Circuit Design

Other volumes in this series:

Volume 2	**Analogue IC Design: The current-mode approach** C. Toumazou, F.J. Lidgey and D.G. Haigh (Editors)
Volume 3	**Analogue–Digital ASICs: Circuit techniques, design tools and applications** R.S. Soin, F. Maloberti and J. France (Editors)
Volume 4	**Algorithmic and Knowledge-based CAD for VLSI** G.E. Taylor and G. Russell (Editors)
Volume 5	**Switched Currents: An analogue technique for digital technology** C. Toumazou, J.B.C. Hughes and N.C. Battersby (Editors)
Volume 6	**High-Frequency Circuit Engineering** F. Nibler *et al.*
Volume 8	**Low-Power High-Frequency Microelectronics: A unified approach** G. Machado (Editor)
Volume 9	**VLSI Testing: Digital and mixed analogue/digital techniques** S.L. Hurst
Volume 10	**Distributed Feedback Semiconductor Lasers** J.E. Carroll, J.E.A. Whiteaway and R.G.S. Plumb
Volume 11	**Selected Topics in Advanced Solid State and Fibre Optic Sensors** S.M. Vaezi-Nejad (Editor)
Volume 12	**Strained Silicon Heterostructures: Materials and devices** C.K. Maiti, N.B. Chakrabarti and S.K. Ray
Volume 13	**RFIC and MMIC Design and Technology** I.D. Robertson and S. Lucyzyn (Editors)
Volume 14	**Design of High Frequency Integrated Analogue Filters** Y. Sun (Editor)
Volume 15	**Foundations of Digital Signal Processing: Theory, algorithms and hardware design** P. Gaydecki
Volume 16	**Wireless Communications Circuits and Systems** Y. Sun (Editor)
Volume 17	**The Switching Function: Analysis of power electronic circuits** C. Marouchos
Volume 18	**System on Chip: Next generation electronics** B. Al-Hashimi (Editor)
Volume 19	**Test and Diagnosis of Analogue, Mixed-Signal and RF Integrated Circuits: The system on chip approach** Y. Sun (Editor)
Volume 20	**Low Power and Low Voltage Circuit Design with the FGMOS Transistor** E. Rodriguez-Villegas
Volume 21	**Technology Computer Aided Design for Si, SiGe and GaAs Integrated Circuits** C.K. Maiti and G.A. Armstrong
Volume 22	**Nanotechnologies** M. Wautelet *et al.*
Volume 23	**Understandable Electric Circuits** M. Wang
Volume 24	**Fundamentals of Electromagnetic Levitation: Engineering sustainability through efficiency** A.J. Sangster
Volume 25	**Optical MEMS for Chemical Analysis and Biomedicine** H. Jiang (Editor)
Volume 26	**High Speed Data Converters** A.M.A. Ali
Volume 27	**Nano-Scaled Semiconductor Devices** E.A. Gutiérrez-D (Editor)
Volume 28	**Security and Privacy for Big Data, Cloud Computing and Applications** L. Wang, W. Ren, K.R. Choo and F. Xhafa (Editors)
Volume 29	**Nano-CMOS and Post-CMOS Electronics: Devices and modelling** Saraju P. Mohanty and Ashok Srivastava
Volume 30	**Nano-CMOS and Post-CMOS Electronics: Circuits and design** Saraju P. Mohanty and Ashok Srivastava
Volume 32	**Oscillator Circuits: Frontiers in design, analysis and applications** Y. Nishio (Editor)
Volume 33	**High Frequency MOSFET Gate Drivers** Z. Zhang and Y. Liu
Volume 34	**RF and Microwave Module Level Design and Integration** M. Almalkawi
Volume 35	**Design of Terahertz CMOS Integrated Circuits for High-Speed Wireless Communication** M. Fujishima and S. Amakawa
Volume 38	**System Design with Memristor Technologies** L. Guckert and E.E. Swartzlander Jr.
Volume 39	**Functionality-Enhanced Devices: An alternative to Moore's law** P.-E. Gaillardon (Editor)
Volume 40	**Digitally Enhanced Mixed Signal Systems** C. Jabbour, P. Desgreys and D. Dallett (Editors)

Volume 43 **Negative Group Delay Devices: From concepts to applications** B. Ravelo (Editor)

Volume 45 **Characterisation and Control of Defects in Semiconductors** F. Tuomisto (Editor)

Volume 47 **Understandable Electric Circuits: Key concepts, 2nd Edition** M. Wang

Volume 53 **VLSI Architectures for Future Video Coding** M. Martina (Editor)

Volume 54 **Advances in High-Power Fiber and Diode Laser Engineering** Ivan Divliansky (Editor)

Volume 58 **Magnetorheological Materials and Their Applications** S. Choi and W. Li (Editors)

Volume 60 **IP Core Protection and Hardware-Assisted Security for Consumer Electronics** A. Sengupta and S. Mohanty

Volume 67 **Frontiers in Securing IP Cores: Forensic detective control and obfuscation techniques** A Sengupta

Volume 68 **High Quality Liquid Crystal Displays and Smart Devices: Vol. 1 and Vol. 2** S. Ishihara, S. Kobayashi and Y. Ukai (Editors)

Volume 69 **Fibre Bragg Gratings in Harsh and Space Environments: Principles and applications** B. Aïssa, E.I. Haddad, R.V. Kruzelecky and W.R. Jamroz

Volume 70 **Self-Healing Materials: From fundamental concepts to advanced space and electronics applications, 2nd Edition** B. Aïssa, E.I. Haddad, R.V. Kruzelecky and W.R. Jamroz

Volume 71 **Radio Frequency and Microwave Power Amplifiers: Vol. 1 and Vol. 2** A. Grebennikov (Editor)

Volume 73 **VLSI and Post-CMOS Electronics Volume 1: VLSI and Post-CMOS Electronics and Volume 2: Materials, devices and interconnects** R. Dhiman and R. Chandel (Editors)

Modelling Methodologies in Analogue Integrated Circuit Design

Edited by
Günhan Dündar and Mustafa Berke Yelten

The Institution of Engineering and Technology

Published by The Institution of Engineering and Technology, London, United Kingdom

The Institution of Engineering and Technology is registered as a Charity in England & Wales (no. 211014) and Scotland (no. SC038698).

First published 2020

The Institution of Engineering and Technology
Michael Faraday House
Six Hills Way, Stevenage
Herts, SG1 2AY, United Kingdom

www.theiet.org

British Library Cataloguing in Publication Data
A catalogue record for this product is available from the British Library

ISBN 978-1-78561-695-2 (hardback)
ISBN 978-1-78561-696-9 (PDF)

Typeset in India by MPS Limited

Contents

About the editors

Günhan Dündar is a full professor in the department of electrical and electronic engineering, Boğaziçi University, Turkey. He has authored and co-authored more than 200 international journal and conference papers in the broad area of circuits and systems, as well as one book and one book chapter. He is also the recipient of several research awards. His research interests lie in the design of, and design methodologies for analogue integrated circuits.

Mustafa Berke Yelten is an assistant professor in the department of Electronics and Communications Engineering, Istanbul Technical University, Turkey. He previously worked as a quality and reliability research engineer in Intel Corporation between 2011 and 2015. As a senior member of IEEE, he has been a technical program committee member of several IEEE conferences. His research interests include the design, optimization, and modelling of nanoscale transistors and the design of analogue/RF integrated circuits.

Chapter 1
Introduction

Günhan Dündar[1] and Mustafa Berke Yelten[2]

This book is intended to fill in a missing link between two well-established disciplines, namely modelling and circuit design. From the perspective of circuit designers, simulation times have increased beyond what is practical with the ever-increasing complexity of circuits, even though computers have also got faster. Modelling systems, especially at higher levels of design abstraction, has become a necessity. On the other hand, the designer is faced with a bewildering array of choices in models. In order to make a feasible choice, the circuit designer has to understand the underlying mathematics of each model as well as its limitations.

From the perspective of model developers, circuit design offers a wide range of applications for which various modelling approaches can be integrated. Models are required not only for simple analogue/digital circuits but also for components, especially as the technologies move into very deep submicron dimensions. Nanometre scale transistors present modelling problems, not only for their nominal behaviour but also for variations and time-dependent drifts. Furthermore, modelling studies of integrated microelectromechanical systems (MEMS)/electronics systems, on-chip inductors, and beyond complementary metal-oxide-semiconductor (CMOS) devices present ample opportunity to develop novel approaches.

In order for a model to be adopted, there are certain specifications to be met. First of all, the models should be accurate enough to capture the experimental results. This requires that the model must account for many details that will lead to increased mathematical complexity. Nevertheless, as the complexity is augmented, the convergence in simulation is mostly impaired, thereby leading to a high computation cost. Another issue arises when the model-representation capability of the underlying physics of a phenomenon is assessed. Empirical models can provide excellent accuracy, yet they cannot be leveraged to provide insights to designers about the underlying causes of an unusual observation. Conversely, a truly-physics-based model will be unable to explain the time- and process-based random variations, which ultimately lead to inaccuracies. Finally, 'the curse of dimensionality' exacerbates all factors described previously: high dimensionality of a problem

[1]Electrical and Electronics Engineering Department, Boğaziçi University, Istanbul, Turkey
[2]Electronics and Communications Engineering Department, Istanbul Technical University, Istanbul, Turkey

requires a plethora of simulations or measurements for high accuracy. In addition, it causes the model to become more convoluted with less intuition to designers, thereby adding up to the computation cost.

This analysis indicates that developing versatile models for analogue integrated circuits can be confined to a delicate balance of trade-offs between accuracy, physical intuition capability, and computational efficiency. On a higher level, a modelling expert should have a vast knowledge and experience in solid-state device physics and microfabrication, analogue circuit design as well as mathematical foundations of modelling methodologies. Often, modelling engineers cannot have full expertise in each of these areas. Nevertheless, resources for modelling approaches where the mechanics of constructing a model meets the practical specifications expected from models in typical applications are strongly desired by integrated circuit designers. This book can be viewed as a response to this expectation.

As the difficulty of both constructing and validating a model is significantly boosted, this book aims to revisit some basic modelling concepts and techniques by concurrently considering the challenges that the community of designers is facing. With an in-depth analysis of different modelling approaches, their applications in analogue integrated circuit design can be better understood. Later, during the discussion of modelling applications, different areas where analogue circuits can be employed will be discussed, where the need for modelling is provided along with possible solutions.

To address the perspective stated previously, this book is divided into two parts. The first part, named as "Fundamentals of Modelling Methodologies", surveys various modelling approaches irrespective of circuit applications. We have tried to have reasonably complete coverage of modelling methodologies with enough theoretical treatises. This part is intended for the readers with a background on circuit design, but not on modelling. It is inevitable that there are some overlaps between chapters, but these have been minimized. Where there are overlaps, we have deemed them necessary for completeness within a chapter, since this book can be utilized as a reference book, where the reader can refer to each chapter individually.

The second part, named as "Applications in Analogue Integrated Circuit Design", deals with various problems in circuit design, where modelling is applied. These problems are by no means complete, and the reader may find many other applications. This second part is intended for readers with a stronger mathematical background who are looking for applications and/or research areas in modelling.

Chapter 2 is about response surface modelling (RSM), which is one of the critical techniques for analysing analogue and mixed-signal circuits. Fitting an RSM to a function involves least-squares regression at its simplest form. However, as circuits become more complicated, the number of variables increases to beyond, thereby leading to an excessive time cost of simulation. Hence, this chapter introduces several different techniques to tackle this growing complexity.

Chapter 3 introduces machine learning. In its most straightforward meaning, machine learning is extracting meaningful information from a large amount of data.

Starting from the fundamentals, this chapter takes the reader quickly towards concepts, such as clustering, learning, and neural networks. During this journey, the theory is as rigorous as possible without bogging down the reader in detail. The final few pages of the chapter deal with a methodological performance assessment of models, which is a subject often overlooked by model developers and users alike.

Chapter 4 concerns itself with data-driven and physics-based modelling. The emphasis in the chapter is on surrogate modelling, where fast replacement models to alleviate the high cost of simulations are developed. Initially, data-driven models are considered, since they are generic and no prior knowledge about the problem is required. In terms of data-driven models, sampling the data properly is of utmost importance, whereby the design of experiments is introduced. Regarding various types of data-driven models, polynomial regression, radial basis function interpolation, kriging, support vector regression, and neural networks are considered as well as some other less well-known approaches. Physics-based models require knowledge about the system to be modelled but result in much more efficient models. These models are based on suitable correction or enhancement of an underlying low-fidelity model. Among the approaches described are variable fidelity models, space mapping, response correction, and feature-based modelling. The chapter concludes with a discussion on model selection and validation.

Chapter 5, which is the last section of the first part, resumes from where the previous chapter left off, namely verification of modelling. This subject has become exceedingly important with more complex systems, where it is virtually impossible to check the equivalence of the system and its model under every scenario possible. Initially, the concept of state space in analogue circuits is defined in the chapter, since it forms the basis of the subsequent discussions. Then, the model validation problem is defined, and the relevant metrics are discussed. The difference between semi-formal and formal model verification is highlighted, and some formal verification techniques are described. Finally, formal modelling methods are introduced and illustrated in an example.

Chapter 6 provides an overview of modelling applied to analogue circuits. As such, it also sets the stage for subsequent chapters where more specific models and/or circuits are considered. A significant portion of the chapter is dedicated to symbolic analysis, which has not been discussed in the previous part as it pertains specifically to analogue circuits, and mostly to linear analysis. However, one can view symbolic analysis as a special case of physics-based models. Next, circuit macromodelling is introduced, which are more ad hoc models developed to fit the behaviour of the system under consideration. Finally, the use of neural networks for modelling the behaviour of analogue circuits is discussed with many examples from the literature.

Chapter 7 deals with machine learning in the modelling of integrated inductors. The integrated inductor is an essential component in radio frequency circuit design, whose modelling has been lacking in accuracy for a long time. The development of the modes follows the design of experiments, data generation, model construction, and model validation steps, as discussed earlier in the book. Kriging models were used as surrogate models for the inductors. However, the chapter does not end here but continues with the utilization of the models in an optimizer to choose the best inductors in that particular technology and finally to design circuits around the inductors.

Chapter 8 considers modelling, but not in the classical sense of describing the nominal behaviour. Modelling of variability and reliability is considered in this chapter. The reliability models developed are physics-based and obtained from the measured data of the ENDURANCE chip. Finally, a circuit reliability simulator based on these models is presented.

Chapter 9 focuses on the impact of modelling in mixed-signal circuits. Particularly, the design of a pipelined analog-to-digital converter (ADC) along with the associated nonidealities has been considered. The chapter starts with a short discussion on the operation of a pipeline ADC. Then, various types of nonidealities stemming from the shortcomings of the operational amplifiers (finite gain and slew rate, input offset voltage), capacitor, and skew mismatches, as well as different non-linearities (charge injection, clock feedthrough, and switch resistance) are explained in conjunction with their simulation models. Finally, simulation results of a designed pipeline ADC, including all modelled nonidealities, are provided.

In Chapter 10, modelling in the context of power systems has been presented with the example of a buck converter. The switching and averaged models of the converter have been separately investigated with their theoretical background. Battery, capacitor, and inductor models are also provided to yield a comprehensive perspective on the realistic operation of a power system. A comparison between switching and averaged models is also made to reflect on their individual advantages and drawbacks.

Chapters 11 and 12 are short chapters highlighting some future applications in modelling. Chapter 11 concerns itself with modelling and design of MEMS devices in an application-specific integrated circuit (ASIC) design environment, whereas Chapter 12 considers spintronic resistive memories. The aim of these two chapters is to give a taste of some future problems, where modelling will be of utmost importance for accurately and precisely describing the operation of novel semiconductor devices and circuits.

In the last chapter of this book, it is aimed to provide an outlook on the future of modelling in analogue integrated circuits by starting from its historical fundamentals. All the chapters of this book are connected to different trends that have been observed throughout the development of modelling in microelectronics. Therefore, in order to forecast the next steps, a better understanding of the current needs and tools, as well as their applications, is mandatory. The book will conclude by stating the possibilities, in which the reader might be interested in investing their time if they decide to continue their work in this research area.

Part I

Fundamentals of modelling methodologies

Chapter 2
Response surface modeling

Jun Tao[1], Xuan Zeng[1], and Xin Li[2]

During the past decades, response surface modeling (RSM) has been extensively studied as one of the most important techniques for analyzing analog and mixed-signal (AMS) circuits. The objective of RSM is to approximate the circuit-level performance as an analytical function of variables of interest (VoIs) (e.g., design variables, tunable knobs, device-level variations and/or environmental conditions) based on a set of simulated/measured data.

The conventional method for RSM is least-squares (LS) regression. However, when modeling modern AMS circuits, LS regression usually suffers from unaffordable modeling cost due to high-dimensional variable space and expensive simulation/measurement cost. To address this cost issue and make RSM of practical utility, several state-of-the-art techniques have been proposed recently. For instance, by exploring the sparsity of model coefficients, feature selection methods, such as orthogonal matching pursuit (OMP), and L_1-norm regularization, are developed to automatically select important basis functions based on a limited number of training samples. Alternatively, another cost-reduction method, referred to as Bayesian model fusion (BMF), attempts to borrow the information collected from an early stage (e.g., schematic design) to facilitate efficient performance modeling at a late stage (e.g., post layout) during today's multistage design flow. In this chapter, we will review these recent advances and highlight their novelty.

2.1 Introduction

As one of the most important techniques for analyzing AMS circuits, RSM has been extensively studied in the literature during the past decades [1–7]. The objective of RSM is to approximate the circuit-level performance (e.g., gain of an analog amplifier) as an analytical (e.g., linear and quadratic) function of design variables (e.g., transistor widths), tunable knobs (e.g., bias current and capacitor array), device-level variations (e.g., oxide thickness and threshold voltage) and/or

[1]State Key Laboratory of ASIC and System, School of Microelectronics, Fudan University, Shanghai, China
[2]Data Science Research Center, Duke Kunshan University, Kunshan, China

environmental conditions (e.g., temperature and power supply). Compared to the expensive transistor-level simulations (e.g., SPICE simulations), such a performance model can evaluate the circuit performances with substantially reduced computational cost. Hence, RSM has been widely adopted to support a number of important applications, e.g., circuit optimization [1,2,8–17], worst-case corner extraction [18,19], and parametric yield estimation [20,21].

Once a response surface model is created, it can be utilized to improve the efficiency of circuit optimization, including pre-silicon sizing and post-silicon tuning. Pre-silicon sizing aims to optimize the design variables before fabrication to satisfy all given specifications [10–16]. Due to large-scale process variations, it is usually difficult to guarantee high parametric yield at advanced technology nodes. For this reason, reconfigurable analog/radio frequency (RF) circuits, including a set of tunable knobs, have been proposed. The objective of post-silicon tuning is to optimize the tunable knob setups and, next, adaptively reconfigure analog/RF circuits after fabrications to satisfy all given specifications over different process corners and environmental conditions [8,9]. In order to solve these circuit optimization problems, both efficiently (e.g., with low computational cost) and robustly (e.g., with guaranteed global optimum), special constraints are often posed for the performance models [16]. For instance, if posynomial performance models are available, circuit optimization can be formulated as a convex geometric programming problem and solved with guaranteed global optimum [10–12]. To afford further modeling flexibility, we can alternatively approximate analog performances by general non-convex polynomial functions. Even though the resulting polynomial optimization is non-convex, its global optimum can be reliably found by semidefinite programming (SDP) [13–16]. To enable hierarchical optimization for large-scale, complex analog systems, we can also apply RSM to extract the analog performance trade-offs, referred to as Pareto optimal fronts (PoFs), at block level [2,17]. Next, block-level performance constraints can be set up for system-level optimization based on these PoF models in order to guarantee that the optimized design is feasible.

Worst-case corner extraction aims to identify the "worst" corners in the device-level variable space (i.e., determining the specific values of device-level variables), at which the performance of a given circuit becomes the worst. Instead of performing expensive Monte Carlo analysis, designers only need to verify or optimize their circuit at these corners and come up with a robust solution [11,22]. Therefore, worst-case corner extraction can facilitate efficient yield optimization. By using RSM, we can cast worst-case corner extraction to a sequential quadratic programming problem [1] or an SDP problem [19] and, next, use existing optimization algorithms [23] to solve them and extract the worst-case corners.

Parametric yield can also be efficiently estimated based on the RSM of circuit performance. If process variations are sufficiently small, we can create a linear regression model to efficiently and accurately approximate a given performance. Without loss of generality, we usually assume that the device-level variations can be represented by a set of independent and standard normal random variables after principal component analysis (PCA). Therefore, the performance will also follow a normal distribution [20]. To capture larger scale process variations, nonlinear response surface models (e.g., quadratic polynomials) must be utilized, and moment-matching method (e.g., APEX [21]) can be used to estimate the unknown probability density function (PDF) or cumulative distribution function of the performance variations.

Given an AMS circuit, most existing RSM methods approximate its performance of interest (PoI) as a linear combination of a set of basis functions (also referred to as feature functions). To calculate the model coefficients, conventional RSM adopts LS fitting [4–7]. First, we generate a set of training samples and run transistor-level simulations to obtain the PoIs over these samples. Next, the unknown model coefficients can be determined by solving an overdetermined linear equation to minimize the sum-of-squares error. In this case, the number of training samples must be equal to or greater than the number of model coefficients (i.e., the number of adopted basis functions) [24]. However, LS usually suffers from unaffordable computational cost when modeling modern AMS circuits because of two reasons [25].

- *High-dimensional variable space:* First, due to the remarkable increase of AMS circuit size, while a typical analog circuit (e.g., OpAmp and current mirror) consists of only 100 devices, an entire AMS system (e.g., phase-locked loop, analog-to-digital converter and RF front-end) may assemble a few hundred such circuits and comprise 10^5 devices or more [2]. To create a performance model for a large-scale AMS system, the number of associated design variables (and also the number of basis functions) will be tremendous (e.g., more than 10^5). Second, with the aggressive scaling of integrated circuit technology, a large number of device-level random variables must be used to model the process variations. For instance, about 40 independent random variables are required to model the device mismatches of a single transistor for a commercial 32 nm complementary metal oxide semiconductor (CMOS) process. If an AMS system contains 10^5 transistors, there are about 4×10^6 random variables in total to capture the corresponding device-level variations, resulting in a high-dimensional variation space [25].
- *Expensive circuit simulation:* The computational cost of circuit simulation substantially increases, as the AMS circuit size becomes increasingly large. For instance, it may take a few days or even a few weeks to run transistor-level simulation of a large AMS circuit, such as phase-locked loop or high-speed link [25].

These recent trends of today's AMS circuits make performance modeling extremely difficult. On one hand, a large number of simulation samples must be generated in order to fit a high-dimensional model. On the other hand, creating a single sample by transistor-level simulation can take a large amount of computational time. The challenge here is how to make performance modeling computationally affordable for today's large-scale AMS circuits [25].

To reduce the modeling cost and make RSM of practical utility for large-scale AMS systems, several state-of-the-art techniques have been proposed recently.

For instance, while numerous basis functions must be used to span the high-dimensional variable space, not all these functions play an important role for a given PoI. In other words, although there are a large number of unknown coefficients, many of them are close to zero, rendering a unique sparse structure. Taking into account this sparse structure as our prior knowledge, a large number of (e.g., 10^4–10^6) model coefficients can be efficiently solved from a small set of (e.g., 10^2–10^3) sampling points without over-fitting by using OMP [17,26] or L_1-norm regularization [27–30]. OMP is a classic algorithm to approximate the optimal solution of L_0-norm regularization [24].

Another alternative method to reduce the modeling cost is referred to as BMF. BMF was proposed on the basis of observation that today's AMS circuits are often designed via a multistage flow. For instance, an AMS design often spans three major stages: (i) schematic design, (ii) layout design and (iii) chip manufacturing and testing [30]. At each stage, simulation or measurement data are collected to validate the circuit design, before moving to the next stage. To build performance models, most conventional RSM techniques only rely on the data at a single stage but completely ignore the data that are generated at other stages. The basic idea of BMF is to reuse the early-stage data when fitting a late-stage performance model. As such, the performance modeling cost can be substantially reduced [25,30].

The remainder of this chapter is organized as follows: in Section 2.2, we describe the problem definition of performance modeling. Next, the conventional LS regression method is summarized in Section 2.3. In Section 2.4, we discuss OMP and L_1-norm regularization in detail and, consequently, present the general idea of regularization in Section 2.5. BMF and its applications are illustrated in Section 2.6. Finally, a brief summary is concluded in Section 2.7.

2.2 Problem formulation

Without loss of generality, let $\mathbf{x} = [x_1\ x_2\ \cdots\ x_N]^T \in \mathcal{R}^N$ denote the VoIs, including design variables, tunable knobs, device-level variations and/or environmental conditions. The PoI y can be approximated by using the linear combination of a set of basis functions $\{b_m(\mathbf{x}); m = 1, 2, \cdots, M\}$:

$$y(\mathbf{x}) = \sum_{m=1}^{M} c_m b_m(\mathbf{x}) \tag{2.1}$$

where $\{c_m; m = 1, 2, \ldots, M\}$ contains all unknown model coefficients, and M is the total number of basis functions. The basis functions $\{b_m(\mathbf{x}); m = 1, 2, \cdots, M\}$ can be chosen as posynomials [10,12], monomials [26,27], polynomials [31,32], etc.

Suppose that we have obtained the transistor-level simulation results or measurement data on N_S training samples, RSM attempts to calculate the unknown model coefficients $\{c_m; m = 1, 2, \cdots, M\}$ in (2.1) by solving the following linear system:

$$\mathbf{B} \cdot \mathbf{c} = \mathbf{y}^S \tag{2.2}$$

where

$$\mathbf{B} = \begin{bmatrix} b_1(\mathbf{x}_1) & b_2(\mathbf{x}_1) & \cdots & b_M(\mathbf{x}_1) \\ b_1(\mathbf{x}_2) & b_2(\mathbf{x}_2) & \cdots & b_M(\mathbf{x}_2) \\ \vdots & \vdots & \vdots & \vdots \\ b_1(\mathbf{x}_{N_S}) & b_2(\mathbf{x}_{N_S}) & \cdots & b_M(\mathbf{x}_{N_S}) \end{bmatrix} \tag{2.3}$$

$$\mathbf{c} = \begin{bmatrix} c_1 & c_2 & \cdots & c_M \end{bmatrix}^T \tag{2.4}$$

$$\mathbf{y}^S = \begin{bmatrix} y_1^S & y_2^S & \cdots & y_{N_S}^S \end{bmatrix}^T \tag{2.5}$$

and $\mathbf{x}_n$ and y_n^S where $n \in \{1, 2, \cdots, N_S\}$ denote the VoIs and PoI for the nth sampling point, respectively. In the following sections, we will describe several state-of-the-art methods to solve the unknown model coefficients from (2.1).

2.3 Least-squares regression

If the number of training samples (i.e., N_S) is greater than the number of unknown model coefficients (i.e., M), (2.2) becomes an overdetermined linear equation and can be solved by adopting the conventional LS regression method. LS regression attempts to determine the model coefficients $\mathbf{c}$ by minimizing a specific error function $E(\mathbf{c})$, where $E(\mathbf{c})$ is defined as the sum-of-squares of the errors between the modeling result $y(\mathbf{x}_n)$ and the corresponding simulated/measured performance value y_n^S over each sampling point $\mathbf{x}_n$:

$$\min_{\mathbf{c}} E(\mathbf{c}) = \sum_{n=1}^{N_S} \left(y(\mathbf{x}_n) - y_n^S\right)^2 = \sum_{n=1}^{N_S} \left(\sum_{m=1}^{M} c_m b_m(\mathbf{x}_n) - y_n^S \right)^2 = \left\|\mathbf{B} \cdot \mathbf{c} - \mathbf{y}^S\right\|_2^2 \tag{2.6}$$

where $\|\bullet\|_2$ stands for the L_2-norm of a vector. Since the performance model in (2.1) is a linear function of the model coefficients $\mathbf{c}$, the cost function $E(\mathbf{c})$ in (2.6) is quadratic in $\mathbf{c}$, and its first-order derivative with respect to the unknown coefficients is also linear over $\mathbf{c}$:

$$\nabla_{\mathbf{c}} E(\mathbf{c}) = \left(\mathbf{B} \cdot \mathbf{c} - \mathbf{y}^S\right)^T \cdot \mathbf{B}. \tag{2.7}$$

By setting this derivative in (2.7) to zero, we can obtain an analytical LS solution (denoted as $\mathbf{c}_{LS}$) for the unknown model coefficients:

$$\mathbf{c}_{LS} = \left(\mathbf{B}^T\mathbf{B}\right)^{-1}\mathbf{B}^T\mathbf{y}^S. \tag{2.8}$$

The aforementioned LS solution can also be derived from maximum likelihood estimation (MLE). Without loss of generality, we can assume that the modeling error follows a zero-mean Gaussian distribution with the variance σ_ε^2:

$$y^S = y(\mathbf{x}) + \varepsilon, \tag{2.9}$$

$$\varepsilon \sim N\left(0, \sigma_\varepsilon^2\right). \tag{2.10}$$

As a result, the distribution of y^S will also be Gaussian:

$$pdf\left(y^S\right) = \frac{1}{\sqrt{2\pi} \cdot \sigma_\varepsilon} \cdot \exp\left[-\frac{\left(y^S - y(\mathbf{x})\right)^2}{2 \cdot \sigma_\varepsilon^2}\right]. \tag{2.11}$$

Since all sampling points are drawn independently and follow the same distribution in (2.11), $\{y_n^S;\ n = 1, 2, \cdots, N_S\}$ is independent and identically distributed (abbreviated to i.i.d.). Consequently, the corresponding likelihood function can be written as a product of the probabilities for each sampling point:

$$pdf\left(\mathbf{y}^S | \mathbf{c}, \sigma_\varepsilon\right) = \prod_{n=1}^{N_S} \frac{1}{\sqrt{2\pi} \cdot \sigma_\varepsilon} \cdot \exp\left[-\frac{\left(y_n^S - y(\mathbf{x}_n)\right)^2}{2 \cdot \sigma_\varepsilon^2}\right]. \tag{2.12}$$

This likelihood function represents the probability of observing all training samples given the model coefficients **c** and the variance σ_ε^2. After taking the natural logarithm on both sides of (2.12) and following the definition of the sum-of-squares error function $E(\mathbf{c})$ in (2.6), we have:

$$\begin{aligned}
\ln\left[pdf\left(\mathbf{y}^S | \mathbf{c}, \sigma_\varepsilon\right)\right] &= \sum_{n=1}^{N_S} \ln\left\{\frac{1}{\sqrt{2\pi} \cdot \sigma_\varepsilon} \cdot \exp\left[-\frac{\left(y_n^S - y(\mathbf{x}_n)\right)^2}{2 \cdot \sigma_\varepsilon^2}\right]\right\} \\
&= -\frac{N_S}{2} \ln(2\pi) - N_S \ln(\sigma_\varepsilon) - \frac{1}{2 \cdot \sigma_\varepsilon^2} \sum_{n=1}^{N_S} \left(y_n^S - y(\mathbf{x}_n)\right)^2 \\
&= -\frac{N_S}{2} \ln(2\pi) - N_S \ln(\sigma_\varepsilon) - \frac{1}{2 \cdot \sigma_\varepsilon^2} E(\mathbf{c}).
\end{aligned} \tag{2.13}$$

Since the natural logarithm function is monotonically increasing, maximizing the likelihood function in (2.12) is equivalent to maximizing its natural logarithm in (2.13). At the right-hand side of (2.13), only the last term depends on **c**, and the positive constant $1/2 \cdot \sigma_\varepsilon^2$ of this term does not affect the optimal solution of the model coefficients **c**. Therefore, the LS regression in (2.6) is equivalent to the MLE problem.

As discussed in the aforementioned section, the LS method has been extensively studied in the literature [4–7,11,33,34]. However, it often becomes computationally unaffordable due to high dimensionality and expensive simulation/measurement cost in practice. When applying LS regression, the number of training samples (i.e., N_S) must be equal to or greater than the number of model coefficients (i.e., M). If M is large (e.g., 10^4–10^6 in the high-dimensional variable space), we must collect more than M samples for a large-scale AMS system, which is intractable, if not impossible. For this reason, the traditional LS method is limited to small-size or medium-size problems (e.g., 10–1,000 model coefficients) [26]. To address this cost issue, we will present two feature selection methods in the following section.

2.4 Feature selection

If the number of training samples (i.e., N_S) is less than the number of model coefficients (i.e., M), i.e., there are fewer equations than the unknowns, the linear equation in (2.2) becomes underdetermined, and the solution for the model coefficients $\mathbf{c}$ is not unique. In this case, directly applying LS regression may result in over-fitting [24].

To address this underdetermined issue, we can explore the sparsity of $\mathbf{c}$ and introduce additional constraints in order to uniquely and accurately determine its value [26–28]. While numerous basis functions must be used to span the high-dimensional and nonlinear variation space, not all of them play an important role in modeling a given PoI. In other words, although the dimensionality of $\mathbf{c}$ maybe huge, many of its elements are close to zero, rendering a unique sparse structure. For instance, for a 65 nm SRAM circuit containing 21,310 independent random variables, only 50 basis functions are required to accurately approximate the delay variation of its critical path [27]. However, we do not know the important basis functions or, equivalently, the exact locations of nonzero coefficients in advance. To automatically select these important basis functions based on a limited number of training samples, L_0-norm regularization can be utilized.

Based on the theory of L_0-norm regularization, we can formulate the following optimization problem to calculate the sparse solution for the model coefficients $\mathbf{c}$:

$$\begin{aligned} \min_{\mathbf{c}} \quad & E(\mathbf{c}) = \sum_{n=1}^{N_S} \left(y(\mathbf{x}_n) - y_n^S \right)^2 = \left\| \mathbf{B} \cdot \mathbf{c} - \mathbf{y}^S \right\|_2^2 \\ s.t. \quad & \left\| \mathbf{c} \right\|_0 \le \lambda, \end{aligned} \tag{2.14}$$

where $\|\bullet\|_0$ stands for the L_0-norm of a vector, i.e., the number of nonzeros in the vector. Therefore, $\|\mathbf{c}\|_0$ measures the sparsity of $\mathbf{c}$. By directly constraining this L_0-norm, the optimization in (2.14) attempts to find a sparse $\mathbf{c}$ that minimizes the sum-of-squares error $E(\mathbf{c})$ [26–28].

The parameter λ in (2.14) explores the trade-off between the sparsity of the model coefficients $\mathbf{c}$ and the minimal value of the cost function $E(\mathbf{c})$. For instance, a large λ will result in a small $E(\mathbf{c})$, but meanwhile, it will increase the number of nonzeros in $\mathbf{c}$. Note that a small cost function does not necessarily mean a small modeling error. Even though the minimal cost function value can be reduced by increasing λ, such a strategy may result in over-fitting especially because (2.2) is underdetermined. In the extreme case, if λ is sufficiently large and the constraint in (2.14) is not active, we can always find a solution for $\mathbf{c}$ to exactly make the cost function equal to zero. However, such a solution is likely to be useless, since it over-fits the given training samples [26,27]. In practice, the optimal value of λ can be automatically determined by cross-validation, as will be discussed in detail in Section 2.4.3.

While the aforementioned L_0-norm regularization can effectively guarantee a sparse solution for the model coefficients $\mathbf{c}$, the optimization in (2.14) is non-deterministic polynomial-time (NP) hard [24,34] and, hence, is extremely difficult to solve. We can approximate its solution by adopting an efficient heuristic algorithm, referred to as OMP, in Section 2.4.1 or, alternatively, relaxing (2.14) to a computationally efficient L_1-norm regularization problem as shown in Section 2.4.2.

2.4.1 Orthogonal matching pursuit

Given the underdetermined linear equation in (2.2), OMP applies a heuristic and iterative algorithm to identify a small set of important basis functions and use them to approximate the performance $y(\mathbf{x})$. For other noncritical basis functions, the corresponding coefficients are set to zero. If the number of selected basis functions (i.e., λ) is substantially less than the total number of basis functions (i.e., M), the resulting solution of model coefficients $\mathbf{c}$ is sparse [26].

To guarantee the fast convergence of OMP, we usually adopt a set of basis functions that are normalized and orthogonal. Namely, the inner production of any two basis functions $b_i(\mathbf{x})$ and $b_j(\mathbf{x})$ must satisfy:

$$\langle b_i(\mathbf{x}), b_j(\mathbf{x})\rangle = \int b_i(\mathbf{x}) \cdot b_j(\mathbf{x}) \cdot pdf(\mathbf{x}) \cdot d\mathbf{x} = \begin{cases} 1 & \text{if } i = j \\ 0 & \text{if } i \neq j \end{cases}. \tag{2.15}$$

For instance, if $\mathbf{x}$ represents a set of independent random VoIs that follow the standard normal distribution (e.g., the device-level variations after PCA), we can adopt high-dimensional Hermite polynomials as basis functions. Generally speaking, for a given $pdf(\mathbf{x})$, a set of normalized and orthogonal basis functions can be created by using the Gram–Schmidt technique [35].

According to (2.1) and (2.15), we can calculate the inner product between $y(\mathbf{x})$ and each basis function $b_m(\mathbf{x})$:

$$\begin{aligned}\langle y(\mathbf{x}), b_m(\mathbf{x})\rangle &= \int y(\mathbf{x}) \cdot b_m(\mathbf{x}) \cdot pdf(\mathbf{x}) \cdot d\mathbf{x} \\ &= \int \left[\sum_{i=1}^{M} c_i b_i(\mathbf{x})\right] \cdot b_m(\mathbf{x}) \cdot pdf(\mathbf{x}) \cdot d\mathbf{x} \\ &= \sum_{i=1}^{M} c_i \cdot \int b_i(\mathbf{x}) \cdot b_m(\mathbf{x}) \cdot pdf(\mathbf{x}) \cdot d\mathbf{x} = c_m.\end{aligned} \tag{2.16}$$

This inner product is equal to the model coefficient c_m. It implies that if $\langle y(\mathbf{x}), b_m(\mathbf{x})\rangle$ (i.e., c_m) is far away from zero, the corresponding $b_m(\mathbf{x})$ must be selected as an important basis function to approximate the performance $y(\mathbf{x})$. Therefore, this inner product $\langle y(\mathbf{x}), b_m(\mathbf{x})\rangle$ can be used as a good criterion to measure the importance of each basis function.

However, we do not know the analytical form of $y(\mathbf{x})$ in advance and, hence, can only numerically approximate the integration in (2.16) from a set of training samples:

$$\langle y(\mathbf{x}), b_m(\mathbf{x})\rangle \approx \frac{1}{N_S} \cdot \sum_{n=1}^{N_S} b_m(\mathbf{x}_n) \cdot y_n^S = \frac{1}{N_S} \cdot \mathbf{b}_m^T \cdot \mathbf{y}^S = \xi_m, \tag{2.17}$$

where $\mathbf{b}_m = [b_m(\mathbf{x}_1)\ b_m(\mathbf{x}_2)\ \cdots\ b_m(\mathbf{x}_{N_S})]^T$ denotes the mth column vector of the matrix $\mathbf{B}$ defined in (2.3) and is also referred to as the mth basis vector. ξ_m in (2.17) gives a statistic estimation for the unknown coefficient c_m. However, since ξ_m is calculated from a set of random sampling data $\{(\mathbf{x}_n, y_n^S); n = 1, 2, \cdots, N_S\}$ that may contain large fluctuations, such an estimation cannot guarantee sufficient accuracy [31,36]. Therefore, instead of directly using (2.17) to determine the value of c_m, OMP applies an iterative algorithm. During each iteration, the inner product in (2.17) is only used to identify one important basis function from all candidates and, next, the model coefficients are calculated by applying LS regression. In what follows, we will describe the OMP algorithm in detail.

We start from a large set of possible basis functions $\mathcal{B}_C = \{b_m(\mathbf{x}); m = 1, 2, \cdots, M\}$ that can be used to approximate the performance function $y(\mathbf{x})$. Initially, without knowing which basis function is important, we use each basis function from the set $\mathcal{B}_C$ (i.e., each column vector of $\mathbf{B}$) to calculate the inner product values $\{\xi_m; m = 1, 2, \cdots, M\}$ based on (2.17). The basis function $b_{m_{S1}}(\mathbf{x})$ that is most correlated with $\mathbf{y}^S$, results in the largest absolute value of the inner product $|\xi_{m_{S1}}|$, is chosen as the first important basis function.

Once $b_{m_{S1}}(\mathbf{x})$ is identified, OMP uses it to approximate $y(\mathbf{x})$ by solving the following LS fitting problem:

$$\min_{c_{m_{S1}}} \sum_{n=1}^{N_S} \left(c_{m_{S1}} b_{m_{S1}}(\mathbf{x}_n) - y_n^S\right)^2 = \left\| c_{m_{S1}} \mathbf{b}_{m_{S1}} - \mathbf{y}^S \right\|_2^2, \tag{2.18}$$

where $\mathbf{b}_{m_{S1}} = [b_{m_{S1}}(\mathbf{x}_1)\ b_{m_{S1}}(\mathbf{x}_2)\ \cdots\ b_{m_{S1}}(\mathbf{x}_{N_S})]^T$. Next, OMP calculates the residual

$$\mathbf{r} = \mathbf{y}^S - c_{m_{S1}}^* \cdot \mathbf{b}_{m_{S1}} \tag{2.19}$$

with the optimal coefficient $c_{m_{S1}}^*$ calculated by (2.18) and removes $b_{m_{S1}}(\mathbf{x})$ from the set of possible basis functions $\mathcal{B}_C$.

Based on (2.19), OMP further identifies the next important basis function $b_{m_{S2}}(\mathbf{x})$ by estimating the inner product values between the residual $\mathbf{r}$ and each basis function included in the current $\mathcal{B}_C$:

$$\xi_m = \frac{1}{N_S} \cdot \sum_{n=1}^{N_S} r_n b_m(\mathbf{x}_n) = \frac{1}{N_S} \cdot \mathbf{b}_m^T \cdot \mathbf{r}, \tag{2.20}$$

where r_n denotes the nth element in $\mathbf{r}$. Once $b_{m_{S2}}(\mathbf{x})$ is known, OMP approximates $\mathbf{y}^S$ in the directions of both $\mathbf{b}_{m_{S1}}$ and $\mathbf{b}_{m_{S2}}$ by solving the following optimization problem:

$$\min_{c_{m_{S1}}, c_{m_{S2}}} \sum_{n=1}^{N_S} \left(c_{m_{S1}} b_{m_{S1}}(\mathbf{x}_n) + c_{m_{S2}} b_{m_{S2}}(\mathbf{x}_n) - y_n^S\right)^2 = \left\| c_{m_{S1}} \mathbf{b}_{m_{S1}} + c_{m_{S2}} \mathbf{b}_{m_{S2}} - \mathbf{y}^S \right\|_2^2. \tag{2.21}$$

It is important to note that the coefficient $c_{m_{S1}}^*$ calculated by (2.18) maybe different from that calculated by (2.21). In other words, every time when a new basis function is selected, OMP recalculates all model coefficients to minimize the sum of squared residuals. This recalculation step is required, because even though the basis functions $\{b_m(\mathbf{x}); m = 1, 2, \cdots, M\}$ are orthogonal as defined in (2.15), the basis vectors $\{\mathbf{b}_m; m = 1, 2, \cdots, M\}$ are not necessarily orthogonal due to random sampling, i.e., $\mathbf{b}_i^T\mathbf{b}_j \neq 0$ when $i \neq j$. Hence, the new basis function selected at the current iteration step may change the model coefficient values calculated at previous iteration steps [26].

The aforementioned iteration for basis function selection and LS regression continues until a sufficient number of (i.e., λ) important basis functions are selected. Algorithm 1 summarizes the major iteration steps of OMP.

Algorithm 1: Orthogonal matching pursuit (OMP)

---

1. Start from the linear (2.2) generated from a set of normalized and orthogonal basis functions and an integer λ representing the total number of basis functions that should be selected.
2. Initialize the residual $\mathbf{r} = \mathbf{y}^S$, the number of selected basis functions $M_S = 0$, the index set of selected basis functions $\mathcal{I}_S = \phi$ and the index set of all possible basis functions $\mathcal{I}_C = \{1, 2, \cdots, M\}$.
3. While $M_S \leq \lambda$.
    4. For each $m \in \mathcal{I}_C$, calculate the inner product ξ_m between the residue $\mathbf{r}$ and the corresponding basis vector $\mathbf{b}_m$ based on (2.20).
    5. Find out the index m_S corresponding to the largest absolute value of the inner product $|\xi_{m_S}|$.
    6. Update $M_S = M_S + 1$, $\mathcal{I}_S = \mathcal{I}_S \cup \{m_S\}$ and remove m_S from $\mathcal{I}_C$.
    7. Solve the following optimization problem:

    $$\min_{\{c_m; m \in \mathcal{I}_S\}} \sum_{n=1}^{N_S} \left(\sum_{m \in \mathcal{I}_S} c_m b_m(\mathbf{x}_n) - y_n^S \right)^2 = \left\| \sum_{m \in \mathcal{I}_S} c_m \mathbf{b}_m - \mathbf{y}^S \right\|_2^2, \quad (2.22)$$

    to determine the optimal model coefficients $\{c_m^*; m \in \mathcal{I}_S\}$ by using LS regression.
    8. Update the residual:

    $$\mathbf{r} = \mathbf{y}^S - \sum_{m \in \mathcal{I}_S} c_m^* \mathbf{b}_m^T. \quad (2.23)$$

9. End while.
10. Set the model coefficients $\{c_m; m \in \mathcal{I}_C\}$ to zero.

It is important to note that even though OMP is a heuristic algorithm to solve the L_0-norm regularization problem in (2.14), the quality of its solution is guaranteed according to several theoretical studies from the statistics community [37]. Roughly speaking, if the M-dimensional vector $\mathbf{c}$ contains λ nonzeros ($\lambda \ll M$) and the linear equation in (2.2) is well conditioned, the actual solution $\mathbf{c}$ can be almost uniquely determined (with a probability nearly equal to one) from N_S sampling points, where N_S is in the order of $O(\lambda \cdot \log M)$ [37]. While this theoretical result does not precisely give the number of required sampling points, it presents an important scaling trend. Namely, N_S (i.e., the number of training samples) is a logarithmic function of M (i.e., the number of unknown coefficients). It, in turn, provides the theoretical foundation that by solving the sparse solution of an underdetermined equation, a large number of model coefficients can be uniquely determined from a small number of training samples.

2.4.2 L_1-norm regularization

Besides OMP, we can also approximate the solution of L_0-norm regularization in (2.14) by relaxing it to a computationally efficient L_1-norm regularization problem [27,28]:

$$\begin{aligned} \min_{\mathbf{c}} \quad & \left\|\mathbf{B} \cdot \mathbf{c} - \mathbf{y}^S\right\|_2^2 \\ s.t. \quad & \left\|\mathbf{c}\right\|_1 \le \lambda, \end{aligned} \tag{2.24}$$

where $\|\mathbf{c}\|_0$ in (2.14) is replaced by $\|\mathbf{c}\|_1$. $\|\mathbf{c}\|_1$ denotes the L_1-norm of the vector $\mathbf{c}$, i.e., the summation of the absolute values of all elements in $\mathbf{c}$:

$$\left\|\mathbf{c}\right\|_1 = |c_1| + |c_2| + \cdots + |c_M|. \tag{2.25}$$

The L_1-norm regularization in (2.24) can easily be reformulated to a convex optimization problem. By introducing a set of slack variables $\boldsymbol{\beta} = \{\beta_m; m = 0, 1, \cdots, M\}$, we can rewrite (2.24) into the following equivalent form [27]:

$$\begin{aligned} \min_{\mathbf{c},\boldsymbol{\beta}} \quad & \left\|\mathbf{B} \cdot \mathbf{c} - \mathbf{y}^S\right\|_2^2 \\ s.t. \quad & \beta_1 + \beta_2 + \cdots + \beta_M \le \lambda \\ & -\beta_m \le c_m \le \beta_m \qquad (m = 1, 2, \cdots, M). \end{aligned} \tag{2.26}$$

In (2.26), the cost function is quadratic and positive semi-definite. Hence, it is convex. All constraints are linear and, therefore, the resulting constraint set is a convex polytope. For these reasons, the L_1-norm regularization in (2.24) is a convex optimization problem and can be solved both efficiently (i.e., with low computational cost) and robustly (i.e., with guaranteed global optimum) by using several state-of-the-art methods, e.g., the interior-point method [23].

To understand the connection between L_1-norm regularization and sparse solution, we consider the two-dimensional example (i.e., $\mathbf{c} = [c_1\ c_2]^T$) in Figure 2.1 [27,28]. Since the cost function $\left\|\mathbf{B} \cdot \mathbf{c} - \mathbf{y}^S\right\|_2^2$ is quadratic and positive semi-definite, its contour lines can be represented by multiple ellipsoids. On the other hand, the constraint $\left\|\mathbf{c}\right\|_1 \le \lambda$ corresponds to a number of rotated squares, associated with different values of λ. For example, two of such squares are shown in Figure 2.1, where $\lambda_1 < \lambda_2$.

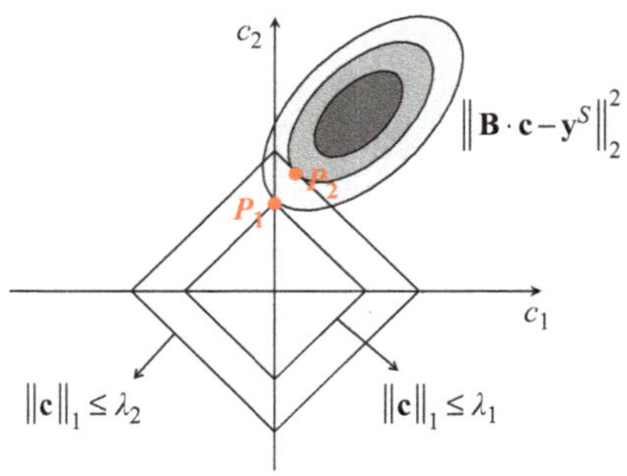

Figure 2.1 The L_1-norm regularization $\|\mathbf{c}\|_1 \le \lambda$ results in a sparse solution (i.e., $c_1 = 0$) if λ is sufficiently small (i.e., $\lambda = \lambda_1$) [27,28]

Studying Figure 2.1, we would notice that if λ is large (e.g., $\lambda = \lambda_2$), both c_1 and c_2 are not zero (e.g., at the red point P_2). However, as λ decreases (e.g., $\lambda = \lambda_1$), the contour of $\|\mathbf{B} \cdot \mathbf{c} - \mathbf{y}^S\|_2^2$ eventually intersects the polytope $\|\mathbf{c}\|_1 \le \lambda$ at one of its vertices (e.g., at the red point P_1). It, in turn, implies that one of the coefficients (e.g., c_1 in this case) becomes exactly zero. From this point of view, by decreasing λ of the L_1-norm regularization in (2.24), we can pose a strong constraint for sparsity and force a sparse solution. This intuitively explains why L_1-norm regularization guarantees sparsity, as is the case for L_0-norm regularization.

In addition, various theoretical studies from the statistics community demonstrate that under some general assumptions, both L_1-norm regularization and L_0-norm regularization result in the same solution [37]. However, the L_1-norm regularization is much more computationally efficient than the L_0-norm regularization that is NP hard. This is the major motivation to replace L_0-norm by L_1-norm.

2.4.3 Cross-validation

To make both the OMP algorithm (i.e., Algorithm 1) and the L_1-norm regularization in (2.24) of practical utility, we must find the optimal value of λ. In practice, λ is not known in advance. The appropriate value of λ must be determined by considering the following two important issues. First, if λ is too small, the aforementioned feature selection methods will not select a sufficient number of important basis functions to approximate the given PoI, thereby leading to large modeling error. On the other hand, if λ is too large and too many basis functions are used to approximate $y(\mathbf{x})$, it will result in over-fitting that again prevents us from extracting an accurate performance model. Hence, in order to achieve the best modeling accuracy, we must accurately estimate the modeling errors for different λ values and find the optimal λ that minimizes the error [26–28].

However, given a limited number of sampling points, accurately estimating modeling error is not a trivial task. To avoid over-fitting, we cannot simply measure the modeling error from the same sampling data that are used to calculate the model coefficients. Instead, modeling error must be measured from an independent data set. Cross-validation is an efficient method for model validation that has been

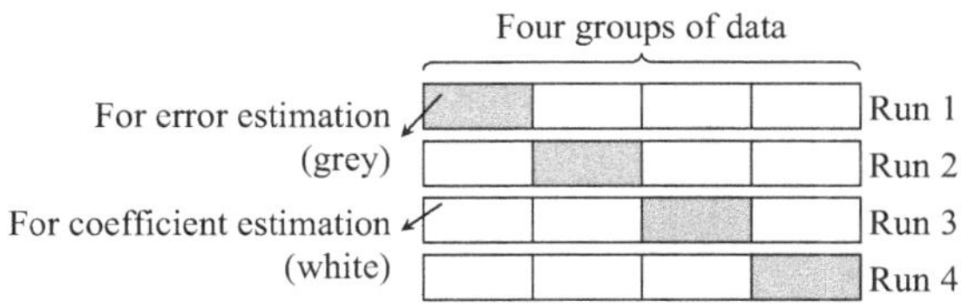

Figure 2.2 A 4-fold cross-validation partitions the data set into four groups and the modeling error is estimated from four independent runs: $\varepsilon = (\varepsilon_1 + \varepsilon_2 + \varepsilon_3 + \varepsilon_4)/4$

widely used in the statistics community [24,34]. A Q-fold cross-validation partitions the entire data set into Q groups, as shown by the example in Figure 2.2 (where $Q = 4$). Modeling error is estimated from Q-independent runs. In each run, one of the Q groups is used to estimate the modeling error and all other groups are used to calculate the model coefficients. Different groups should be selected for error estimation in different runs. As such, each run results in an error value ε_q (where $q \in \{1, 2, \cdots, Q\}$) that is measured from a unique group of sampling points. In addition, when a model is trained and tested in each run, non-overlapped data sets are used so that over-fitting can be easily detected. The final modeling error is computed as the average of $\{\varepsilon_q; q = 1, 2, \cdots, Q\}$, i.e., $\varepsilon = (\varepsilon_1 + \varepsilon_2 + \cdots + \varepsilon_Q)/Q$ [26–28].

For our application, OMP or L_1-norm regularization is used to identify important basis functions and calculate model coefficients for different λ values during each cross-validation run. Next, the modeling error associated with each run is estimated, resulting in $\{\varepsilon_q(\lambda); q = 1, 2, \cdots, Q\}$. Note that ε_q is not simply a value but a 1-D function of λ. Once all cross-validation runs are complete, the final modeling error is calculated as $\varepsilon(\lambda) = \left[\varepsilon_1(\lambda) + \varepsilon_2(\lambda) + \cdots + \varepsilon_Q(\lambda)\right]/Q$, which is again a 1-D function of λ. The optimal λ is determined by finding the minimal value of $\varepsilon(\lambda)$.

The major drawback of cross-validation is the need to repeatedly extract the model coefficients for Q times. However, for our circuit modeling application, the overall modeling cost is often dominated by the simulation and/or measurement cost that is required to generate sampling data. Hence, the computational overhead by cross-validation is almost negligible.

2.4.4 Least angle regression

As discussed in the previous subsection, we need to adopt a two-step approach to implement the L_1-norm regularization in (2.24) to automatically determine the optimal value of λ: (i) repeatedly solve the convex programming problem in (2.26) by using the interior-point method [23] for different λ's and (ii) select the optimal λ by cross-validation as discussed in Section 2.4.3. This approach, however, is computationally expensive, as we must run a convex solver for many times in order to visit a sufficient number of possible values of λ. Instead of applying the interior-point method, we can adopt an efficient algorithm, i.e., least angle regression (LAR), to accomplish these two steps and further reduce the computational cost [27,28].

According to (2.24), the sparsity of the solution **c** depends on the value of λ. In the extreme case, if λ is zero, all coefficients in **c** are equal to zero. As λ gradually increases, more and more coefficients in **c** become nonzero. In fact, it can be proven that the solution **c** of (2.24) is a piece-wise linear function of λ [38]. As a result, we do not have to repeatedly solve the L_1-norm regularization at many different λ's. Instead, we only need to estimate the local linear function in each interval $[\lambda_i, \lambda_{i+1}]$. This property allows us to find the entire solution trajectory $\mathbf{c}(\lambda)$ with low computational cost. Next, we will show an iterative algorithm based on LAR [27] to efficiently find the solution trajectory $\mathbf{c}(\lambda)$.

Without loss of generality, we first normalize each basis vector by performing:

$$\mathbf{b}_m = \frac{\mathbf{b}_m}{\|\mathbf{b}_m\|_2}, \tag{2.27}$$

where $m \in \{1, 2, \cdots, M\}$. Next, we start from the extreme case where λ is zero. In this case, the solution of (2.24) is trivial, i.e., $\mathbf{c} = 0$. As λ increases from zero, we calculate the correlation between $\mathbf{y}^S$ and each normalized $\mathbf{b}_m$:

$$cor_m = \left|\mathbf{b}_m^T \cdot \mathbf{y}^S\right|, \tag{2.28}$$

and find the vector $\mathbf{b}_{mS1}$ that is most correlated with $\mathbf{y}^S$, i.e., cor_{mS1} takes the largest value [38]. By using $\mathbf{b}_{mS1}$ to approximate $\mathbf{y}^S$, we can express the modeling residue as:

$$\mathbf{r} = \mathbf{y}^S - \gamma_1 \mathbf{b}_{mS1}, \tag{2.29}$$

where γ_1 is an unknown coefficient to be determined.

To intuitively understand the LAR algorithm, we consider the two-dimensional example shown in Figure 2.3. In this example, the vector $\mathbf{b}_2$ has a higher correlation with $\mathbf{y}^S$ than the vector $\mathbf{b}_1$. Hence, $\mathbf{b}_2$ is first selected to approximate $\mathbf{y}^S$. From the geometrical point of view, finding the largest correlation is equivalent to finding the least angle between the normalized vectors $\{\mathbf{b}_m;\ m = 1, 2, \cdots, M\}$ and the

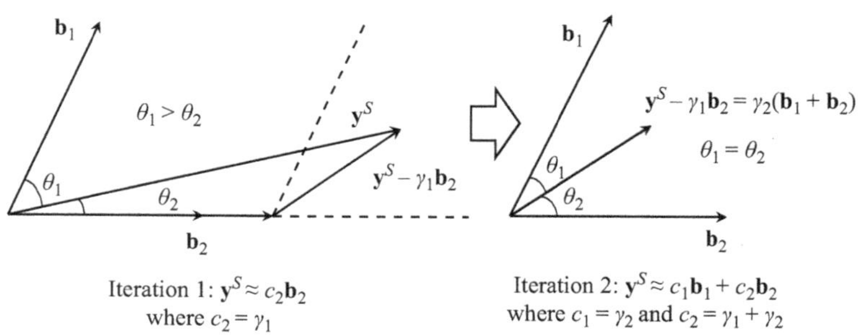

Figure 2.3 LAR calculates the solution trajectory $\mathbf{c}(\lambda)$ of a two-dimensional example $\mathbf{y}^S = c_1\mathbf{b}_1 + c_2\mathbf{b}_2$ [27]

performance $\mathbf{y}^S$. Therefore, the aforementioned algorithm is referred to as LAR in [38].

As $|\gamma_1|$ increases, the correlation between the vector $\mathbf{b}_{m_{S1}}$ (e.g., $\mathbf{b}_2$ in Figure 2.3) and the residual $\mathbf{r}$ defined in (2.29) (e.g., $\mathbf{r} = \mathbf{y}^S - \gamma_1 \mathbf{b}_2$ in Figure 2.3) decreases. LAR uses an efficient algorithm to compute the maximal value of $|\gamma_1|$ at which the correlation between $\mathbf{b}_{m_{S1}}$ and $\mathbf{r} = \mathbf{y}^S - \gamma_1 \mathbf{b}_{m_{S1}}$ is no longer dominant. In other words, there is another vector $\mathbf{b}_{m_{S1}}$ (e.g., $\mathbf{b}_1$ in Figure 2.3) that has the same correlation with the current residual $\mathbf{r}$:

$$\left|\mathbf{b}_{m_{S1}} \cdot \left(\mathbf{y}^S - \gamma_1 \mathbf{b}_{m_{S1}}\right)\right| = \left|\mathbf{b}_{m_{S2}} \cdot \left(\mathbf{y}^S - \gamma_1 \mathbf{b}_{m_{S1}}\right)\right|. \tag{2.30}$$

In this case, instead of continuing along $\mathbf{b}_{m_{S1}}$, LAR proceeds in a direction equiangular between $\mathbf{b}_{m_{S1}}$ and $\mathbf{b}_{m_{S2}}$. Namely, it approximates $\mathbf{y}^S$ by the linear combination of $\mathbf{b}_{m_{S1}}$ and $\mathbf{b}_{m_{S2}}$:

$$\mathbf{y}^S \approx \gamma_1 \mathbf{b}_{m_{S1}} + \gamma_2 (\mathbf{b}_{m_{S1}} + \mathbf{b}_{m_{S2}}), \tag{2.31}$$

where γ_1 is fixed and γ_2 is unknown at this second iteration.

Taking Figure 2.3 as an example, the residual $\mathbf{y}^S - \gamma_1 \mathbf{b}_2$ is approximated by $\gamma_2 \cdot (\mathbf{b}_1 + \mathbf{b}_2)$. If $|\gamma_2|$ is sufficiently large, $\mathbf{y}^S$ will be exactly equal to $\mathbf{y}^S = \gamma_1 \mathbf{b}_2 + \gamma_2 (\mathbf{b}_1 + \mathbf{b}_2) = \gamma_2 \mathbf{b}_1 + (\gamma_1 + \gamma_2) \mathbf{b}_2$. In this example, because only two basis functions $b_1(\mathbf{x})$ and $b_2(\mathbf{x})$ (i.e., two basis vectors $\mathbf{b}_1$ and $\mathbf{b}_2$) are used, LAR stops at the second iteration step. If more than two basis functions are involved, LAR will keep increasing $|\gamma_2|$ until a third vector $\mathbf{b}_{m_{S3}}$ earns its way into the "most correlated" set, and so on. Algorithm 2 summarizes the major iteration steps of LAR.

Algorithm 2: Least angle regression (LAR)

---

1. Start from the simulated/measured performance $\mathbf{y}^S$ defined in (2.5) and the normalized vectors $\{\mathbf{b}_m; m = 1, 2, \cdots, M\}$.
2. Initialize the residual $\mathbf{r}_{cur} = \mathbf{y}^S$, the index set of selected basis functions $\mathcal{I}_S = \phi$, the index set of all possible basis functions $\mathcal{I}_C = \{1, 2, \cdots, M\}$ and the iteration index $p = 1$.
3. Calculate the correlation between $\mathbf{y}^S$ and each possible basis vector $\mathbf{b}_m$, where $m \in \{1, 2, \cdots, M\}$, based on (2.28).
4. Select the vector $\mathbf{b}_{m_S}$ that has the largest correlation cor_{m_S}.
5. While $\mathbf{r}_{cur} \neq 0$.
    6. Update $p = p + 1$, $\mathcal{I}_S = \mathcal{I}_S \cup \{m_S\}$ and remove m_S from $\mathcal{I}_C$.
    7. Use the algorithm in [38] to determine the maximal $|\gamma_p|$ such that either the resulted residual

$$\mathbf{r} = \mathbf{r}_{cur} - \gamma_p \cdot \sum_{m \in \mathcal{I}_S} \mathbf{b}_m. \tag{2.32}$$

is equal to 0 or another basis vector $\mathbf{b}_{m_S}$ ($m_S \in \mathcal{I}_C$) has as much correlation with the resulted residual:

$$\left|\mathbf{b}_{m_S}^T \cdot \mathbf{r}\right| = \left|\mathbf{b}_m^T \cdot \mathbf{r}\right| \quad (\forall m \in \mathcal{I}_S) \tag{2.33}$$

8. Update $\mathbf{r}_{cur} = \mathbf{r}$.
9. End while.

It can be proven that with several small modifications, LAR will generate the entire piece-wise linear solution trajectory $\mathbf{c}(\lambda)$ for the L_1-norm regularization in (2.24) [38]. Based on $\mathbf{c}(\lambda)$, the optimal λ can efficiently be determined by cross-validation. First, we partition the entire data set into Q groups and use LAR to extract the model coefficient trajectory $\mathbf{c}_q(\lambda)$ (where $q \in \{1, 2, \cdots, Q\}$) during each cross-validation run. Next, instead of repeatedly solving (2.26) for different λ's, the modeling error $\varepsilon_q(\lambda)$ associated with each run can directly be estimated on the basis of $\mathbf{c}_q(\lambda)$. Once all cross-validation runs are complete, the final modeling error is calculated as $\varepsilon(\lambda) = \left[\varepsilon_1(\lambda) + \varepsilon_2(\lambda) + \cdots + \varepsilon_Q(\lambda)\right]/Q$, and the optimal λ is determined by minimizing $\varepsilon(\lambda)$ [27].

The computational cost of LAR is similar to that of applying the interior-point method to solve a single convex optimization in (2.24) with a fixed λ value. Therefore, compared to the simple L_1-norm regularization approach, LAR typically achieves orders of magnitude more efficiency, as is demonstrated in [38].

2.4.5 Numerical experiments

In this subsection, we compare the efficacy of several aforementioned RSM methods by using a two-stage operational amplifier (OpAmp) designed in a commercial 65 nm process [26]. We aim to build linear performance models for 4 PoIs, i.e., gain, bandwidth, power and offset, considering 630 independent random variables to capture both inter-die and intra-die variations of metal oxide semiconductor (MOS) transistors and layout parasitics.

Figure 2.4 shows the relative modeling error for three different modeling techniques (i.e., LS fitting, OMP, and LAR) as a function of the number of training samples. To achieve the same accuracy, OMP and LAR require much fewer training samples than LS, because they solve the unknown model coefficients from an underdetermined equation by exploiting the underlying sparsity of model coefficients. In this example, such a sparse structure exists, since the variability of each circuit-level performance metric is dominated by a few device-level variation sources. For instance, the offset of the OpAmp is mainly determined by the device mismatches of the input differential pair.

Studying Figure 2.4, we find that OMP and LAR yield different modeling accuracy, given the same number of training samples. Even though both OMP and LAR build sparse performance models, they rely on different algorithms to select the important basis functions and/or determine the model coefficients. Compared to LAR, OMP shows slightly improved modeling accuracy in most cases. However, there are a few examples where LAR outperforms OMP, as shown in Figure 2.4(b).

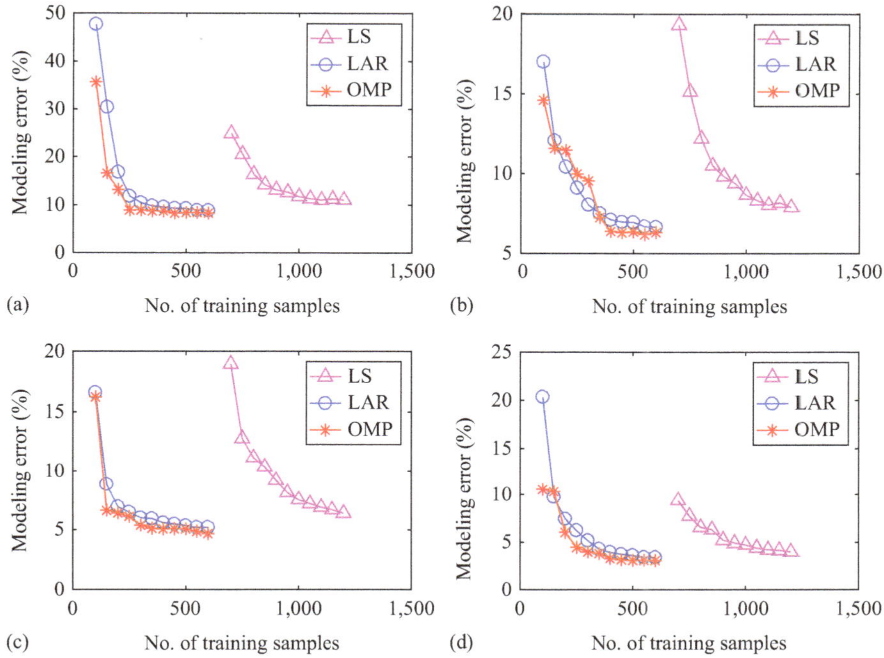

Figure 2.4 Performance modeling errors of different RSM approaches are shown for four performance metrics with different numbers of training samples: (a) gain, (b) bandwidth, (c) power, and (d) offset [26]

To the best of our knowledge, there is no theoretical evidence to prove that one method is always better than the other.

2.5 Regularization

In the literature, the general objective of regularization is to avoid over-fitting when using RSM to approximate a given PoI. Over-fitting usually occurs due to the limited number of available training data (similar to the case discussed in Section 2.4), the unavoidable measurement noise and/or a highly complicated model template (e.g., a highly nonlinear model template with a large number of unknown model coefficients). In these cases, by carefully tuning the model coefficients, we may optimize the model to exactly match the training data; however, the resulting model is not able to accurately predict the data points outside the training set. Whenever over-fitting occurs, the magnitude of the obtained model coefficients often becomes extremely large [24].

To intuitively understand the aforementioned over-fitting issue, we take a simple curve-fitting example for illustration purpose [24]. Suppose that we have obtained a set of training samples $\{(x_n, y_n^S); n = 1, 2, \cdots, 11\}$, where the dimension of VoIs (i.e., N) is equal to 1 and the number of training samples (i.e., N_S) is equal to 11 as shown in Figure 2.5. These samples are randomly generated from a sinusoid function $y = \sin(2\pi x)$ within the region $x \in [0\ 1]$ by adding a small Gaussian noise. In this example, the modeling objective is to approximate the sinusoid function $y = \sin(2\pi x)$ from these 11 training samples by using a set of monomial basis functions $\{b_m(x) = x^{m-1};\ m = 1, 2, \cdots, M\}$.

By setting M to 4 and 11, respectively, we can calculate the corresponding coefficients $\mathbf{c}$ based on (2.8) and, next, obtain the predictions over the entire region $x \in [0\ 1]$ shown as the red curves in Figure 2.5(b) and (c). Table 2.1 summarizes the estimated values of model coefficients for different M. According to Figure 2.5(b) and (c) and Table 2.1, we can find that when more basis functions are adopted, the magnitude of model coefficients gets larger. When $M = 11$, the fitted model exactly matches each training sample with extremely large coefficients; however, it exhibits large-scale oscillations due to over-fitting as shown in Figure 2.5(c). In this example, the model template with $M = 11$ is prohibitively complicated given $N_S = 11$ training samples and the model coefficients are over-fitted to incorrectly capture the random noise [24].

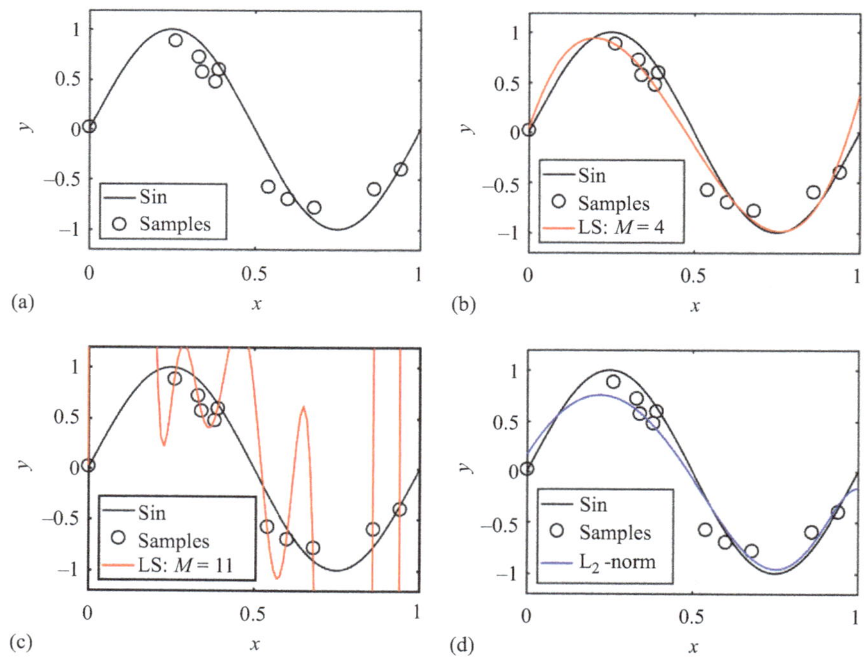

Figure 2.5 The modeling results for a sinusoid function $y = \sin(2\pi x)$ in (a) by using different methods and different numbers of basis functions (i.e., M), (b) LS regression with $M = 4$, (c) LS regression with $M = 11$, and (d) L_2-norm regularization with $M = 11$

Table 2.1 The estimated model coefficients to approximate the sinusoid function $y = \sin(2\pi x)$ by using different methods

	LS: $M = 4$	LS: $M = 10$	L_2-norm regularization
c_1	0.038	0.022	0.164
c_2	9.924	1.683×10^4	5.237
c_3	−31.447	-3.619×10^5	−10.871
c_4	21.858	3.348×10^6	−4.162
c_5		-1.750×10^7	2.277
c_6		5.694×10^7	4.958
c_7		-1.198×10^8	4.756
c_8		1.630×10^8	2.982
c_9		-1.385×10^8	0.579
c_{10}		6.673×10^7	−1.902
c_{11}		-1.391×10^7	−4.180

To prevent the model coefficients from reaching large values and, hence, avoid over-fitting, we can introduce prior knowledge (e.g., an appropriate prior distribution) for the model coefficients **c**. For simplicity, let us first assume that all coefficients are independent and follow a zero-mean Gaussian distribution:

$$pdf(\mathbf{c}) = \prod_{m=1}^{M} \frac{1}{\sqrt{2\pi} \cdot \sigma_G} \cdot \exp\left(-\frac{c_m^2}{2 \cdot \sigma_G^2}\right), \tag{2.34}$$

where all elements in **c** share the same standard deviation σ_G. This standard deviation controls the prior distribution of all model coefficients. As shown in Figure 2.6, if σ_G is small (e.g., $\sigma_G = 0.1$), almost all model coefficients are constrained to small values (i.e., close to zero). On the other side, if σ_G is large (e.g., $\sigma_G = 1$), there may exist several coefficients with large magnitude (i.e., located far away from zero).

According to Bayes' theorem, given a set of training data $\mathbf{y}^S$, the posterior distribution of **c** is proportional to the product of its prior distribution and the likelihood function:

$$pdf\left(\mathbf{c}|\mathbf{y}^S\right) \propto pdf(\mathbf{c}) \cdot pdf\left(\mathbf{y}^S|\mathbf{c}\right). \tag{2.35}$$

Next, the model coefficients **c** can be estimated by applying the maximum a posterior (MAP) method, i.e., maximizing the posterior PDF $pdf(\mathbf{c}|\mathbf{y}^S)$ in (2.35). Taking the natural logarithm of (2.35) and combining the prior PDF in (2.34) and the likelihood in (2.12), maximizing the posterior $pdf(\mathbf{c}|\mathbf{y}^S)$ is equivalent to maximizing:

$$\begin{aligned}\ln\left[pdf(\mathbf{c}) \cdot pdf\left(\mathbf{y}^S|\mathbf{c}\right)\right] = &-\frac{M + N_S}{2} \ln(2\pi) - N_S \ln(\sigma_\varepsilon) - M \ln(\sigma_G) \\ &- \frac{1}{2 \cdot \sigma_\varepsilon^2} \cdot \left\|\mathbf{B} \cdot \mathbf{c} - \mathbf{y}^S\right\|_2^2 - \frac{1}{2 \cdot \sigma_G^2} \cdot \left\|\mathbf{c}\right\|_2^2.\end{aligned} \tag{2.36}$$

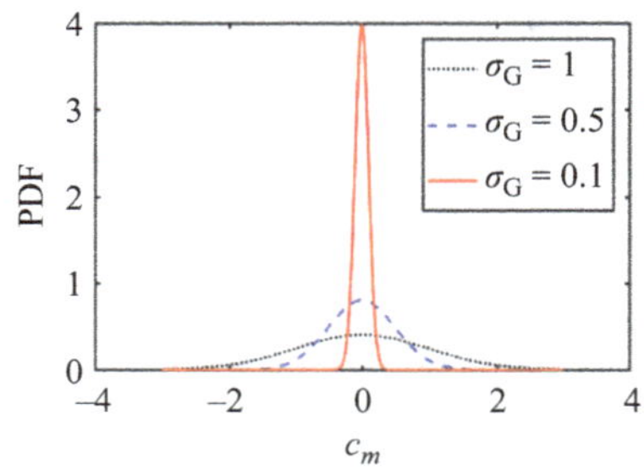

Figure 2.6 The zero-mean Gaussian PDF for c_m with different σ_G

At the right-hand side of (2.36), only the last two terms depend on **c**. Therefore, by defining a regularization parameter:

$$\rho = \frac{\sigma_\varepsilon^2}{\sigma_G^2}, \tag{2.37}$$

maximizing (2.36) can be rewritten as minimizing the error function $E(\mathbf{c})$ plus a regularization term over **c**:

$$\min_{\mathbf{c}} \quad \|\mathbf{B}\cdot\mathbf{c} - \mathbf{y}^S\|_2^2 + \rho\cdot\|\mathbf{c}\|_2^2 = E(\mathbf{c}) + \rho\cdot\|\mathbf{c}\|_2^2, \tag{2.38}$$

where the hyper-parameter ρ is used to balance the trade-off between $E(\mathbf{c})$ and the regularization term $\rho\cdot\|\mathbf{c}\|^2$. By using cross-validation to find an appropriate value of ρ, we can approach the optimal solution with minimum modeling error while avoiding over-fitting. Based on the theory of Lagrange multipliers [24], the optimization problem in (2.38) is equivalent to the following constrained optimization under Karush-Kuhn-Tucher (KKT) conditions:

$$\begin{aligned} &\min_{\mathbf{c}} \quad \|\mathbf{B}\cdot\mathbf{c} - \mathbf{y}^S\|_2^2 \\ &s.t. \quad \|\mathbf{c}\|_2^2 \le \lambda. \end{aligned} \tag{2.39}$$

The aforementioned discussions demonstrate that we can adopt L_2-norm regularization in (2.38) or (2.39) to avoid over-fitting by introducing a prior Gaussian distribution in (2.34).

Let us return to the curve fitting example in Figure 2.5. By solving L_2-norm regularization in (2.38) with $\rho = 0.001$ and $M = 11$ (i.e., 11 basis functions), we can solve the optimal coefficients shown in Table 2.1 and the fitted model is shown as the blue curve in Figure 2.5(d). Compared to the LS solution with $M = 11$, the magnitudes of our model coefficients calculated by L_2-norm regularization are much smaller and, as a result, over-fitting is avoided.

Generally speaking, by introducing different prior knowledge (i.e., different prior distributions) over the model coefficients, we can derive different regularization methods. For instance, if we assume that the elements in **c** are correlated with Gaussian variables, the resulting regularization term must contain their covariance matrix [30]. Alternatively, if we assume that all model coefficients are independent and follow the same Laplace distribution, RSM will be cast to the L_1-norm regularization problem discussed in Section 2.4.2.

2.6 Bayesian model fusion

In this section, we discuss the mathematical formulation of BMF and its applications in detail. Unlike the conventional RSM approaches (e.g., LS regression and regularization methods described in the previous sections) that fit the performance model based on the simulation or measurement data at a single stage only (e.g., post-layout simulation data), BMF attempts to identify the underlying pattern of the unknown model coefficients by reusing the early-stage data (e.g., schematic-level simulation data) in order to efficiently fit a late-stage (e.g., post-layout) performance model. By fusing the early-stage and late-stage data together through Bayesian inference, the simulation and/or measurement cost can be significantly reduced. In particular, BMF consists of the following two major steps: (i) statistically defining the prior knowledge for the unknown model coefficients based on the early-stage simulation data and (ii) optimally determining the late-stage performance model by combining the prior knowledge and very few late-stage samples [8,25,30].

To intuitively explain the key idea of BMF, let us consider two different performance models: the early-stage model $y_E(\mathbf{x})$ and the late-stage model $y_L(\mathbf{x})$:

$$y_E(\mathbf{x}) = \sum_{m=1}^{M} c_{E,m} b_m(\mathbf{x}), \tag{2.40}$$

$$y_L(\mathbf{x}) = \sum_{m=1}^{M} c_{L,m} b_m(\mathbf{x}), \tag{2.41}$$

where $\{c_{E,m}; m = 1, 2, \cdots, M\}$ and $\{c_{L,m}; m = 1, 2, \cdots, M\}$ represent the early-stage and late-stage model coefficients, respectively. For simplicity, we assume that the early-stage model $y_E(\mathbf{x})$ and the late-stage model $y_L(\mathbf{x})$ share the same basis functions. Other special cases where $y_L(\mathbf{x})$ contains additional basis functions or variables that are not found from $y_E(\mathbf{x})$ can also be appropriately handled [25,30].

The early-stage model $y_E(\mathbf{x})$ is fitted from the early-stage simulation data. In practice, the early-stage simulation data are collected to validate the early-stage design, before we move to the next stage. For this reason, we should already know the early-stage model $y_E(\mathbf{x})$ before fitting the late-stage model $y_L(\mathbf{x})$. Namely, we can assume that the early-stage model coefficients $\{c_{E,m}; m = 1, 2, \cdots, M\}$ are provided as the input to the BMF method for late-stage performance modeling.

Given the early-stage model $y_E(\mathbf{x})$, we have to extract the prior knowledge that can be used to facilitate efficient late-stage modeling. To this end, BMF attempts to learn the underlying pattern of the late-stage model coefficients $\{c_{L,m}; m = 1, 2, \cdots, M\}$ based on the early-stage model coefficients $\{c_{E,m}; m = 1, 2, \cdots, M\}$.

Remember that both the early-stage and late-stage models are fitted for the same PoI of the same circuit. Their model coefficients should be similar. We statistically represent such prior knowledge as a prior PDF [24] for the late-stage model coefficients $\{c_{L,m};\ m = 1, 2, \cdots, M\}$. In particular, we consider two different cases to define the prior distribution: zero-mean prior distribution and nonzero-mean prior distribution [25].

2.6.1 Zero-mean prior distribution

If the early-stage model coefficient $c_{E,m}$ has a large (or small) magnitude, it is likely that the late-stage model coefficient $c_{L,m}$ also has the same. Such prior knowledge can be mathematically encoded as a zero-mean Gaussian distribution:

$$c_{L,m} \sim N\left(0, \sigma_m^2\right), \tag{2.42}$$

where $m \in \{1, 2, \cdots, M\}$, and the standard deviation σ_m is a parameter that encodes the magnitude information of the model coefficient $c_{L,m}$. If the standard deviation σ_m is small, the prior distribution $pdf(c_{L,m})$ is narrowly peaked around zero, implying that the coefficient $c_{L,m}$ is possibly close to zero. Otherwise, if σ_m is large, the prior distribution $pdf(c_{L,m})$ widely spreads over a large range and the coefficient $c_{L,m}$ can possibly take a value that is far away from zero [25,30]. Different from (2.34) where we assume that all model coefficients share the same standard deviation σ_G, in (2.42), the values of σ_m for different model coefficients (i.e., different values of m) are different.

Given (2.42), we need to appropriately determine the standard deviation σ_m to fully specify the prior distribution $pdf(c_{L,m})$. The value of σ_m should be optimized so that the probability distribution $pdf(c_{L,m})$ correctly represents the prior knowledge. In other words, by appropriately choosing the value of σ_m, the prior distribution $pdf(c_{L,m})$ should take a large value (i.e., a high probability) at the location where the actual late-stage model coefficient $c_{L,m}$ occurs. However, we only know the early-stage model coefficient $c_{E,m}$, instead of the late-stage model coefficient $c_{L,m}$, at this moment. Remember that $c_{E,m}$ and $c_{L,m}$ are expected to be similar. Hence, the prior distribution $pdf(c_{L,m})$ should also take a large value at $c_{L,m} = c_{E,m}$. Based on this criterion, the optimal prior distribution $pdf(c_{L,m})$ can be found by maximizing the probability for $c_{E,m}$ to occur:

$$\max_{\sigma_m} \quad pdf\left(c_{L,m} = c_{E,m}\right). \tag{2.43}$$

Namely, given the early-stage model coefficient $c_{E,m}$, the optimal standard deviation σ_m is determined by solving the MLE in (2.43).

To solve σ_m from (2.43), we consider the following first-order optimality condition [25,30]:

$$\frac{d}{d\sigma_m} pdf\left(c_{L,m} = c_{E,m}\right) = \frac{1}{\sqrt{2\pi} \cdot \sigma_m} \cdot \exp\left(-\frac{c_{E,m}^2}{2 \cdot \sigma_m^2}\right) \cdot \left(\frac{c_{E,m}^2}{\sigma_m^3} - \frac{1}{\sigma_m}\right) = 0. \tag{2.44}$$

Hence, the optimal value of σ_m is equal to:

$$\sigma_m = \left|c_{E,m}\right|. \tag{2.45}$$

Equation (2.45) reveals an important fact that the optimal standard deviation σ_m is simply equal to the absolute value of the early-stage model coefficient $|c_{E,m}|$. This observation is consistent with our intuition. Namely, if the early-stage model coefficient $c_{E,m}$ has a large (or small) magnitude, the late-stage model coefficient $c_{L,m}$ should also have the same and, hence, the standard deviation σ_m should be large (or small).

To complete the definition of the prior distribution for all late-stage model coefficients $\{c_{L,m};\ m = 1, 2, \cdots, M\}$, we further assume that these coefficients are statistically independent and, hence, their joint distribution is:

$$pdf(\mathbf{c}_L) = \frac{1}{\left(\sqrt{2\pi}\right)^M \cdot \prod_{m=1}^{M} \sigma_m} \cdot \exp\left(-\sum_{m=1}^{M} \frac{c_{L,m}^2}{2 \cdot \sigma_m^2}\right), \tag{2.46}$$

where $\mathbf{c}_L = \left[c_{L,1}\ c_{L,2}\ \cdots\ c_{L,M}\right]^T$ contains all late-stage model coefficients. The independence assumption in (2.46) simply implies that we do not know the correlation information among these coefficients as our prior knowledge. The correlation information will be learned from the late-stage simulation data when the posterior distribution is calculated by MAP [25,30].

Given a set of N_L late-stage simulation or measurement samples $\{(\mathbf{x}_n, y_n^S);\ n = 1, 2, \cdots, N_L\}$, the objective of MAP is to calculate the model coefficients $\mathbf{c}_L$ by maximizing the posterior distribution $pdf(\mathbf{c}_L|\mathbf{y}_L^S)$, where $\mathbf{y}_L^S = \left[y_{L,1}^S\ y_{L,2}^S\ \cdots\ y_{L,N_L}^S\right]^T$. Based on Bayes' theorem, the posterior distribution $pdf(\mathbf{c}_L|\mathbf{y}_L^S)$ is proportional to the product of the prior distribution $pdf(\mathbf{c}_L)$ in (2.46) and the likelihood function $pdf(\mathbf{y}_L^S \mid \mathbf{c}_L)$:

$$pdf\left(\mathbf{c}_L\middle|\mathbf{y}_L^S\right) \propto pdf(\mathbf{c}_L) \cdot pdf\left(\mathbf{y}_L^S\middle|\mathbf{c}_L\right). \tag{2.47}$$

Similar to (2.9) and (2.10), to derive the likelihood function $pdf(\mathbf{y}_L^S \mid \mathbf{c}_L)$, we assume that the error for the late-stage performance modeling follows a zero-mean Gaussian distribution with the variance $\sigma_{\varepsilon_L}^2$:

$$y_L^S = y_L(\mathbf{x}) + \varepsilon_L, \tag{2.48}$$

$$\varepsilon_L \sim N\left(0, \sigma_{\varepsilon_L}^2\right), \tag{2.49}$$

where the value of σ_{ε_L} can be optimally determined by cross-validation. As a result, the likelihood function $pdf(\mathbf{y}_L^S \mid \mathbf{c}_L)$ is a multivariate Gaussian distribution [25].

Combining (2.46), (2.47), and (2.48), it is straightforward to prove that the posterior distribution $pdf(\mathbf{c}_L|\mathbf{y}_L^S)$ is Gaussian [25,30]:

$$\mathbf{c}_L\middle|\mathbf{y}_L^S \sim N\left(\boldsymbol{\mu}_L^0, \boldsymbol{\Sigma}_L^0\right), \tag{2.50}$$

where:

$$\mu_L^0 = \sigma_{\varepsilon_L}^{-2} \cdot \Sigma_L^0 \cdot \mathbf{B}_L^T \cdot \mathbf{y}_L^S, \tag{2.51}$$

$$\Sigma_L^0 = \left[\sigma_{\varepsilon_L}^{-2} \cdot \mathbf{B}_L^T \cdot \mathbf{B}_L + diag\left(\sigma_1^{-2}, \sigma_2^{-2}, \cdots, \sigma_M^{-2}\right)\right]^{-1}, \tag{2.52}$$

$$\mathbf{B}_L = \begin{bmatrix} b_1(\mathbf{x}_1) & b_2(\mathbf{x}_1) & \cdots & b_M(\mathbf{x}_1) \\ b_1(\mathbf{x}_2) & b_2(\mathbf{x}_2) & \cdots & b_M(\mathbf{x}_2) \\ \vdots & \vdots & \vdots & \vdots \\ b_1(\mathbf{x}_{N_L}) & b_2(\mathbf{x}_{N_L}) & \cdots & b_M(\mathbf{x}_{N_L}) \end{bmatrix}, \tag{2.53}$$

and *diag*(•) represents an operator to construct a diagonal matrix. Since the Gaussian PDF $pdf(\mathbf{c}_L|\mathbf{y}_L^S)$ reaches the maximum at its mean value, the MAP solution for the model coefficients $\mathbf{c}_L$ is equal to the mean vector μ_L^0:

$$\mathbf{c}_{L,MAP} = \mu_L^0 = \sigma_{\varepsilon_L}^{-2} \cdot \Sigma_L^0 \cdot \mathbf{B}_L^T \cdot \mathbf{y}_L^S, \tag{2.54}$$

where σ_{ε_L} is a hyper-parameter that can be determined by cross-validation.

2.6.2 Nonzero-mean prior distribution

An alternative prior definition is to construct a nonzero-mean Gaussian distribution for each late-stage model coefficient $c_{L,m}$:

$$c_{L,m} \sim N\left(c_{E,m}, \lambda^2 \cdot c_{E,m}^2\right), \tag{2.55}$$

where λ is a hyper-parameter that can be determined by cross-validation. Figure 2.7 shows a simple example of nonzero-mean prior distribution for two model

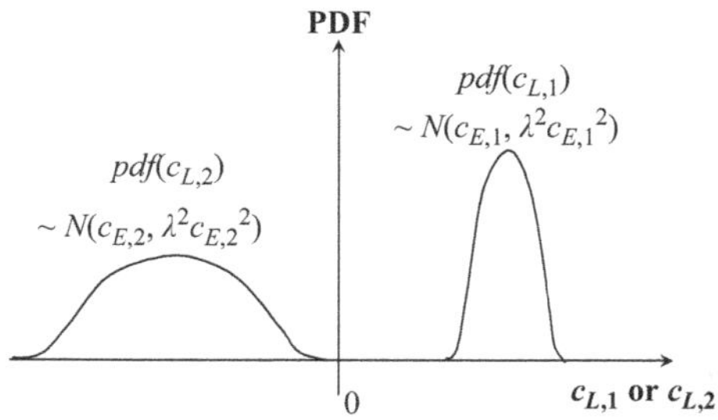

Figure 2.7 A simple example of nonzero-mean prior distribution is shown for two model coefficients $c_{L,1}$ and $c_{L,2}$. The coefficient $c_{L,1}$ possibly takes a small magnitude, since its prior distribution is narrowly peaked around a small value. The coefficient $c_{L,2}$ possibly takes a large magnitude, since its prior distribution widely spreads around a large value [25]

coefficients $c_{L,1}$ and $c_{L,2}$, where the absolute value of $c_{E,1}$ is small and that of $c_{E,2}$ is large [25].

The prior distribution in (2.55) has a 2-fold meaning. First, the Gaussian distribution $pdf(c_{L,m})$ is peaked at its mean value $c_{L,m} = c_{E,m}$, implying that the early-stage coefficient $c_{E,m}$ and the late-stage coefficient $c_{L,m}$ are likely to be similar. In other words, since the Gaussian distribution $pdf(c_{L,m})$ exponentially decays with $(c_{L,m}-c_{E,m})^2$, it is unlikely to observe a late-stage coefficient $c_{L,m}$ that is extremely different from the early-stage coefficient $c_{E,m}$. Second, the standard deviation of the prior distribution $pdf(c_{L,m})$ is proportional to $|c_{E,m}|$. It means that the absolute difference between the late-stage coefficient $c_{L,m}$ and the early-stage coefficient $c_{E,m}$ can be large (or small), if the magnitude of the early-stage coefficient $|c_{E,m}|$ is large (or small). Restating in words, each late-stage coefficient $c_{L,m}$ (where $m \in \{1, 2, \cdots, M\}$) has been provided with a relatively equal opportunity to deviate from the corresponding early-stage coefficient $c_{E,m}$ [25]. Similar to (2.46), we again assume that all late-stage model coefficients $\{c_{L,m};\ m = 1, 2, \cdots, M\}$ are statistically independent and their joint distribution is:

$$
\begin{aligned}
pdf(\mathbf{c}_L) &= \prod_{m=1}^{M} pdf(c_{L,m}) \\
&= \frac{1}{\left(\sqrt{2\pi}\cdot\lambda\right)^M \cdot \prod_{m=1}^{M} |c_{E,m}|} \cdot \exp\left(-\sum_{m=1}^{M} \frac{\left(c_{L,m} - c_{E,m}\right)^2}{2\cdot\lambda^2\cdot c_{E,m}^2}\right).
\end{aligned} \tag{2.56}
$$

To derive the posterior distribution of $\mathbf{c}_L$, we assume that the error for the late-stage performance model follows the zero-mean Gaussian distribution in (2.48). Next, combining (2.47), (2.48), and (2.56), we find that the posterior distribution $pdf(\mathbf{c}_L|\mathbf{y}_L^S)$ is Gaussian [25]:

$$
\mathbf{y}_L^S | \mathbf{c}_L \sim N\left(\boldsymbol{\mu}_L^{no0}, \boldsymbol{\Sigma}_L^{no0}\right), \tag{2.57}
$$

where

$$
\boldsymbol{\mu}_L^{no0} = \boldsymbol{\Sigma}_L^{no0} \cdot \left[\eta \cdot diag\left(c_{E,1}^{-2}, c_{E,2}^{-2}, \cdots, c_{E,M}^{-2}\right) \cdot \mathbf{c}_E + \mathbf{B}_L^T \cdot \mathbf{y}_L^S\right], \tag{2.58}
$$

$$
\boldsymbol{\Sigma}_L^{no0} = \left[\eta \cdot diag\left(c_{E,1}^{-2}, c_{E,2}^{-2}, \cdots, c_{E,M}^{-2}\right) + \mathbf{B}_L^T \cdot \mathbf{B}_L\right]^{-1}, \tag{2.59}
$$

$$
\mathbf{c}_E = \left[c_{E,1}\ \ c_{E,2}\ \ \cdots\ \ c_{E,M}\right]^T, \text{ and}
$$

$$
\eta = \frac{\sigma_{\varepsilon_L}^2}{\lambda^2}. \tag{2.60}
$$

Similar to (2.54), the MAP solution for $\mathbf{c}_L$ is equal to the mean vector $\boldsymbol{\mu}_L^{no0}$:

$$
\mathbf{c}_{L,MAP} = \boldsymbol{\mu}_L^{no0} = \boldsymbol{\Sigma}_L^{no0} \cdot \left[\eta \cdot diag\left(c_{E,1}^{-2}, c_{E,2}^{-2}, \cdots, c_{E,M}^{-2}\right) \cdot \mathbf{c}_E + \mathbf{B}_L^T \cdot \mathbf{y}_L^S\right]. \tag{2.61}
$$

Studying (2.61) reveals an important observation that we only need to determine η, instead of the individual parameters σ_{ε_L} and λ, in order to find the MAP solution $\mathbf{c}_{L,MAP}$. Similar to the case of zero-mean prior distribution, the hyper-parameter η can be optimally determined by using the cross-validation technique discussed in Section 2.4.3.

For a given performance modeling problem by using BMF, it is important to determine whether a nonzero-mean or zero-mean prior distribution is preferred. Intuitively, a nonzero-mean prior distribution provides stronger prior knowledge than a zero-mean prior distribution. The nonzero-mean prior distribution encodes both the sign and the magnitude information about the late-stage model coefficients, while the zero-mean prior distribution encodes the magnitude information only. From this point of view, a nonzero-mean prior distribution is preferred, if the early-stage and late-stage model coefficients are extremely close and, hence, the prior knowledge learned from the early-stage model coefficients is highly accurate. On the other hand, if the early-stage and late-stage model coefficients are substantially different, we should not pose an overly strong prior distribution and, hence, a zero-mean prior distribution is preferred in this case [25,30].

To solve a broad range of practical problems and further reduce the modeling cost for large-scale AMS circuits, various efficient methods based on the BMF scheme have been developed during the past several years. For instance, to reduce the number of required physical samples, co-learning BMF [39] trains an extremely low-complexity model to generate pseudo samples for fitting a high-complexity model with high accuracy. To further reduce the late-stage modeling (e.g., the post-silicon performance modeling) cost by taking into account multiple prior knowledge (e.g., the pre-silicon simulated results and the post-silicon measurement data), dual-prior BMF [40] is derived from the Bayesian inference that can be represented as a graphical model [24]. To facilitate the performance modeling of tunable AMS circuits, correlated BMF [41] encodes the correlation information for both the model template and the coefficient magnitude among different knob configurations by using a unified prior distribution. The details for these extended versions of BMF can be found from the literature [39–41].

2.6.3 Numerical experiments

In this subsection, we attempt to demonstrate the efficacy of BMF by using a ring oscillator example [30] with 7,177 independent random variables to model device-level process variations, including both inter-die variations and random mismatches at the post-layout stage. The schematic stage is considered the early stage and the post-layout stage is considered the late stage. Our objective is to approximate three post-layout performance metrics (i.e., power, phase noise and frequency) as linear functions of these random variables.

For testing and comparison purposes, three different performance modeling techniques are implemented: (i) OMP, (ii) the BMF method with zero-mean prior distribution (BMF-ZM), and (iii) the BMF method with nonzero-mean prior distribution (BMF-NZM). The OMP algorithm does not consider any prior

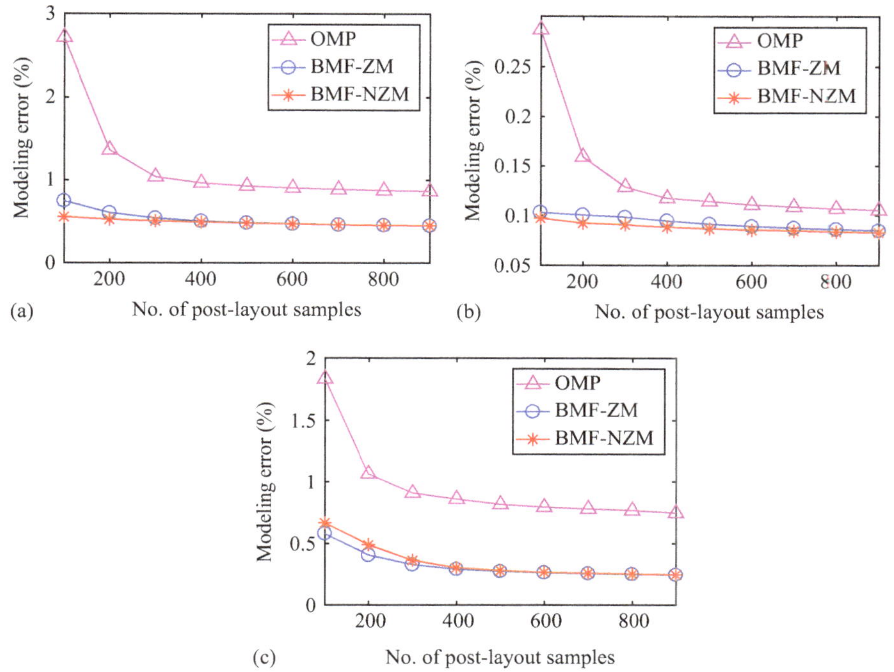

Figure 2.8 Performance modeling errors for three PoIs of different methods with different numbers of training samples: (a) power, (b) phase noise, and (c) frequency

information from the schematic stage. When applying BMF-ZM or BMF-NZM, we use the schematic-level performance model to define our prior knowledge for post-layout performance modeling.

Figure 2.8 summarizes the relative modeling error as a function of the number of post-layout training samples. Studying Figure 2.8 reveals two important observations. First, the modeling error decreases as the number of simulation samples increases. Given the same number of samples, both BMF-ZM and BMF-NZM achieve significantly higher accuracy than OMP. Second, BMF-ZM is less accurate than BMF-NZM for power but is more accurate than BMF-NZM for frequency. In other words, the optimal prior distribution can vary from case to case in practice. Since the overall modeling cost is dominated by post-layout transistor-level simulations, both BMF methods achieve 9× runtime speed-up over OMP with superior accuracy in this example.

2.7 Summary

In this chapter, we discuss several state-of-the-art RSM methods for performance modeling of analog and AMS circuits. RSM aims to approximate a given PoI by the linear combination of a set of basis functions. If the number of training samples is much larger than the number of adopted basis functions, the model coefficients can be accurately estimated by using LS regression. To reduce the number of required training samples and, hence, the modeling cost, we can explore the sparsity of model coefficients and, next, cast performance modeling to an L_0-norm regularization problem. Both OMP and L_1-norm regularization can be used to efficiently approximate the sparse solution of L_0-norm regularization. Alternatively, based on the observation that today's AMS circuits are often designed via a multistage flow, BMF attempts to reduce the modeling cost by fusing the early-stage and late-stage data together through Bayesian inference. As an important aspect of future research, a number of recently developed machine learning techniques, such as deep learning, maybe further adopted for RSM for AMS applications.

Acknowledgments

This work is supported partly by National Natural Science Foundation of China (NSFC) research projects 61874032, 61574046, and 61774045, partly by the project 2018MS005 from the State Key Laboratory of ASIC and System at Fudan University.

References

[1] G. Gielen and R. Rutenbar, "Computer-aided design of analog and mixed-signal integrated circuits," *Proceedings of the IEEE*, vol. 88, no. 18, pp. 1825–1854, 2000.

[2] R. Rutenbar, G. Gielen and J. Roychowdhury, "Hierarchical modeling, optimization, and synthesis for system-level analog and RF designs," *Proceedings of the IEEE*, vol. 95, no. 3, pp. 640–669, 2007.

[3] X. Li, J. Le and L. Pileggi, *Statistical Performance Modeling and Optimization*, Boston, USA: Now Publishers, 2007.

[4] A. Singhee and R. Rutenbar, "Beyond low-order statistical response surfaces: latent variable regression for efficient, highly nonlinear fitting," *Design Automation Conference (DAC)*, San Diego, CA, USA, pp. 256–261. 2007.

[5] X. Li, J. Le, L. Pileggi and A. Strojwas, "Projection-based performance modeling for inter/intra-die variations," *International Conference on Computer-Aided Design (ICCAD)*, San Jose, CA, USA, pp. 721–727, 2005.

[6] Z. Feng and P. Li, "Performance-oriented statistical parameter reduction of parameterized systems via reduced rank regression," *International Conference on Computer-Aided Design (ICCAD)*, San Jose, CA, USA, pp. 868–875, 2006.

[7] A. Mitev, M. Marefat, D. Ma and J. Wang, "Principle Hessian direction based parameter reduction for interconnect networks with process variation," *International Conference on Computer-Aided Design (ICCAD)*, San Jose, CA, USA, pp. 632–637, 2007.

[8] X. Li, F. Wang, S. Sun and C. Gu, "Bayesian model fusion: a statistical framework for efficient pre-silicon validation and post-silicon tuning of complex analog and mixed-signal circuits," *International Conference on Computer-Aided Design* (*ICCAD*), San Jose, CA, USA, pp. 795–802, 2013.

[9] J. Plouchart, F. Wang, X. Li, *et al*., "Adaptive circuit design methodology and test applied to millimeter-wave circuits," *IEEE Design & Test*, vol. 31, no. 6, pp. 8–18, 2014.

[10] W. Daems, G. Gielen and W. Sansen, "An efficient optimization-based technique to generate posynomial performance models for analog integrated circuits," *Design Automation Conference* (*DAC*), New Orleans, Louisiana, USA, pp. 431–436, 2002.

[11] X. Li, P. Gopalakrishnan, Y. Xu and L. Pileggi, "Robust analog/RF circuit design with projection-based posynomial modeling," *International Conference on Computer-Aided Design* (*ICCAD*), San Jose, CA, USA, pp. 855–862, 2004.

[12] M. Hershenson, S. Boyd and T. Lee, "Optimal design of a CMOS op-amp via geometric programming," *IEEE Transactions on Computer-Aided Design of Integrated Circuits and Systems* (*TCAD*), vol. 20, no. 1, pp. 1–21, 2001.

[13] Y. Wang, M. Orshansky and C. Caramanis, "Enabling efficient analog synthesis by coupling sparse regression and polynomial optimization," *Design Automation Conference* (*DAC*), San Francisco, CA, USA, 2014.

[14] F. Wang, S. Yin, M. Jun, *et al*., "Re-thinking polynomial optimization: efficient programming of reconfigurable radio frequency (RF) systems by convexification," *IEEE Asia and South Pacific Design Automation Conference* (*ASPDAC*), Macau, China, pp. 545–550, 2016.

[15] Y. Wang, C. Caramanis and M. Orshansky, "PolyGP: improving GP-based analog optimization through accurate high-order monomials and semi-definite relaxation," *IEEE Design, Automation & Test in Europe Conference* (*DATE*), Dresden, Germany, pp. 1423–1428, 2016.

[16] J. Tao, Y. Su, D. Zhou, X. Zeng and X. Li, "Graph-constrained sparse performance modeling for analog circuit optimization via SDP relaxation," *IEEE Transactions on Computer-Aided Design of Integrated Circuits and Systems* (*TCAD*), vol. 38, no. 8, pp. 1385–1398, 2018.

[17] J. Tao, C. Liao, X. Zeng and X. Li, "Harvesting design knowledge from internet: high-dimensional performance trade-off modeling for large-scale analog circuits," *IEEE Transactions on Computer-Aided Design of Integrated Circuits and Systems* (*TCAD*), vol. 35, no. 1, pp. 23–36, 2016.

[18] M. Sengupta, S. Saxena, L. Daldoss, G. Kramer, S. Minehane and J. Cheng, "Application-specific worst case corners using response surfaces and statistical models," *IEEE Transactions on Computer-Aided Design of Integrated Circuits and Systems* (*TCAD*), vol. 24, no. 9, pp. 1372–1380, 2005.

[19] H. Zhang, T. Chen, M. Ting and X. Li, "Efficient design-specific worst case corner extraction for integrated circuits," *Design Automation Conference* (*DAC*), San Francisco, CA, USA, pp. 386–389, 2009.

[20] S. Nassif, "Modeling and analysis of manufacturing variations," *IEEE Custom Integrated Circuits Conference* (*CICC*), San Diego, CA, USA, pp. 223–228, 2001.

[21] X. Li, J. Le, P. Gopalakrishnan and L. Pileggi, "Asymptotic probability extraction for nonnormal performance distributions," *IEEE Transactions on Computer-Aided Design of Integrated Circuits and Systems* (*TCAD*), vol. 26, no. 1, pp. 16–37, 2007.

[22] A. Dharchoudhury and S. Kang, "Worse-case analysis and optimization of VLSI circuit performance," *IEEE Transactions on Computer-Aided Design of Integrated Circuits and Systems* (*TCAD*), vol. 14, no. 4, pp. 481–492, 1995.

[23] S. Boyd and L. Vandenberghe, *Convex Optimization*, Cambridge, UK: Cambridge University Press, 2004.

[24] C. Bishop, *Pattern Recognition and Machine Learning*. New York, NY, USA: Springer, 2006.

[25] F. Wang, P. Cachecho, W. Zhang, *et al.*, "Bayesian model fusion: large-scale performance modeling of analog and mixed-signal circuits by reusing early-stage data," *IEEE Transactions on Computer-Aided Design of Integrated Circuits and Systems* (*TCAD*), vol. 35, no. 8, pp. 1255–1268, 2016.

[26] X. Li, "Finding deterministic solution from underdetermined equation: large-scale performance modeling of analog/RF circuits," *IEEE Transactions on Computer-Aided Design of Integrated Circuits and Systems* (*TCAD*), vol. 29, no. 11, pp. 1661–1668, 2010.

[27] X. Li, "Finding deterministic solution from underdetermined equation: large-scale performance modeling by least angle regression," *Design Automation Conference* (*DAC*), San Francisco, CA, USA, pp. 364–369, 2009.

[28] X. Li, W. Zhang and F. Wang, "Large-scale statistical performance modeling of analog and mixed-signal circuits," *IEEE Custom Integrated Circuits Conference* (*CICC*), San Jose, CA, USA, 2012.

[29] C. Fang, Q. Huang, F. Yang, X. Zeng, D. Zhou and X. Li, "Efficient performance modeling of integrated circuits via kernel density based sparse regression," *Design Automation Conference* (*DAC*), Austin, TX, USA, 2016.

[30] F. Wang, W. Zhang, S. Sun, X. Li and C. Gu, "Bayesian model fusion: large-scale performance modeling of analog and mixed-signal circuits by reusing early-stage data," *Design Automation Conference* (*DAC*), Austin, TX, USA, 2013.

[31] X. Li and H. Liu, "Statistical regression for efficient high-dimensional modeling of analog and mixed-signal performance variations," *Design Automation Conference* (*DAC*), Anaheim, CA, USA, pp. 38–43, 2008.

[32] C. Liao, J. Tao, H. Yu, *et al.*, "Efficient hybrid performance modeling for analog circuits using hierarchical shrinkage priors," *IEEE Transactions on Computer-Aided Design of Integrated Circuits and Systems* (*TCAD*), vol. 35, no. 12, pp. 2148–2152, 2016.

[33] T. McConaghy and G. Gielen, "Template-free symbolic performance modeling of analog circuits via canonical-form functions and genetic

programming," *IEEE Transactions on Computer-Aided Design of Integrated Circuits and Systems (TCAD)*, vol. 28, no. 8, pp. 1162–1175, 2009.

[34] T. Hastie, R. Tibshirani and J. Friedman, *The Elements of Statistical Learning*, Berlin, Germany: Springer, 2003.

[35] G. Golub and C. van Loan, *Matrix Computations*, Baltimore, MA: Johns Hopkins University Press, 2012.

[36] C. Robert and G. Casella, *Monte Carlo Statistical Methods* Berlin, Germany: Springer, 2005.

[37] J. Tropp and A. Gilbert, "Signal recovery from random measurements via orthogonal matching pursuit," *IEEE Transactions on Information Theory (TIT)*, vol. 53, no. 12, pp. 4655–4666, 2007.

[38] B. Efron, T. Hastie and I. Johnstone, "Least angle regression," *The Annals of Statistics*, vol. 32, no. 2, pp. 407–499, 2004.

[39] F. Wang, M. Zaheer, X. Li, J. Plouchart and A. Valdes-Garcia, "Co-learning Bayesian model fusion: efficient performance modeling of analog and mixed-signal circuits using side information," *IEEE/ACM International Conference on Computer-Aided Design (ICCAD)*, Austin, TX, USA, pp. 575–582, 2015.

[40] Q. Huang, C. Fang, F. Yang, X. Zeng, D. Zhou and X. Li, "Efficient performance modeling via dual-prior Bayesian model fusion for analog and mixed-signal circuits," *IEEE/ACM Design Automation Conference (DAC)*, Austin, TX, USA, 2016.

[41] F. Wang and X. Li, "Correlated Bayesian model fusion: efficient performance modeling of large-scale tunable analog/RF integrated circuits," *Design Automation Conference (DAC)*, Austin, TX, USA, 2016.

Chapter 3
Machine learning

Olcay Taner Yıldız[1]

3.1 Introduction

Today, with the advances in the hardware technologies, it is possible to store, process, and output large amount of data. With the increase in the size of data, explaining it, that is, extracting meaningful information from it, becomes a bottleneck. Machine learning, i.e., the science of extracting useful information from data comes as an aid. Empowered with concepts from mathematics, statistics, and computer science, machine learning is arguably the solution for all of our information extraction problems.

Machine learning is about *curve fitting*. In regression problems, we try to fit a linear/nonlinear function to data. For example, in support vector machine regression, current disposition of the data is not enough and we transform our data by a kernel function to a high-dimensional space and fit the curve in that high-dimensional space [1]. In K-class classification problems, we try to fit a linear/nonlinear function to separate classes. For example, in naïve Bayes, we assume that the data are Gaussian-distributed, and the separating function is easily calculated [2].

Machine learning is about *optimization*. In learning via optimization, we define an error function (loss function) on the model and try to optimize it, i.e., try to find the optimal parameters of the model by optimization techniques borrowed from optimization literature. For example, in multivariate linear regression, the error function is the sum of squared errors, we take the partial derivatives of the error function with respect to the weights, equalize the derivatives to zero to get the linear equations, and solve the equations to extract the weights of the model. Yet in neural-network-based classification, the error function is cross-entropy, we take the partial derivatives of the error function with respect to the weights of the network to get the update equations, and using those update equations we train the network [3].

Machine learning is about *algorithms*. In decision/regression trees, we need to write a recursive-learning algorithm to classify the data arriving at each decision node, and we also need to write a recursive code to generate the decision tree structure [4]. In Bayesian networks, we use graph algorithms to calculate conditional independences between variables, or to calculate marginal probabilities [5,6].

[1]Department of Computer Engineering, Işık University, Istanbul, Turkey

Yet in hidden Markov models, we use Viterbi algorithm, a clever application of dynamic programming, to calculate the most probable state sequence given an observation sequence [7].

Machine learning is about *statistics*. We usually assume Gaussian noise on the data, we assume multivariate normal distribution on class covariance matrices in quadratic discriminant analysis, and we use cross-validation or bootstrapping to generate multiple training sets. We also use hypothesis testing to decide which classifier is better than the other based on their classification performance on several test sets [8,9].

Machine learning is about *models*. In decision trees, the data structure is a binary/L-ary tree depending on the type of the features and the decision function used [10]. In rule learning, the model is an ordered set of rules [11]. In Bayesian networks, hidden Markov models, and neural networks, the most complex model, i.e., graphs are used. In naïve Bayes, linear discriminant analysis (LDA), we use a multivariate polynomial to represent the data.

Machine learning is about *performance metrics*. In classification, we use accuracy if we want to get a crude estimate on the performance of the classifier. If we need more details on pairwise classes, confusion matrix comes in handy. If the dataset has two classes, more metrics follow: precision, recall, true positive rate, false positive rate, F-measure, etc. [12]. In regression, we use mean square error for a general estimate on the performance of the regressor. There are also other performance metrics, such as hinge-loss for specific algorithms, i.e., support vector machines [13].

So what is machine learning about? Like defining an elephant, we need the combination of all of these topics to define machine learning: we start machine learning by assuming a certain *model* on the data, use *algorithmic* and/or *optimization* and/or *statistical* techniques to learn the parameters of that model (which is sometimes defined as *curve fitting*), and use *performance metrics* to evaluate our model/algorithm/classifier.

This chapter is organized as follows: everything begins with data. We will talk about data representation in Section 3.2. Dimension reduction techniques are reviewed in Section 3.3. We discuss basic clustering algorithms in Section 3.4. Classification and regression algorithms are discussed in Section 3.5. Lastly, we conclude with performance assessment of algorithms in Section 3.6.

3.2 Data

In any machine-learning application, we start with data, i.e., measurements/calculations/observations made by the user. The multivariate data are represented by a 2D matrix as:

$$\mathbf{X} = \begin{bmatrix} x_1^{(1)} & x_2^{(1)} & x_3^{(1)} & \dots & x_d^{(1)} & y^{(1)} \\ x_1^{(2)} & x_2^{(2)} & x_3^{(2)} & \dots & x_d^{(2)} & y^{(2)} \\ \dots & & & & & \\ x_1^{(N)} & x_2^{(N)} & x_3^{(N)} & \dots & x_d^{(N)} & y^{(N)} \end{bmatrix}$$

where d represents the number of features (variables, attributes, and inputs) and N represents the number of instances (observations and examples) in the dataset. The instances are assumed independent and identically distributed. Each row in the matrix corresponds to a single instance, whereas each column in the matrix corresponds to a single feature. The variable y represents the output feature and (i) takes discrete values for classification and (ii) continuous values for regression problems. Datasets are usually grouped into three: sparse datasets, where $d \gg N$ (i.e., genome datasets, image datasets); *big datasets*, where $N \gg d$ (i.e., text datasets and geo datasets); and standard datasets.

Each feature can be continuous (an integer, a real number) or discrete (taking L distinct values) depending on the application. As an example, let say we have a dataset about $N = 1{,}000$ students.

$$\mathbf{X} = \begin{bmatrix} 1976 & \text{male} & 2.34 & \text{amber} \\ 1984 & \text{female} & 3.12 & \text{brown} \\ \ldots & & & \\ 2000 & \text{female} & 1.23 & \text{blue} \end{bmatrix}$$

where we have four input features, namely, birth year of the student (continuous), sex of the student (discrete with $L = 2$), GPA of the student (continuous), and eye color of the student (discrete with $L = 4$). See Figure 3.1 for the famous Iris dataset plotted with respect to third and fourth features.

Many machine-learning algorithms cannot process discrete features; nevertheless, discrete features can be converted to continuous features via 1-of-L encoding.

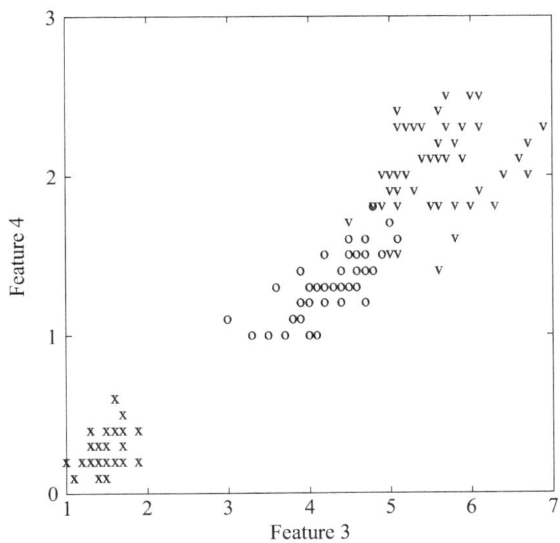

Figure 3.1 Iris dataset

In 1-of-L encoding, each discrete feature is encoded by an extra L feature, where only one of the new features is 1 (corresponding to the value of that feature), and the remaining are 0. As extra features are added, the size of the dataset grows and application of some algorithms may get impractical. For instance, if we convert the previous dataset, we get the following matrix:

$$\mathbf{X} = \begin{bmatrix} 1976 & 1 & 0 & 2.34 & 1 & 0 & 0 & 0 \\ 1984 & 0 & 1 & 3.12 & 0 & 1 & 0 & 0 \\ \cdots \\ 2000 & 0 & 1 & 1.23 & 0 & 0 & 0 & 1 \end{bmatrix}$$

Yet other machine-learning algorithms require abstraction of continuous data, whereby estimation of mean vector:

$$\mathbf{m} = [m_1 m_2 \cdots m_d]$$

and covariance matrix:

$$\mathbf{S} = \begin{bmatrix} s_1^2 & s_{12} & \cdots & s_{1d} \\ s_{21} & s_2^2 & \cdots & s_{2d} \\ \cdots \\ s_{d1} & s_{d2} & \cdots & s_d^2 \end{bmatrix}$$

is required. In the covariance matrix, the diagonal terms are the variances and the off-diagonal terms are the covariances. If the data are normally distributed, we write $\mathbf{X} \sim \mathcal{N}$ $(\mathbf{m}, \mathbf{S})$.

3.3 Dimension reduction

As a preprocessing step before classification or regression, we may introduce a dimension reduction step, whereby we reduce the number of dimensions of the dataset. Some of the reasons for dimension reduction are as follows: (i) in many learning algorithms the complexity of the algorithm depends on d, and if we reduce d, the training complexity of our algorithm decreases; (ii) "Occam's Razor" principle: if we can explain the same data with a *model* having fewer parameters, we must choose that model; and (iii) if some input features can be removed, they are unnecessary, so we reduce the cost of extracting those features.

Dimension reduction algorithms can be grouped into two. *Feature-selection* algorithms, where we select a subset of the original features (k features out of d features) [14,15]; *feature-extraction* algorithms, where we generate a new set of k features which are some linear/nonlinear combination of the original features.

3.3.1 Feature selection

In feature selection, in our case subset selection, our objective is to find the best subset of the original features, where the best is determined by the *performance metric*. Since there are 2^d possible feature subsets, exhaustive search is impossible and we use ourselves into ingenious search algorithms. Depending on the path on the number of selected features k, subset selection algorithms are categorized into three types: forward, backward, and floating, where k increases, decreases, and sometimes increases sometimes decreases, respectively. All three algorithms are actually state space search algorithms, where a state is a subset of features, and search corresponds to generating a new subset from the current best subset.

In forward selection, we start by empty feature subset $F = \emptyset$. At each step, we produce a candidate subset list FL, which includes the subsets obtained by adding one possible feature to F $(FL = \{1\}, \{2\}, \ldots, \{d\})$. The current best subset F is selected from the candidate subset list according to the performance metric. The algorithm continues by selecting a new F at each step and terminates when the performance metric does not improve.

In backward selection, we start by the original feature subset $F = \{1, 2, \ldots, d\}$. At each step, we again produce a candidate subset list, which includes the subsets obtained by removing one possible feature from F $(FL = \{2, 3, \ldots, d\}, \{1, 3, \ldots, d\}, \{1, 2, \ldots, d-1\})$. The current best subset F is selected from the candidate subset list according to the performance metric. The algorithm continues by selecting a new F at each step and terminates when the performance metric does not improve.

Floating selection is the combination of the previous two algorithms. At each step, when producing SL, we take into account of both adding and removing features [16]. Floating selection can start both from empty and original feature subset.

The best feature-selection algorithm depends on the number of features k in the optimal feature subset. If k is near to 0, forward selection is more hopeful to find the optimal subset, whereas if k is near to d, backward can be lucky. Although time complexity of floating selection is more than the other two, it could be very helpful if we do not know k.

3.3.2 Feature extraction

In feature extraction, we generate a totally new set of features from the original feature set. The feature-extraction algorithms map the data to a new space, where the mapping can be linear or nonlinear, and the mapping can be supervised or unsupervised (will use output y or will not). In this chapter, we will talk about principal component analysis and LDA as feature-extraction algorithms, where the aim is to find the projection matrix $\mathbf{W}$ to get the new dataset via the following equation:

$$\mathbf{X} = \mathbf{X}\mathbf{W} \tag{3.1}$$

Principal component analysis (PCA) is an unsupervised linear feature-extraction algorithm, which maximizes the variance throughout the projected features, where each projection vector $\mathbf{w}_i$ is orthogonal to other projection vectors [2]. So, the problem reduces into an *optimization* problem:

$$\begin{aligned} \underset{\mathbf{W}}{\text{maximize}} \quad & \mathbf{W}\mathbf{S}\mathbf{W}^T \\ \text{subject to} \quad & \|\mathbf{w}_i\| = 1, \quad i = 1, \ldots, d, \\ & \mathbf{w}_i^T \mathbf{w}_j = 0, \quad i,j = 1, \ldots, d; \quad i \neq j, \end{aligned}$$

whereby the eigenvectors of the covariance matrix $\mathbf{S}$ are the solution to the optimization problem. The eigenvectors of $\mathbf{S}$ are sorted according to their corresponding eigenvalues λ and the most important k eigenvectors are the ones having the largest k eigenvalues. Figure 3.2(a) shows the application of PCA on *Iris* dataset. We can also control the dimension reduction (value of k) according to the proportion of variance explained.

$$\frac{\sum_{i=1}^{k} \lambda_i}{\sum_{i=1}^{d} \lambda_i}$$

LDA is a supervised linear feature-extraction algorithm which separates instances from different classes as much as possible and brings together instances from the same class as much as possible [2]. Between-class scatter matrix $\mathbf{S}_B = \sum_{i=1}^{K} N_i(\mathbf{m}_i - \mathbf{m})(\mathbf{m}_i - \mathbf{m})^T$ and within-class scatter matrix $\mathbf{S}_W = \sum_{i=1}^{K} \mathbf{S}_i$ of a K-class dataset quantifies the separability between and within classes, respectively. If we write the problem as an *optimization* problem:

$$\begin{aligned} &\underset{\mathbf{W}}{\text{maximize}} && \frac{|\mathbf{W}\mathbf{S}_B\mathbf{W}^T|}{|\mathbf{W}\mathbf{S}_W\mathbf{W}^T|} \\ &\text{subject to} && \|\mathbf{w}_i\| = 1, \quad i = 1, \ldots, d, \\ &&& \mathbf{w}_i^T\mathbf{w}_j = 0, i, j = 1, \ldots, d; i \neq j \end{aligned}$$

then the eigenvectors of the matrix $\mathbf{S}_W^{-1}\ \mathbf{S}_B$ will be the solution to this optimization problem. In LDA, the maximum rank of $\mathbf{S}_B$ is $K-1$, therefore $k \leq K-1$. Figure 3.2(b) shows the application of LDA on *Iris* dataset.

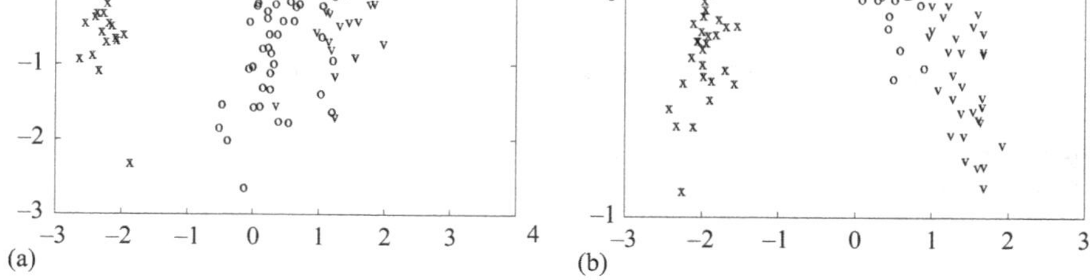

Figure 3.2 Feature extraction on Iris dataset: (a) PCA and (b) LDA

3.4 Clustering

Before delving into the supervised algorithms, we will briefly introduce one of the most important topics in unsupervised machine learning, namely, clustering. In clustering, the aim is to divide $\mathbf{X}$ into a set of groups (clusters, sets), where the groups are disjoint and possibly represent meaningful divisions. After clustering, the groups can be named by experts in the respective domains, which further simplify the descriptive *model* of the data. In this chapter, we will talk about K-means clustering and hierarchical clustering [17,18].

3.4.1 K-Means clustering

In K-means clustering, we are interested in splitting the data $\mathbf{X}$ into K disjoint groups $\mathbf{X}_i$, where the group means $\mathbf{m}_i$ form the representers of the groups. In this problem, there are two sets of unknowns. The first set of unknowns is the elements of the group membership matrix $\mathbf{G}$, where $g_i^{(t)}$ is 1, if the instance $\mathbf{x}^{(t)}$ belongs to the group i, 0 otherwise. The second set of unknowns is group means $\mathbf{m}_i$. The important and interesting thing is, these two sets of unknowns are bound to each other. If we know one set of unknowns, we can optimally determine the other set of unknowns based on the optimization of the reconstruction error.

$$E(\mathbf{m}_i, \mathbf{G}|\mathbf{X}) = \sum_t \sum_i g_i^{(t)} \|\mathbf{x}^{(t)} - \mathbf{m}_i\|^2 \tag{3.2}$$

If we know $\mathbf{G}$, we can calculate group means as:

$$\mathbf{m}_i = \frac{\sum_t g_i^{(t)} \mathbf{x}^{(t)}}{\sum_t g_i^{(t)}} \tag{3.3}$$

On the other hand, if we know $\mathbf{m}_i$, we can extract the group membership of $\mathbf{x}^{(t)}$, because the Euclidean distance of an instance $\mathbf{x}^{(t)}$ to the group means determines the group membership as:

$$g_i^{(t)} = \begin{cases} 1 & \text{if } \|\mathbf{x}^{(t)} - \mathbf{m}_i\| = \min_j \|\mathbf{x}^{(t)} - \mathbf{m}_j\| \\ 0 & \text{otherwise} \end{cases}. \tag{3.4}$$

To solve for these two sets of unknowns, we will use a special version of expectation maximization optimization technique and proceed with an iterative procedure. First, we initialize the mean vectors $\mathbf{m}_i$ randomly. Then, at each iteration, we first use (3.4) to calculate the group membership matrix $\mathbf{G}$. Then, once we have calculated $\mathbf{G}$, we can easily calculate group means using (3.3). These two steps are repeated (calculate $\mathbf{G} \rightarrow$ calculate $\mathbf{m}_i \rightarrow$ calculate $\mathbf{G} \rightarrow$ calculate $\mathbf{m}_i \rightarrow$ etc.) until the group membership matrix $\mathbf{G}$ stabilizes.

The most important disadvantage of this iterative procedure is its initialization. Since the optimization technique is a local search procedure, final group means are heavily dependent on the initial group means. Possible techniques, such as randomly selecting K instances as initial mean vectors, are proposed in the literature to overcome this drawback [2].

K-means clustering can be generalized to soft clustering by:

- Assigning a probability value to $g_i^{(t)}$. Again $\sum_i g_i^{(t)} = 1$, but this way an instance $\mathbf{x}^{(t)}$ will belong to more than one cluster, and the cluster membership decision will be soft (a probability value between 0 and 1) instead of hard (only 0 and 1).
- Assuming the groups that are Gaussian distributed with group covariances $\mathbf{S}_i$, i.e., $\mathbf{X}_i \sim \mathcal{N}(\mathbf{m}_i, \mathbf{S}_i)$.
- Optimizing the log-likelihood instead of the reconstruction error. The expectation maximization procedure appears here again [19,20], and with a similar iterative procedure, we optimize the parameters $\mathbf{m}_i$, $\mathbf{S}_i$, and $\mathbf{G}$.

3.4.2 *Hierarchical clustering*

Instead of clustering instances with respect to the cluster means, we can cluster instances with respect to their similarities to other instances. In other words, instance to instance distance matters, but not the distance of instance to cluster mean. The most distant the instances, the mere dissimilar they are.

Hierarchical clustering is a type of agglomerative clustering, where we start with N clusters (where each instance is actually a cluster) and merge clusters one by one until only one cluster remains. At each iteration, we select the two closest clusters to merge and get a single merged cluster. But how will we decide on the closest clusters, or in other words, how will we compute the distance between clusters, where until now we have only talked about distances between instances.

In single-link clustering, the distance between two clusters is the smallest distance between all possible pairs of instances of the two groups. So, single-link clustering has an optimistic point of view, if there is only a single pair of instances which are very near, then single-link clustering thinks these two clusters as very close. In complete-link clustering, the distance between two clusters is the largest distance between all possible pairs of instances of the two groups. Yet, complete-link clustering has a pessimistic point of view, if there is only a single pair of instances which are very far, then complete-link clustering expects these two clusters as very distant.

In average-link clustering, the distance between two clusters is the average of all distances between all possible pairs of instances of the two groups. In centroid-link clustering, the distance between two clusters is the distance between the cluster means.

3.5 Supervised learning algorithms

In this section, we will briefly review supervised learning algorithms used in machine learning. Supervised algorithms can be categorized in multiple ways. First, supervised algorithms can be parametric vs nonparametric. In the parametric algorithms, the algorithm makes certain assumptions on the underlying distribution of the data (usually called *bias*) and fits the model based on those assumptions.

Nevertheless, in the nonparametric case, the situation is reverse. The algorithm does not make any assumption on the data, and the model is actually the data itself.

Second, supervised algorithms can be discriminative vs generative. In the discriminative algorithms, the algorithm's aim is to find such a discriminant that it will separate the classes as best as it gets. Therefore, only the instances near the discriminant are important. On the other hand, generative algorithms make certain distribution assumptions on the data and use that assumption to produce a discriminator. For this case, since all data are required for the satisfaction of the assumption, all instances are equally important.

Third, supervised algorithms can be classification vs regression. In classification, the output feature **y** takes discrete values. In regression, the output feature **y** takes continuous values. For classification, only counting the mismatches in the output is enough, therefore the loss function is accuracy; whereas for regression, there is actually no mismatch but a real valued absolute discrepancy in the output value, therefore the loss function is squared error.

3.5.1 Simplest algorithm: prior

Before introducing more meaningful supervised algorithms, we must first talk about the simplest algorithm ever, the base case, or the so-called prior. According to Occam's Razor' principle, if we can explain the same data with a *model* having fewer parameters, we must choose that model. Leading that principle, the simplest model about the data is the prior distribution of the data. For classification, prior distribution corresponds to the prior probabilities of the classes, so the prior classifier labels all the test data with the most probable class label; whereas for regression, prior distribution corresponds to the mean value of the output vector **y**; and the prior regressor assigns the mean value to all test data.

If we propose a new classification/regression algorithm on a specific dataset, we must be ensured that it must beat the prior. Note also that according to the *no free lunch theorem* [21], there does not exist any algorithm that can beat every algorithm on all datasets. So, there can be some datasets, where prior is the superior.

3.5.2 A simple but effective algorithm: nearest neighbor

The most commonly used representative of nonparametric algorithms is the nearest neighbor. The assumption of nearest neighbor is simple, the world does not change much, i.e., similar things perform similarly. Therefore, we only need to store the dataset itself and make the decision on the test instance based on the similarity of it to the instances in the dataset. In other words, the class label (regression value) of an instance is strongly influenced by its *nearby* instances.

More formally, in k-nearest neighbor classification, the class-label of a test instance is obtained by taking the majority vote of the k-nearest instances in the training set [22]. Ties are broken arbitrarily, where k is usually taken as an odd number to get rid of ties mostly. In k-nearest neighbor regression, the output value of a test instance is calculated by taking the average of the output values of the k-nearest examples in the training set. Euclidean distance is usually used to calculate the distance between two instances.

In machine-learning literature, nearest neighbor techniques are also called instance-based or memory-based learning algorithms, because they simply store the training set in a table and look up the output values of the nearest k instances [23]. They are also called lazy learning algorithms, since they do not do anything in the training phase of the algorithm, but start processing while doing testing phase. Although the training complexity of nearest neighbor is as negligible as prior, the test complexity is one of the largest in all learning algorithms ($\mathcal{O}(N)$ for a single test instance). In spite of this drawback, due to its simplicity, k-nearest neighbor is one of the widely used and one of the successful learning algorithms in practice. Theoretically, it has been shown that for larger datasets ($N \to \infty$), the risk of 1-nearest neighbor is never worse than the best risk that can be achieved [24].

3.5.3 Parametric methods: five shades of complexity

Starting from the Bayes' rule, we have:

$$p\{C_i|\mathbf{x}) = \frac{p(\mathbf{x}|C_i)p(C_i)}{\sum_{j=1}^{K} p(\mathbf{x}|C_j)p(C_j)} \tag{3.5}$$

where $p(C_i)$, $p(C_i|\mathbf{x})$, and $p(\mathbf{x}|C_i)$ represent the prior probability, the posterior probability, and the class distribution of the class C_i, respectively. As we can see, if all class distributions are equal, choosing the maximum posterior probability reduces to the choosing the maximum prior probability, i.e., *prior algorithm*.

Since the denominator is the same for all classes, we can simplify the posterior probability and get the discriminant function as $f_i(\mathbf{x}) = \log p(\mathbf{x}|C_i) + \log p(C_i)$. If the class distributions are assumed to follow Gaussian density, we have $p(\mathbf{x}|C_i) \sim \mathcal{N}(\mathbf{m}_i, \mathbf{S}_i)$, and the discriminant function reduces to:

$$f_i(\mathbf{x}) = -\frac{d \log 2\pi}{2} - \frac{\log |\mathbf{S}_i|}{2} - \frac{(\mathbf{x}-\mathbf{m}_i)^T \mathbf{S}_i^{-1}(\mathbf{x}-\mathbf{m}_i)}{2} + \log p(C_i) \tag{3.6}$$

The first part is the same for all classes, and if we get rid of it, we obtain our first parametric classifier, namely, *quadratic discriminant*.

$$f_i(\mathbf{x}) = -\frac{\log |\mathbf{S}_i| + \mathbf{x}^T \mathbf{S}_i^{-1}\mathbf{x} - 2\mathbf{x}^T \mathbf{S}_i^{-1}\mathbf{m}_i + \mathbf{m}_i^T \mathbf{S}_i^{-1}\mathbf{m}_i}{2} + \log p(C_i) \tag{3.7}$$

The number of parameters, i.e., the model complexity is $Kd + Kd(d+1)/2$, where the first part is for class means and the second part for class covariance matrices.

For complexity reduction, we can assume a single shared covariance matrix for all classes. In this case, first and second parts of the equation are the same for all classes and simplifying it reduces $f_i(\mathbf{x})$ to our second classifier, namely, *linear discriminant*.

$$f_i(\mathbf{x}) = \mathbf{x}^T \mathbf{S}_i^{-1}\mathbf{m}_i - \frac{\mathbf{m}_i^T \mathbf{S}_i^{-1}\mathbf{m}_i}{2} + \log p(C_i) \tag{3.8}$$

The model complexity is $Kd + d(d+1)/2$, where the first part is for class means and the second part for shared covariance matrix.

For more reduction, we assume all off-diagonals of the shared covariance matrix are zero. For this case, first and second parts of the equation simplify to vector multiplications instead of matrix multiplications, and we get the *naïve Bayes* classifier. The model complexity is $Kd + d$, where the first part is for class means and the second part for the diagonal of the shared covariance matrix.

As the fifth algorithm, we further reduce by taking priors equal and a single covariance value s. In this case, we get the *nearest mean* classifier and the model complexity is only $Kd + 1$.

Depending on the application, we can choose our complexity level (as bias) and report the test error. Yet, another possibility may well be applying all algorithms and make a selection considering both the test error and model complexity [25].

3.5.4 Decision trees

Decision trees are significantly different than the previous models we have discussed. First, they have a tree-based structure where each non-leaf node m implements a decision function, $f_m(\mathbf{x})$, and each leaf node corresponds to a class decision. Second, they are one of most interpretable learning algorithms available. When written as a set of IF-THEN rules, the decision tree can be transformed into a human-readable format, which then can be modified and/or validated by human experts in their corresponding domains.

Decision trees are categorized depending on the type of the decision function. In univariate decision trees [4] (Figure 3.3(a)), the decision function $f_m(\mathbf{x})$ uses only one feature x_i and depending on the type of that feature, each non-leaf node of a univariate decision tree will have two or L children. If x_i is a continuous feature, the decision function is in the form of $x_i < \theta$, and each non-leaf node has two children, where the instances satisfying (not satisfying) the decision function follow the left (right) child, respectively. If x_i is a discrete feature, the decision function takes on

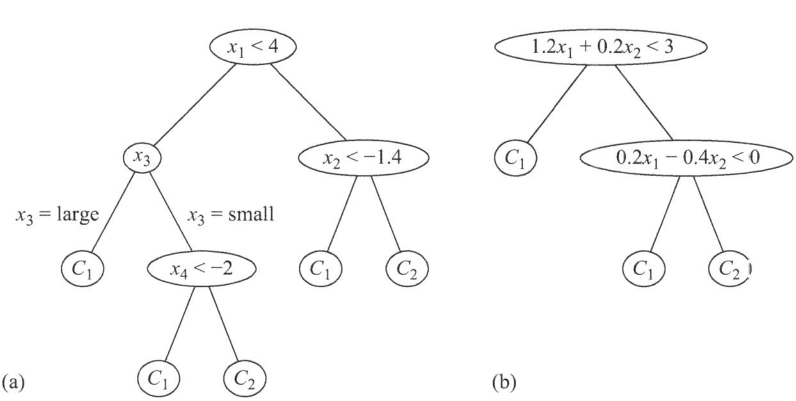

Figure 3.3 Decision trees: (a) univariate and (b) multivariate

the forms $x_i = v_1, \ x_i = v_2, \ \ldots, \ x_i = v_L$, where $v_1, v_2, \ldots, v_L$ correspond to all possible values of the discrete feature x_i.

In multivariate decision trees (Figure 3.3(b)), the decision function $f_m(\mathbf{x})$ uses all features of **x** and each non-leaf node has two children. Depending on the type of the multivariate function, we have (i) multivariate linear tree [26,27], where the decision function is in the form of $\mathbf{w}^T\mathbf{x} < \theta$, (ii) multivariate nonlinear tree [28], where the decision function is a nonlinear function in terms of **x**, and (iii) omnivariate tree [10], where the decision function can be any of the previous three.

Tree induction algorithms are recursive and at each decision node m, starting from the root node (we have the complete training set), we look for the best decision function $f_m(\mathbf{x})$. When the best decision function is found, the training data are split according to $f_m(\mathbf{x})$, and the learning continues recursively with the children of m. We continue splitting until there is no need for the same, where the instances are all from the same class.

For decision trees, the *loss function*, i.e., the criterion for comparing the candidate decision functions is quantified by an impurity measure. Several impurity measures are proposed in the literature such as Gini index [29], entropy.

$$\text{Entropy} = \sum_{i=1}^{K} p(C_i) \log p(C_i) \tag{3.9}$$

where $p(C_i)$ is the probability of class C_i at node m [30].

3.5.5 Kernel machines

Kernel machines, in other words, support vector machines [1,31], are maximum margin methods, where the model is written as a weighted sum of *support vectors*. Kernel machines are discriminative methods, i.e., they are only interested in the instances across the class boundaries in classification, or instances across the regressor in regression. For obtaining the optimal separating hyperplane, kernel machines try to maximize separability, or *margin*, and write the problem as a quadratic optimization problem, whose solution gives us support vectors.

Kernel functions improve our notion of instance model, and not only we can define kernel functions on instances but we can also define on networks, graphs, words, sentences, or trees. This great range of applicability of kernel functions made kernel functions popular across many domains, including natural language processing, bioinformatics [32], and robotics.

3.5.5.1 Separable case: optimal separating hyperplane

For a linearly separable two-class problem (see Figure 3.4(a)), we define the output $y^{(t)}$ as $+1/-1$ and the aim is to find the optimal separating hyperplane **w** and bias w_0 which satisfies the following condition:

$$y^{(t)}(\mathbf{w}^T\mathbf{x}^{(t)} + w_0) \geq 1 \tag{3.10}$$

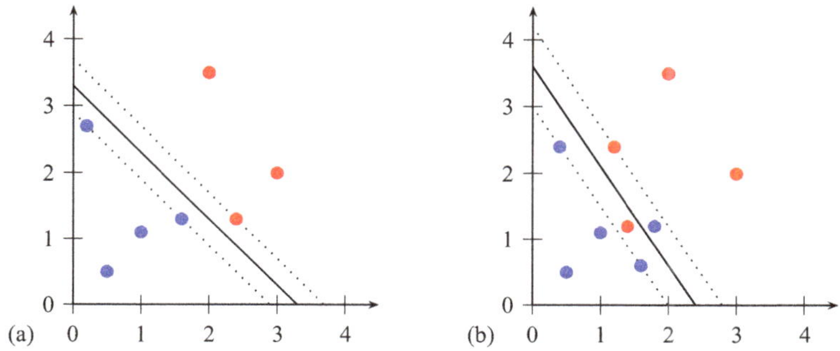

Figure 3.4 For a two-class problem the separating hyperplane and support vectors: (a) separable case and (b) nonseparable case

For the instances of positive class ($y^{(t)} = +1$), we want to see them on the positive side of the hyperplane. Similarly, for the instances of negative class ($y^{(t)} = -1$), we want to see them on the negative side of the hyperplane. Note that we also want the instances some distance away from the hyperplane, which we call *margin*, which we also want to maximize for the best generalization. If we write the margin maximization problem as an optimization problem:

$$\begin{aligned} &\underset{\mathbf{w}}{\text{minimize}} \quad \frac{\|\mathbf{w}\|^2}{2} \\ &\text{subject to} \quad y^{(t)}(\mathbf{w}^T\mathbf{x}^{(t)} + w_0) \geq 1, \quad t = 1, \ldots, N \end{aligned} \tag{3.11}$$

This standard quadratic optimization problem can be converted to its dual formulation using Lagrange multipliers $\alpha^{(t)}$ as:

$$L_p = \frac{\|\mathbf{w}\|^2}{2} - \sum_t \alpha^{(t)}[y^{(t)}(\mathbf{w}^T\mathbf{x}^{(t)} + w_0) - 1] \tag{3.12}$$

With Karush–Kuhn–Tucker conditions $\partial L_p/\partial \mathbf{w} = 0$ and $\partial L_p/\partial w_0 = 0$, we get the following dual formulation:

$$\begin{aligned} &\underset{\alpha^{(t)}}{\text{maximize}} \quad \sum_t \alpha^{(t)} - \frac{\sum_t \sum_s \alpha^{(t)}\alpha^{(s)}y^{(t)}y^{(s)}(\mathbf{x}^{(t)})^T\mathbf{x}^{(s)}}{2} \\ &\text{subject to} \quad \sum_t \alpha^{(t)}y^{(t)} = 0 \\ &\qquad\qquad\quad \alpha^{(t)} \geq 0, \quad t = 1, \ldots, N. \end{aligned} \tag{3.13}$$

Once we solve the previous dual quadratic optimization problem, we obtain a set of instances for which $\alpha^{(t)} > 0$. These instances are called support vectors and they lie on the margin.

3.5.5.2 Nonseparable case: soft margin hyperplane

For a linearly nonseparable two-class problem (see Figure 3.4(b)), we look for the hyperplane which acquires the minimum error. Since some of the constraints will not be satisfied, we add slack variables $\xi^{(t)} \geq 0$, which store the deviation from the margin. Now the aim is to find the optimal $\mathbf{w}$ and bias w_0 which satisfies the following condition:

$$y^{(t)}(\mathbf{w}^T\mathbf{x}^{(t)} + w_0) \geq 1 - \xi^{(t)} \tag{3.14}$$

Since each slack variable contributes to error, the total error is $\sum_t \xi^{(t)}$ and including this error as a weighted penalty in the objective function transforms the margin maximization problem to:

$$\begin{aligned} &\underset{\mathbf{w}}{\text{minimize}} && \frac{\|\mathbf{w}\|^2}{2} + C\sum_t \xi^{(t)} \\ &\text{subject to} && y^{(t)}(\mathbf{w}^T\mathbf{x}^{(t)} + w_0) \geq 1 - \xi^{(t)}, \quad t = 1, \ldots, N. \\ &&& \xi^{(t)} \geq 0, \quad t = 1, \ldots, N \end{aligned} \tag{3.15}$$

Again, we can convert this problem into its dual by introducing Lagrange multipliers $\alpha^{(t)}$ and $\beta^{(t)}$, and obtain:

$$L_p = \frac{\|\mathbf{w}\|^2}{2} + C\sum_t \xi^{(t)} - \sum_t \alpha^{(t)}[y^{(t)}(\mathbf{w}^T\mathbf{x}^{(t)} + w_0) - 1 + \xi^{(t)}] - \sum_t \beta^{(t)}\xi^{(t)} \tag{3.16}$$

With Karush–Kuhn–Tucker conditions solved, we get the following dual formulation:

$$\begin{aligned} &\underset{\alpha^{(t)}}{\text{maximize}} && \sum_t \alpha^{(t)} - \frac{\sum_t\sum_s \alpha^{(t)}\alpha^{(s)}y^{(t)}y^{(s)}(\mathbf{x}^{(t)})^T\mathbf{x}^{(s)}}{2} \\ &\text{subject to} && \sum_t \alpha^{(t)}y^{(t)} = 0 \\ &&& 0 \leq \alpha^{(t)} \leq C, \quad t = 1, \ldots, N \end{aligned} \tag{3.17}$$

Again, once we solve the quadratic problem, the instances having the corresponding value of $\alpha^{(t)} > 0$ are support vectors.

C is the regularization hyperparameter of the support vector machines and is usually tuned between 2^{-10} and 2^{10} with multipliers of 2. The upper bound for the training time complexity is $\mathcal{O}(N^3)$.

When there are more than two classes $(K > 2)$, we resort to multi class to binary class conversion techniques. In one-vs-all strategy, we define K subproblems, and in each subproblem, we try to separate class i from all other classes. On the other hand, in one-vs-one strategy, we define $K(K-1)/2$ subproblems, and in each subproblem, we try to separate class i from class j.

3.5.5.3 Kernel trick

If we had stopped here, support vector machines would not have been so popular. In order to solve a nonlinear problem, we do not fit a nonlinear model but map the original input space into a high-dimensional space and solve the problem linearly in that space. So instead of linearly combining the instances, we linearly combine the projected instances.

Again, the aim is to find the optimal **w** which satisfies the following condition:

$$y^{(t)}\mathbf{w}^T\Phi(\mathbf{x}^{(t)}) \geq 1 - \xi^{(t)} \tag{3.18}$$

where $\Phi(\mathbf{x}^{(t)})$ projects the instance $\mathbf{x}^{(t)}$ into a high-dimensional space. Including the error as a weighted penalty in the objective function transforms the margin maximization problem to:

$$\begin{aligned} &\underset{\mathbf{w}}{\text{minimize}} && \frac{\|\mathbf{w}\|^2}{2} + C\sum_t \xi^{(t)} \\ &\text{subject to} && y^{(t)}\mathbf{w}^T\Phi(\mathbf{x}^{(t)}) \geq 1 - \xi^{(t)}, \quad t = 1, \ldots, N, \\ &&& \xi^{(t)} \geq 0, \quad t = 1, \ldots, N. \end{aligned} \tag{3.19}$$

Yet again we can convert this problem into its dual by introducing Lagrange multipliers $\alpha^{(t)}$ and $\beta^{(t)}$, and obtain:

$$L_p = \frac{\|\mathbf{w}\|^2}{2} + C\sum_t \xi^{(t)} - \sum_t \alpha^{(t)}[y^{(t)}\mathbf{w}^T\Phi(\mathbf{x}^{(t)}) - 1 + \xi^{(t)}] - \sum_t \beta^{(t)}\xi^{(t)}. \tag{3.20}$$

With Karush–Kuhn–Tucker conditions solved, and replacing the inner product of $\Phi(\mathbf{x}^{(t)})^T\Phi(\mathbf{x}^{(s)})$ with kernel function $K(\mathbf{x}^{(t)}, \mathbf{x}^{(s)})$, we get the following dual formulation:

$$\begin{aligned} &\underset{\alpha^{(t)}}{\text{maximize}} && \sum_t \alpha^{(t)} - \frac{\sum_t\sum_s \alpha^{(t)}\alpha^{(s)}y^{(t)}y^{(s)}K(\mathbf{x}^{(t)}, \mathbf{x}^{(s)})}{2} \\ &\text{subject to} && \sum_t \alpha^{(t)}y^{(t)} = 0 \\ &&& 0 \leq \alpha^{(t)} \leq C, \quad t = 1, \ldots, N. \end{aligned} \tag{3.21}$$

So instead of mapping two instances to a high-dimensional space and applying dot product there, we can apply the kernel function directly in the original space. This is the main idea behind the *kernel trick*. Possible kernel functions used in the literature are:

- Polynomial kernel of degree n: $K(\mathbf{x}^{(t)}, \mathbf{x}^{(s)}) = (\mathbf{x}^{(t)}\mathbf{x}^{(s)} + 1)^n$
- Radial basis function: $K(\mathbf{x}^{(t)}, \mathbf{x}^{(s)}) = \exp\left(\frac{-\|\mathbf{x}^{(t)} - \mathbf{x}^{(s)}\|^2}{2s^2}\right)$
- Sigmoidal function: $K(\mathbf{x}^{(t)}, \mathbf{x}^{(s)}) = \tanh(2\mathbf{x}^{(t)}\mathbf{x}^{(s)} + 1)$

3.5.6 Neural networks

Artificial neural networks (ANN) take their inspiration from the brain. The brain consists of billions of neurons and these neurons are interconnected and work in parallel, which makes the brain a powerful computing machine. Each neuron is connected through synapses to thousands of neurons and the *firing* of a neuron depends on those synapses. Research on ANN started with the invention of perceptron [33] but came to its first halt with its limitation on the XOR problem [34]. After 15 years of standby, the resurrection of ANN came with the paper [35].

3.5.6.1 Neurons (units)

There are three types of neurons (units) in ANN. Each unit except the input unit takes an input and calculates an output. Input units represent a single input feature x_i or the bias $+1$.

Hidden units calculate an intermediate output from its inputs. They first combine their inputs linearly as $\mathbf{w}^T\mathbf{x}$ and then use nonlinear activation functions to map that linear combination to a nonlinear space. Well-known activation functions are sigmoid $z(x) = 1/(1+e^{-x})$, hyperbolic tangent $z(x) = (e^x - e^{-x})/(e^x + e^x)$, inverse of tangent $z(x) = \tan^{-1}(x)$, and rectified linear unit $z(x) = \max(0, x)$.

Output units calculate the output of the ANN. For regression problems, the output unit only calculates linear combination of its inputs (see Figure 3.5(a) and (c)). For two-class classification problems, the output unit uses sigmoid function $1/(1+e^{-x})$ to map the output to a probability value (see Figure 3.5(a) and (c)). For K-class classification problems (see Figure 3.5(b) and (d)), there are K output units, and each output unit i uses softmax function $e^{o_i}/\sum_{j=1}^{K} e^{o_j}$ to map its output to a probability value, so that the sum of all output units is 1.

3.5.6.2 Models and forward-propagation

In this chapter, we are interested in four types of neural networks models (see Figure 3.5), (i) perceptron for regression and two-class classification, (ii) perceptron for K-class classification, (iii) multilayer perceptron for regression and two-class classification, and (iv) multilayer perceptron for K-class classification.

In linear perceptron or simply perceptron, there are two layers, (i) input layer consisting of $d+1$ input units (including bias unit) and (ii) output layer consisting of one output for regression (K outputs for K-class classification). The weights are represented by $\mathbf{W}$, where w_{ij} represents the weight of the connection between the output unit i and input unit j.

Forward-propagation in linear perceptron calculates the values of output units. For regression, the output is $o = \sum_{j=0}^{d} w_{0j}x_j$. For two-class classification, the output is $o = 1/(1+e^{-z})$, where $z = \sum_{j=0}^{d} w_{0j}x_j$. For K-class classification, the output unit i is $o_i = e^{z_i}/\sum_{k=1}^{K} e^{z_k}$ where $z_i = \sum_{j=0}^{d} w_{ij}x_j$. In multilayer perceptron, there are three layers, (i) input layer consisting of $d+1$ input units (including bias unit), (ii) hidden layer consisting of H hidden units and one bias unit, and (iii) output layer consisting of one output for regression (K outputs for K-class classification).

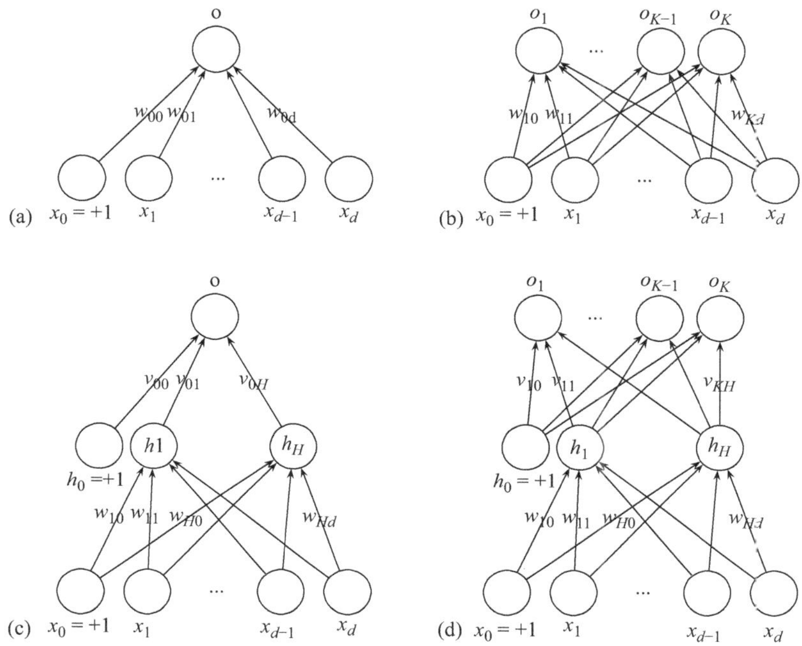

Figure 3.5 Neural network models: (a) perceptron for regression and two-class classification, (b) perceptron for K-class classification, (c) multilayer perceptron for regression and two-class classification, (d) multilayer perceptron for K-class classification

The weights between input and hidden units are represented by $\mathbf{W}$, where w_{ij} represents the weight of the connection between the hidden unit i and input unit j. On the other hand, the weights between hidden units and output units are represented by $\mathbf{V}$, where v_{ij} represents the weight of the connection between the output unit i and hidden unit j.

Forward-propagation in multilayer perceptron calculates the values of hidden and output units. For all models, if the activation function is sigmoid, the value of hidden unit i is $h_i = 1/(1 + e^{-g_i})$, where $g_i = \sum_{j=0}^{d} w_{ij}x_j$. For regression, the output is $o = \sum_{j=0}^{H} v_{0j}h_j$. For two-class classification, the output is $o = 1/(1 + e^{-z})$, where $z = \sum_{j=0}^{H} v_{0j}h_j$. For K-class classification, the output unit i is $o_i = e^{z_i}/\sum_{k=1}^{K} e^{z_k}$, where $z_i = \sum_{j=0}^{H} v_{ij}h_j$.

3.5.6.3 Backward-propagation

Before delving into learning in ANN, we need to verify two notions; what is to be learned and what is to be optimized. In regression, we optimize the mean square error:

$$E = (y^{(t)} - o^{(t)})^2, \tag{3.22}$$

in two-class classification we optimize the cross-entropy:

$$E = -y^{(t)} \log o^{(t)} - (1 - y^{(t)})\log(1 - o^{(t)}), \tag{3.23}$$

in K-class classification, again we optimize cross-entropy but calculated on K classes:

$$E = -\sum_{k=0}^{K} y_k^{(t)} \log o_k^{(t)} \tag{3.24}$$

For all of the networks, the aim is to learn the parameters of the network, which are simply the weight matrix $\mathbf{W}$ (and $\mathbf{V}$ for multilayer perceptron) [3].

Since all the functions in ANN are differentiable and not easily solvable, one resorts to gradient-descent style optimization. In gradient-descent optimization, we take the partial derivatives of to be optimized function with respect to be learned parameters $((\partial E/\partial \mathbf{W}), (\partial E/\partial \mathbf{V}))$ and use those partial derivatives to calculate the update rules for those parameters.

For regression in linear perceptron, the update rule is:

$$\Delta w_{0j} = -\eta \frac{\partial E}{\partial w_{0j}} = -\eta \frac{\partial E}{\partial o} \frac{\partial o}{\partial w_{0j}} = \eta(y^{(t)} - o^{(t)})x_j. \tag{3.25}$$

For two-class classification in linear perceptron, the update rule is:

$$\Delta w_{0j} = -\eta \frac{\partial E}{\partial w_{0j}} = -\eta \frac{\partial E}{\partial o} \frac{\partial o}{\partial z} \frac{\partial z}{\partial w_{0j}} = \eta(y^{(t)} - o^{(t)})x_j. \tag{3.26}$$

For K-class classification in linear perceptron, the update rule is:

$$\Delta w_{ij} = -\eta \frac{\partial E}{\partial w_{ij}} = -\eta \frac{\partial E}{\partial o_i} \frac{\partial o_i}{\partial z_i} \frac{\partial z_i}{\partial w_{ij}} = \eta(y^{(t)} - o^{(t)})x_j. \tag{3.27}$$

For regression in multilayer perceptron, the update rules are:

$$\Delta v_{0j} = -\eta \frac{\partial E}{\partial v_{0j}} = -\eta \frac{\partial E}{\partial o} \frac{\partial o}{\partial v_{0j}} = \eta(y^{(t)} - o^{(t)})h_j \tag{3.28}$$

$$\begin{aligned} \Delta w_{ij} &= -\eta \frac{\partial E}{\partial w_{ij}} = -\eta \frac{\partial E}{\partial o} \frac{\partial o}{\partial h_i} \frac{\partial h_i}{\partial g_i} \frac{\partial g_i}{\partial w_{ij}} \\ &= \eta(y^{(t)} - o^{(t)})v_{0j}h_i(1 - h_i)x_j \end{aligned} \tag{3.29}$$

For two-class classification in multilayer perceptron, the update rules are:

$$\Delta v_{0j} = -\eta \frac{\partial E}{\partial v_{0j}} = -\eta \frac{\partial E}{\partial o} \frac{\partial o}{\partial z} \frac{\partial z}{\partial v_{0j}} = \eta(y^{(t)} - o^{(t)})h_j \tag{3.30}$$

$$\begin{aligned} \Delta w_{ij} &= -\eta \frac{\partial E}{\partial w_{ij}} = -\eta \frac{\partial E}{\partial o} \frac{\partial o}{\partial z} \frac{\partial z}{\partial h_i} \frac{\partial h_i}{\partial g_i} \frac{\partial g_i}{\partial w_{ij}} \\ &= \eta(y^{(t)} - o^{(t)})v_{0j}h_i(1 - h_i)x_j \end{aligned} \tag{3.31}$$

For K-class classification in multilayer perceptron, the update rules are:

$$\Delta v_{ij} = -\eta \frac{\partial E}{\partial v_{ij}} = -\eta \frac{\partial E}{\partial o_i} \frac{\partial o_i}{\partial z_i} \frac{\partial z_i}{\partial v_{ij}} = \eta(y^{(t)} - o^{(t)})h_j \tag{3.32}$$

$$\begin{aligned} \Delta w_{ij} &= -\eta \frac{\partial E}{\partial w_{ij}} = -\eta \frac{\partial E}{\partial o_k} \frac{\partial o_k}{\partial z_k} \frac{\partial z_k}{\partial h_i} \frac{\partial h_i}{\partial g_i} \frac{\partial g_i}{\partial w_{ij}} \\ &= \eta \left(\sum_{k=0}^{K} (y_k^{(t)} - o_k^{(t)}) v_{ij} \right) h_i (1 - h_i) x_j \end{aligned} \tag{3.33}$$

3.6 Performance assessment and comparison of algorithms

Machine learning experiments are usually done to assess the performance of learning algorithms. If we have more than one candidate algorithm to experiment, we face with the second objective: comparing the performance of algorithms.

The first thing we need to know is, we cannot compare algorithms based on the training error. Training errors are optimistic and overly biased. The algorithm tries its best to learn the dataset and therefore is highly prone to recognize (remember) instances instead of learning the relationships among them. This is called *overfitting* in the machine-learning literature and is the reverse of *underfitting*, where we have not learned enough. To overcome overfitting, we need a separate dataset other than training set, on which we will compare learning algorithms. This set is called *test set*. We also need a third set *validation set*, on which we tune the hyperparameters of the algorithm. For example, in k-nearest neighbor, k is a hyperparameter and training on different ks produce different training errors, and in order to finalize the learner and test in on the test set, we need a separate set to differentiate between different ks.

Another important factor in learning is the impact of chance. Maybe there were one or more mislabeled instances in the training/validation/test set, or there were some outliers in the given training/validation pair, or there was noise while obtaining the features on the training/validation/test set. Yet another important possibility is the factor of chance on the training of the algorithm. Many learning algorithms (i.e., neural networks, K-means clustering) have an iterative type of optimization, where initially we start with a random solution. The initial solution may have a well impact on the end result. So, whatever possible reason is, training and validation on a single training/validation set pair is not healthy. We need to run our learning algorithm(s) on multiple sets, so we need to resample the dataset to obtain multiple training and validation sets.

Last but not least, remember that, error is not the only one criterion on which we compare algorithms. There are also other criteria in real life, which may get our attention more than the error, such as training time or space complexity, testing time or space complexity, interpretability, and risks when errors are generalized to other loss functions [36].

3.6.1 Sensitivity analysis

The aim of sensitivity analysis is to find the parameter/a set of parameters which has/have the greatest impact on the output of the model/algorithm [37]. It provides important insight into the model, whereby the most effective parameter/set of parameters is/are determined. In other words, sensitivity analysis helps the experiment designer to understand the input–output relationship; determines how the uncertainty in parameters affects the actual output of the system; and guides experiment designer for future experiment designs. There are two important types of sensitivity analysis: (i) local and (ii) global.

3.6.1.1 Local sensitivity analysis

In local sensitivity analysis, one evaluates the change in the model output with respect to a change in a single parameter. Only small variations on one parameter are applied, and the effect of this simple variation is obtained via local sensitivity indices. Local sensitivity indices are usually calculated as the partial derivatives of the model output with respect to that local parameter.

The main limitation of the local sensitivity analysis is that it evaluates the effect of one parameter at a time, and therefore does not allow the evaluation of simultaneous effects of changes in multiple parameters, and does not allow calculations of interactions between multiple parameters.

3.6.1.2 Global sensitivity analysis

In global sensitivity analysis, all parameters of the system are changed at the same time over the whole parameter space, which enables us to calculate the respective contributions of each parameter and the interactions between those multiple parameters. The interactions between the model parameters identify the model output variance.

Well-known global sensitivity analysis methodologies are

- Weighted average of local sensitivity analysis, which calculates the local sensitivity analysis indices at different random values of the input parameters. The weighted average of the local indices is then used to calculate the global sensitivity.
- Partial rank correlation coefficient, which uses the Pearson correlation coefficients (between -1 and 1) to identify the important parameters.
- Sobol method, which is based on variance decomposition technique to obtain the corresponding contributions of the input parameters on the model output variance. Sobol's method also can identify the interactions on the input parameters on the overall output variance.

3.6.2 Resampling

In this section, we will talk about how to generate K training/validation set pairs. We also want to hold back as much as from overlapping the validation and training set pairs.

3.6.2.1 *K*-Fold cross-validation

In K-fold cross-validation, the aim is to generate K training/validation set pair, where training and validation sets on fold i do no overlap. First, we divide the dataset $\mathbf{X}$ into K parts as $\mathbf{X}_1, \mathbf{X}_2, \ldots, \mathbf{X}_K$. Then for each fold i, we use $\mathbf{X}_i$ as the validation set and $\mathbf{X} - \mathbf{X}_i$ as the training set. So, the training and validation sets are

$$\begin{aligned} \mathbf{T}_1 &= \mathbf{X}_2 \cup \mathbf{X}_3 \cup \ldots \mathbf{X}_K \quad \mathbf{V}_1 = \mathbf{X}_1 \\ \mathbf{T}_2 &= \mathbf{X}_1 \cup \mathbf{X}_3 \cup \ldots \mathbf{X}_K \quad \mathbf{V}_2 = \mathbf{X}_2 \\ &\ldots \\ \mathbf{T}_K &= \mathbf{X}_1 \cup \mathbf{X}_2 \cup \ldots \mathbf{X}_{K-1} \; \mathbf{V}_K = \mathbf{X}_K \end{aligned}$$

Possible values of K are 10 or 30. One extreme case of K-fold cross-validation is leave-one-out, where $K = N$ and each validation set has only one instance.

If we have more computation power, we can have multiple runs of K-fold cross-validation, such as 10×10 cross-validation [38] or 5×2 cross-validation [8,39].

3.6.2.2 Bootstrapping

If we have very small datasets, we do not insist on the non-overlap of training and validation sets. In bootstrapping, we generate K multiple training sets, where each training set contains N examples (like the original dataset). To get N examples, we draw examples with replacement. For the validation set, we use the original dataset. The drawback of bootstrapping is that the bootstrap samples overlap more than the cross-validation sample, hence they are more dependent.

3.6.3 Comparison of algorithms

In this section, we will talk about statistical tests comparing two classification algorithms, and test whether those two algorithms have the same error rate.

3.6.3.1 *K*-Fold cv paired *t*-test

In K-fold cv paired t-test, we assume that we sampled the training set with K-fold cross-validation and the difference of error rates of two algorithms on fold i are calculated as $e_i = f_i - s_i$, where f_i and s_i represent the error rates of the first and second algorithms on fold i, respectively.

The null hypothesis of the K-fold cv paired t-test is that the distribution of e_i has 0 mean. The test statistic:

$$\tau_K = \frac{\sqrt{K}m}{S} \tag{3.34}$$

is t-distributed with $K - 1$ degrees of freedom, where m and S are the mean and the variance of e_i's. The test rejects the null hypothesis at significance level α, if τ_K is outside the interval $(-t_{\alpha/2,K-1}, t_{\alpha/2,K-1})$.

3.6.3.2 5 × 2 cv paired *t*-test

In 5 × 2 cv paired *t*-test [39], we assume that we sampled the training set with 5 × 2-fold cross-validation and the difference of error rates of two algorithms on fold j of replication i are calculated as $e_{ij} = f_{ij} - s_{ij}$, where f_{ij} and s_{ij} represent the error rates of the first and second algorithms on fold j of replication i, respectively.

Under the null hypothesis that the two classifications have the same error rate, the test statistic

$$\tau_t = \frac{e_{11}}{\sqrt{\sum_{i=1}^{5} s_i^2/5}} \tag{3.35}$$

is t-distributed with 5 degrees of freedom, where s_i^2 is the variance of the error differences on replication i. The test rejects the null hypothesis at significance level α, if τ_t is outside the interval $(-t_{\alpha/2,5}, t_{\alpha/2,5})$.

3.6.3.3 Combined 5 × 2 cv *F*-test

In combined 5 × 2 cv *F*-test [8], the assumptions are the same as 5 × 2 cv paired *t*-test.

Under the null hypothesis that two classifications have the same error rate, the test statistic:

$$\tau_F = \frac{\sum_{i=1}^{5}\sum_{j=1}^{2} e_{ij}^2}{2\sum_{i=1}^{5} s_i^2} \tag{3.36}$$

is F-distributed with 10 and 5 degrees of freedoms. The test rejects the null hypothesis at significance level α, if τ_F is greater than $F_{\alpha,10,5}$.

3.6.3.4 Combined 5 × 2 cv *t*-test

In combined 5 × 2 cv *t*-test [9], the assumptions are the same as 5 × 2 cv paired *t*-test.

Under the null hypothesis that two classifications have the same error rate, the test statistic:

$$\tau_C = \frac{\sum_{i=1}^{5}\sum_{j=1}^{2} e_{ij}}{\sqrt{\sum_{i=1}^{5}\sum_{j=1}^{2} e_{ij}^2 - 2\sum_{i=1}^{5} e_{i1}e_{i2}}} \tag{3.37}$$

is t-distributed with 5 degrees of freedom. The test rejects the null hypothesis at significance level α, if τ_C is outside the interval $(-t_{\alpha/2,5}, t_{\alpha/2,5})$.

References

[1] Vapnik V. The Nature of Statistical Learning Theory. New York, NY: Springer Verlag; 1995.

[2] Alpaydın E. Introduction to Machine Learning. Cambridge, MA: The MIT Press; 2010.

[3] Bishop CM. Neural Networks for Pattern Recognition. Oxford, England: Oxford University Press; 1995.

[4] Quinlan JR. C4.5: Programs for Machine Learning. San Meteo, CA: Morgan Kaufmann; 1993.

[5] Pearl J. Causality: Models, Reasoning, and Inference. Cambridge, England: Cambridge University Press; 2000.

[6] Jordan MI. An Introduction to Probabilistic Graphical Models. Cambridge, MT: The MIT Press; 2009.

[7] Rabiner LR. A Tutorial on Hidden Markov Models and Selected Applications in Speech Recognition. Proceedings of the IEEE. 1989;77:257–286.

[8] Alpaydın E. Combined 5 $\times$ 2 cv F Test for Comparing Supervised Classification Learning Classifiers. Neural Computation. 1999;11: 1975–1982.

[9] Yıldız OT. Omnivariate Rule Induction Using a Novel Pairwise Statistical Test. IEEE Transactions on Knowledge and Data Engineering. 2013;25: 2105–2118.

[10] Yıldız OT, and Alpaydın E. Omnivariate Decision Trees. IEEE Transactions on Neural Networks. 2001;12(6):1539–1546.

[11] Fürnkranz J. Separate-and-Conquer Learning. Artificial Intelligence Review. 1999;13:3–54.

[12] Fawcett T. An Introduction to ROC Analysis. Pattern Recognition Letters. 2006;27:861–874.

[13] Cherkassky V, and Mulier F. Learning From Data. Hoboken, NJ: John Wiley and Sons; 1998.

[14] Devjner PA, and Kittler J. Pattern Recognition: A Statistical Approach. New York, NY: Prentice-Hall; 1982.

[15] Guyon I, and Elisseeff A. An Introduction to Variable and Feature Selection. Journal of Machine Learning Research. 2003;3:1157–1182.

[16] Pudil P, Novovicova J, and Kittler J. Floating Search Methods in Feature Selection. Pattern Recognition Letters. 1994;15:1119–1125.

[17] Jain AK, and Dubes RC. Algorithms for Clustering Data. New York, NY: Prentice-Hall; 1988.

[18] Jain AK, Murthy MN, and Flynn PJ. Data Clustering: A Review. ACM Computing Surveys. 1999;31:264–323.

[19] Dempster AP, Laird NM, and Rubin DB. Maximum Likelihood from Incomplete Data via the EM Algorithm. Journal of Royal Statistical Society B. 1977;39:1–38.

[20] Redner RA, and Walker HF. Mixture Densities, Maximum Likelihood and the EM Algorithm. SIAM Review. 1984;26:195–239.

[21] Wolpert DH. The Relationship between PCA, the Statistical Physics Framework, the Bayesian Framework, and the VC Framework. In: Wolpert DH, editor. The Mathematics of Generalization. Cambridge, MA: Addison Wesley; 1995. p. 117–214.

[22] Aha DW, Kibler D, and Albert MK. Instance-Based Learning Algorithm. Machine Learning. 1991;6:37–66.

[23] Aha DW. Special Issue on Lazy Learning. Artificial Intelligence Review. 1997;11(1–5):7–423.

[24] Duda RO, Hart PE, and Stork DG. Pattern Classification. Hoboken, NJ: John Wiley and Sons; 2001.

[25] Yıldız OT, and Alpaydın E. Ordering and Finding the Best of $K > 2$ Supervised Learning Algorithms. IEEE Transactions on Pattern Analysis Machine Intelligence. 2006;28(3):392–402.

[26] Murthy SK. Automatic Construction of Decision Trees from Data: A Multi-Disciplinary Survey. Data Mining and Knowledge Discovery. 1998;2(4): 345–389.

[27] Yıldız OT, and Alpaydın E. Linear Discriminant Trees. In: 17th International Conference on Machine Learning. Morgan Kaufmann; 2000. p. 1175–1182.

[28] Guo H, and Gelfand SB. Classification Trees With Neural Network Feature Extraction. IEEE Transactions on Neural Networks. 1992;3:923–933.

[29] Breiman L, Friedman JH, Olshen RA, *et al.* Classification and Regression Trees. Hoboken, NJ: John Wiley and Sons; 1984.

[30] Quinlan JR. Induction of Decision Trees. Machine Learning. 1986;1:81–106.

[31] Vapnik V. Statistical Learning Theory. New York, NY: Wiley; 1998.

[32] Schölkopf B, Tsuda K, and Vert JP, editors. Kernel Methods in Computational Biology. Cambridge, MA: The MIT Press; 2004.

[33] Rosenblatt F. Principles of Neurodynamics: Perceptrons and the Theory of Brain Mechanisms. New York, NY: Spartan; 1962.

[34] Minsky ML, and Papert SA. Perceptrons. Cambridge, MA: MIT Press; 1969.

[35] Hopfield JJ. Neural Networks and Physical Systems with Emergent Collective Computational Abilities. Proceedings of the National Academy of Sciences. 1982;79:2554–2558.

[36] Turney PD. Types of Cost in Inductive Concept Learning. In: Workshop on Cost-Sensitive Learning in 17th International Conference on Machine Learning. Stanford University, CA, USA; 2000. p. 15–21.

[37] Saltelli A, Chan K, and Scott EM. Sensitivity Analysis: Wiley Series in Probability and Statistics. New York, NY: Wiley; 2000.

[38] Bouckaert RR, and Frank E. Evaluating the Replicability of Significance Tests for Comparing Learning Algorithms. In: Advances in Knowledge Discovery and Data Mining, LNCS. vol. 3056. Sydney, Australia: Springer; 2004. p. 3–12.

[39] Dietterich TG. Approximate Statistical Tests for Comparing Supervised Classification Learning Classifiers. Neural Computation. 1998;10:1895–1923.

Chapter 4
Data-driven and physics-based modeling

Slawomir Koziel[1]

Utilization of computer simulations has become ubiquitous in contemporary engineering design. Accurate simulations can provide a reliable assessment of components and devices, thereby replace the need for prototyping, reduce the cost of the design cycle, as well as its cost. At the same time, the computational cost of computer simulations (e.g., full-wave electromagnetic (EM) analysis in microwave or antenna engineering) maybe considerable or even unmanageable, particularly for complex structures. This is usually not a problem for design verification or even simple simulation-driven design procedures based on parameter sweeping; however, it becomes problematic for procedures that require numerous analyses, such as parametric optimization, statistical analysis, or tolerance-aware design.

The difficulties related to the high cost of simulation can be alleviated to a certain extent by utilization of fast replacement models, also referred to as surrogates. The popularity and importance of surrogate models has been steadily growing over the years along with the development of various modeling techniques and an increasing number of applications in engineering and science. In this chapter, we give a brief introduction to surrogate modeling, starting with model classification, explaining the modeling flow as well as discussing the details of modeling stages. These include design of experiments (DoEs), data acquisition, model identification as well as model validation. We also discuss the most popular modeling techniques, including—among others—polynomial regression, kriging, neural networks (NNs), space mapping (SM), and response correction techniques.

4.1 Model classification

Surrogate models can be classified in different ways. Here, we distinguish the two major classes of models, referred to as data-driven models [1] (also named functional or approximation ones) and physics-based models [2]. In this section, the generic properties of both classes are briefly discussed. Particular techniques are described in further sections of this chapter.

[1]School of Science and Engineering, Reykjavik University, Reykjavik, Iceland

Data-driven models are by far the most popular type of surrogates with a large variety of modeling methods available, e.g., [3–8]. There are certain features that are common to all types of approximation models. These can be identified as follows:

- The models are constructed solely from sampled training data so that no problem-specific knowledge is required.
- They are generic, therefore, applicable to a wide range of problems and easily transferrable between various areas.
- They are typically based on explicit analytical formulas.
- They are cheap to evaluate.

Due to their low evaluation cost, data-driven models are easy to optimize and can be conveniently used to solve design tasks that require numerous system/device simulations. Unfortunately, data-driven surrogates have an important disadvantage, which is a large number of training data points necessary to ensure good predictive power of the model. This number rapidly grows with the number of parameters of the system at hand, a phenomenon often referred to as the curse of dimensionality [5,9]. Furthermore, the number of necessary data points quickly grows with the parameter ranges. The latter is particularly important from the point of view of practical usefulness of the surrogate in the design process. Depending on the type and nonlinearity of the system responses, a practical limit for the parameter space dimensionality maybe as low as a few (four to six) for highly nonlinear systems (e.g., multiband antennas or antenna arrays) to over 20 (e.g., for aerodynamic profiles, such as airfoils). On the other hand, local approximation models can be quite useful as auxiliary optimization tools.

The models of the second class, physics-based ones, utilize a different principle. They are constructed by suitable correction or enhancement of an underlying low-fidelity model. In the case of electronics engineering, the original (or high-fidelity) models are typically evaluated using circuit or full-wave EM analysis [10], the latter being particularly expensive. The low-fidelity models can be obtained as equivalent circuits (e.g., for microwave filters or other components [11]) or coarse-mesh EM simulation (e.g., for antennas or antenna arrays [12]). In some cases, analytical models are used as well if available. The correction process aims at improving the alignment between the low-fidelity model and the high-fidelity model, either locally or across the entire design space.

Because the low-fidelity model embeds certain knowledge about the system of interest, the low- and high-fidelity models are normally well correlated. Consequently, only a limited amount of high-fidelity data is normally sufficient to ensure reasonable accuracy of the surrogate model. For the same reason, physics-based surrogates typically feature good generalization. At the same time, the effect of curse of dimensionality is not as pronounced as for the data-driven surrogates. A certain disadvantage of this class of models is their limited generality: each type of problem requires careful preparation of the low-fidelity model the quality of which maybe critical for the quality of the surrogate. It should also be mentioned that hybridization of both classes of surrogates is possible (e.g., SM enhanced by data-driven correction layer [13]).

4.2 Modeling flow

In this section, a typical modeling flow is described. It is mostly pertinent to data-driven models, although it also applies to physics-based surrogates to a certain extent. The overall flowchart of the surrogate modeling process is shown in Figure 4.1. There are four stages that can be distinguished:

- DoE. DoE refers to a process of allocating a given number of training samples in the design space using a selected strategy. In reality, the available computational budget limits the number of samples that can be used. In the past, factorial type of designs has been popular due to the fact that the majority of data was coming from physical measurements of the system. Nowadays, as the training data comes from computer simulations, space-filling DoEs are usually preferred [3]. An outline of popular DoE techniques is provided in Section 4.3.
- Data acquisition. This stage consists of acquiring the training data at the points allocated by DoE. Data acquisition is normally the bottleneck of the modeling process particularly when the system evaluation requires expensive simulations (e.g., full-wave EM analysis).
- Model identification. This step employs the extraction of the parameters of the approximation model of choice. In most cases (e.g., kriging [15] or NNs [16]), determination of the surrogate model parameters requires solving a suitably posed minimization problem. However, in some situations, the model parameters

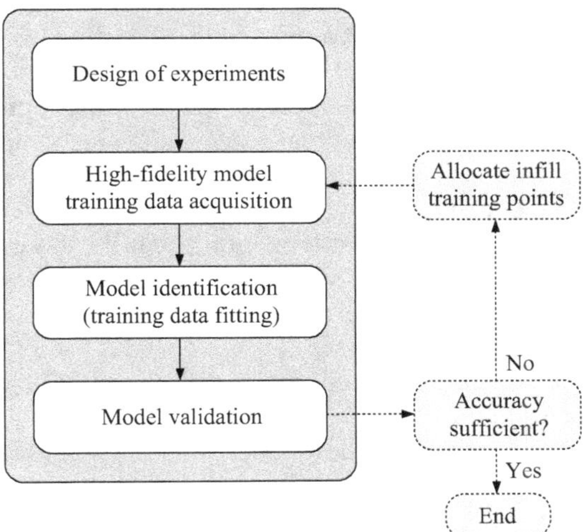

Figure 4.1 A generic surrogate model construction flowchart [14]. The shaded box shows the main flow, whereas the dashed lines indicate an optional iterative procedure in which additional training data points are added (using a suitable infill strategy) and the model is updated accordingly

can be found using explicit formulas by solving an appropriate regression problem (e.g., polynomial approximation [1]). Sections 4.4 and 4.5 discuss selected data-driven and physics-based modeling techniques, respectively.

- Model validation. This step is used to verify the model accuracy. Normally, a generalization capability of the surrogate is of major concern, i.e., estimating the predictive power at the points (designs) not seen during the surrogate identification stage. Consequently, the model testing should involve a separate set of testing samples. Maintaining a proper balance between surrogate approximation and generalization (accuracy at the known and at the unknown data sets) is also of interest. A few comments on model validation have been provided in Section 4.6.

As the validation stage may indicate insufficient model accuracy, the surrogate model construction maybe—in practice—an iterative process so that the steps shown in the core of the diagram shown in Figure 4.1 constitute just a single iteration. In case the model needs to be improved, a new set of samples and the corresponding high-fidelity model evaluations maybe added to the training set upon model validation and utilized to reidentify the model. This adaptive sampling process is then continued until the accuracy goals are met or the computational budget is exceeded. In the optimization context, the surrogate update maybe oriented more toward finding better designs rather than toward ensuring global accuracy of the model [5].

4.3 Design of experiments

As explained before, DoEs [17–19] is a strategy for allocating the training points in the parameter space. In most practical cases, the space is defined by the lower and upper bounds for the parameters so that DoE can be performed on the unity hypercube and later remapped to the actual space. Following sample allocation, the high-fidelity model data is acquired at the selected locations and utilized to construct the surrogate model. Obviously, there is a trade-off between the number of training samples and the amount of information about the system of interest that can be extracted from these points. The estimated model accuracy can be fed back to the sampling algorithm to allocate more points in the regions that exhibit more nonlinear behavior [6,20].

Figure 4.2(a) shows an example of a factorial design [17], which consists of allocating the samples in the corners, edges, and/or faces of the design space. This is a traditional DoE approach that allows estimating the main effects and interactions between design variables without using too many samples. The rationale behind spreading the samples is that it minimizes possible errors of estimating the main trends as well as interactions between design variables, in case the data about the system was coming from physical measurements. Nowadays, certain factorial designs are used because they are "economical" in terms of the number of samples, which is important if the computational budget for data acquiring is very limited. An example is a so-called star distribution (cf. Figure 4.2(a)) which is often used in conjunction with SM [21].

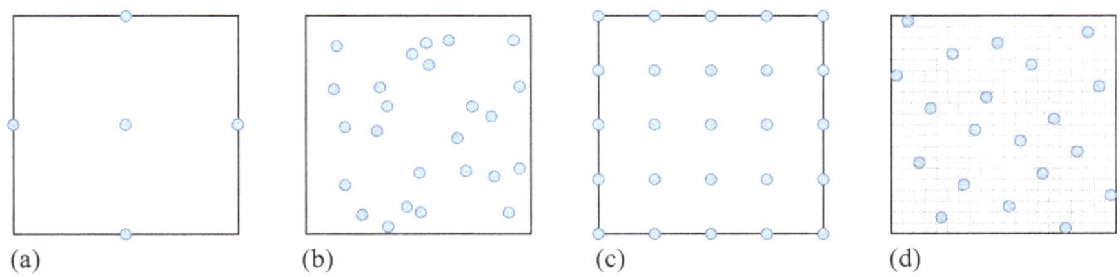

Figure 4.2 Popular DoE techniques [14]: (a) factorial designs (here, star distribution); (b) random sampling; (c) uniform grid sampling; (d) Latin hypercube sampling (LHS)

Vast majority of modern DoE procedures are space-filling designs attempting to allocate training points uniformly within the design space [1]. This is especially useful for constructing an initial surrogate model when the knowledge about the system is limited. A number of space-filling DoEs are available. The simple ones include pseudorandom sampling [17], cf. Figure 4.2(b), or uniform grid sampling (Figure 4.2(c)). The limitation of the former is poor uniformity, the latter is only practical for low-dimensional spaces as the number of samples is restricted to $N_1 N_2 \ldots N_n$, where N_j is the number of samples along jth axis of the design space. Arguably, the most popular DoE for uniform sample distribution is Latin hypercube sampling (LHS) [22]. In order to allocate p samples with LHS, the range for each parameter is divided into p bins, which for n design variables, yields a total number of p^n bins in the design space. The samples are randomly selected in the design space so that (i) each sample is randomly placed inside a bin and (ii) for all one-dimensional projections of the p samples and bins, there is exactly one sample in each bin. An example LHS distribution for $p = 20$ in two-dimensional space has been shown in Figure 4.2(d). LHS can be improved to provide more uniform sampling distributions [23–26]. Other uniform sampling techniques that are commonly used include orthogonal array sampling [1], quasi-Monte Carlo sampling [17], or Hammersley sampling [17]. Space-filling sample distribution can also be posed as an optimization problem, specifically, as minimization of a suitably defined nonuniformity measure, e.g., $\sum_{i=1,\ldots,p}\sum_{j=i=1,\ldots,p} d_{ij}^{-2}$ [24], where d_{ij} is the Euclidean distance between samples i and j.

4.4 Data-driven models

The number of available data-driven modeling techniques is quite considerable. This section provides a brief outline of the selected methods. More details can be found in the literature, e.g., [3]. The common notation utilized throughout is the following. The training data samples will be denoted as $\{\boldsymbol{x}^{(i)}\}$, $i = 1, \ldots, p$ and the corresponding high-fidelity model evaluations as $f(\boldsymbol{x}^{(i)})$. The surrogate model is constructed by approximating the data pairs $\{\boldsymbol{x}^{(i)}, f(\boldsymbol{x}^{(i)})\}$. Here, f is considered to be a scalar function; however, generalization of most methods for vector-values cases is rather straightforward.

4.4.1 Polynomial regression

Polynomial regression is one of the simplest approximation techniques [1]. The surrogate is defined as a linear combination of the basis functions v_j:

$$s(x) = \sum_{j=1}^{K} \beta_j v_j(x), \tag{4.1}$$

where β_j are unknown coefficients. The model parameters can be found as a least-square solution to the linear system:

$$f = X\beta, \tag{4.2}$$

where $f = \left[f\left(x^{(1)}\right)\ f\left(x^{(2)}\right) \cdots f\left(x^{(p)}\right)\right]^T$, X is a $p \times K$ matrix containing the basis functions evaluated at the sample points, and $\beta = [\beta_1\ \beta_2 \cdots \beta_K]^T$. In order to ensure the uniqueness of the solution to (4.2), one needs to ensure consistency between the numbers of data points and basis functions (typically, $p \geq K$).

If the sample points and basis functions are taken arbitrarily, some columns of X maybe linearly dependent. For $p \geq K$ and rank(X) = K, a solution to (4.2) can be computed through X^+, the pseudoinverse of X [27], i.e., $\beta = X^+ f = (X^T X)^{-1} X^T f$.

A simple yet useful example of a regression model is a second-order polynomial:

$$s(x) = s\left(\left[x_1\ x_2 \cdots x_n\right]^T\right) = \beta_0 + \sum_{j=1}^{n} \beta_j x_j + \sum_{i=1}^{n} \sum_{j \leq i}^{n} \beta_{ij} x_i x_j, \tag{4.3}$$

with the basis functions being monomials: 1, x_j, and $x_i x_j$.

4.4.2 Radial basis function interpolation

Radial basis function model interpolation/approximation [5,28] is a regression model where the surrogate is defined as a linear combination of K radially symmetric functions ϕ:

$$s(x) = \sum_{j=1}^{K} \lambda_j \phi\left(\left\|x - c^{(j)}\right\|\right), \tag{4.4}$$

in which $\lambda = [\lambda_1\ \lambda_2 \cdots \lambda_K]^T$ is the vector of model parameters and $c^{(j)}, j = 1, \ldots, K$, are the (known) basis function centers; $\|\ \|$ stands for the L_2-norm. The model coefficients can be found as $\lambda = \Phi^+ f = (\Phi^T \Phi)^{-1} \Phi^T f$, where, $f = \left[f\left(x^{(1)}\right)\ f\left(x^{(2)}\right) \cdots f\left(x^{(p)}\right)\right]^T$ and the $p \times K$ matrix $\Phi = [\Phi_{kl}]_{k=1,\ldots,p;\ l=1,\ldots,K}$, with the entries defined as:

$$\Phi_{kl} = \phi\left(\left\|x^{(k)} - c^{(l)}\right\|\right). \tag{4.5}$$

In a special case, where the number of basis functions is equal to the number of samples, i.e., $p = K$, the centers of the basis functions coincide with the data points and are all different, Φ is a regular square matrix. Then, $\lambda = \Phi^{-1} f$. A popular choice of the basis function is a Gaussian, $\phi(r) = \exp(-r^2/2\sigma^2)$, where σ is the scaling parameter.

4.4.3 Kriging

Kriging belongs to the most popular data-driven modeling techniques today [3,15]. Kriging is a Gaussian-process-based modeling method, which is compact and cheap to evaluate [29]. In its basic formulation (e.g., [3]), kriging assumes that the function of interest is of the following form:

$$\boldsymbol{f}(\boldsymbol{x}) = \boldsymbol{g}(\boldsymbol{x})^T\boldsymbol{\beta} + Z(\boldsymbol{x}), \tag{4.6}$$

where $\boldsymbol{g}(\boldsymbol{x}) = [g_1(\boldsymbol{x})\ \ g_2(\boldsymbol{x})\cdots g_K(\boldsymbol{x})]^T$ are known (e.g., constant) functions, $\boldsymbol{\beta} = [\beta_1\ \ \beta_2\cdots\beta_K]^T$ are the unknown model parameters (hyperparameters), and $Z(\boldsymbol{x})$ is a realization of a normally distributed Gaussian random process with zero mean and variance σ^2. The regression part $\boldsymbol{g}(\boldsymbol{x})^T\beta$ is a trend function for f, and $Z(\boldsymbol{x})$ takes into account localized variations. The covariance matrix of $Z(\boldsymbol{x})$ is given as:

$$Cov\left(Z\left(\boldsymbol{x}^{(i)}\right)Z\left(\boldsymbol{x}^{(j)}\right)\right) = \sigma^2\boldsymbol{R}\left[R\left(\boldsymbol{x}^{(i)},\boldsymbol{x}^{(j)}\right)\right], \tag{4.7}$$

where $\boldsymbol{R}$ is a $p \times p$ correlation matrix with $R_{ij} = R(\boldsymbol{x}^{(i)},\boldsymbol{x}^{(j)})$. Here, $R(\boldsymbol{x}^{(i)},\boldsymbol{x}^{(j)})$ is the correlation function between sampled data points $\boldsymbol{x}^{(i)}$ and $\boldsymbol{x}^{(j)}$. The most popular choice is the Gaussian correlation function:

$$R(\boldsymbol{x},\boldsymbol{y}) = \exp\left[-\sum_{k=1}^{n}\theta_k|x_k - y_k|^2\right], \tag{4.8}$$

where θ_k are the unknown correlation parameters, and x_k and y_k are the kth components of the vectors $\boldsymbol{x}$ and $\boldsymbol{y}$, respectively. The kriging predictor is defined as:

$$s(\boldsymbol{x}) = \boldsymbol{g}(\boldsymbol{x})^T\boldsymbol{\beta} + \boldsymbol{r}^T(\boldsymbol{x})\boldsymbol{R}^{-1}(\boldsymbol{f} - \boldsymbol{G}\boldsymbol{\beta}), \tag{4.9}$$

where $\boldsymbol{r}(\boldsymbol{x}) = \left[R\left(\boldsymbol{x},\boldsymbol{x}^{(1)}\right)\cdots R\left(\boldsymbol{x},\boldsymbol{x}^{(p)}\right)\right]^T, \boldsymbol{f} = \left[f\left(\boldsymbol{x}^{(1)}\right)\ \ f\left(\boldsymbol{x}^{(2)}\right)\cdots f\left(\boldsymbol{x}^{(p)}\right)\right]^T$, and $\boldsymbol{G}$ is a $p\times K$ matrix with $G_{ij} = g_j(\boldsymbol{x}^{(i)})$. The vector of model parameters β can be computed as $\beta = (\boldsymbol{G}^T\boldsymbol{R}^{-1}\boldsymbol{G})^{-1}\boldsymbol{G}^T\boldsymbol{R}^{-1}\boldsymbol{f}$. Model fitting is accomplished by maximum likelihood for θ_k [30]. One of practically important features of kriging is that the random process $Z(\boldsymbol{x})$ gives information on the approximation error that can be used for improving the surrogate, e.g., by allocating additional training samples at the locations, where the estimated model error is the highest [5]. This feature is also utilized in various global optimization methods (e.g., [6]). Figure 4.3 shows an example function of two variables and its kriging model obtained using 20, 50, and 100 samples allocated using LHS.

4.4.4 Support vector regression

Support vector regression (SVR) [31] exhibits good generalization capability [32] and convenient training through quadratic programming [33]. SVR exploits the

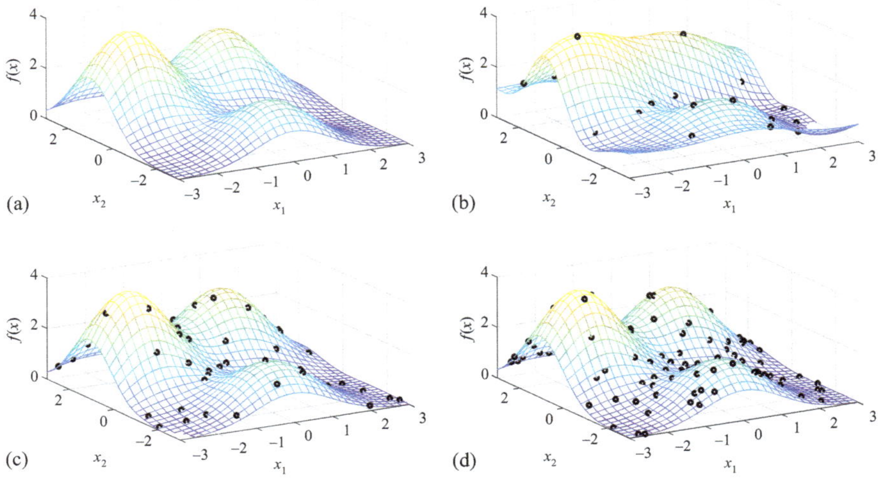

Figure 4.3 Kriging model of a two-parameter scalar function: (a) functional landscape, (b) training data set (o) and kriging model for p = 20 samples, (c) kriging model for p = 50 samples, (d) kriging model for p = 100 samples

structural risk minimization principle, arguably superior [31] to traditional empirical risk minimization principle, employed by, e.g., NNs. SVR has been gaining popularity in various areas, including electrical engineering and aerodynamic design [34–36].

Let $r^k = f(\boldsymbol{x}^k)$, $k = 1,2,\ldots, N$ denote the sampled high-fidelity model responses (here, SVR is formulated for vector-valued f). SVR is to approximate r^k at all base points $\boldsymbol{x}^k$, $k = 1,2,\ldots, N$. Let $r^k = \left[r_1^k\ r_2^k \cdots r_m^k\right]^T$ denote components of the vector r^k. For linear regression, one aims at approximating a training data set, here, the pairs $D_j = \left\{\left(\boldsymbol{x}^1, r_j^1\right), \ldots, \left(\boldsymbol{x}^N, r_j^N\right)\right\}$, $j = 1, 2, \ldots, m$, by a linear function $f_j(\boldsymbol{x}) = w_j^T\boldsymbol{x} + b_j$. The optimal regression function is given by the minimum of the following functional [33]:

$$\Phi_j(\boldsymbol{w}, \xi) = \frac{1}{2}||\boldsymbol{w}_j||^2 + C_j \sum_{i=1}^{N} \left(\xi_{j\cdot i}^{+} + \xi_{j\cdot i}^{-}\right), \tag{4.10}$$

where C_j is a user-defined value, and $\xi_{j\cdot i}^{+}$ and $\xi_{j\cdot i}^{-}$ are the slack variables representing upper and lower constraints on the output of the system. The typical cost function used in SVR is an ε-insensitive loss function defined as:

$$L_\varepsilon(y) = \begin{cases} 0 & \text{for } |f_j(\boldsymbol{x}) - y| < \varepsilon \\ |f_j(\boldsymbol{x}) - y| & \text{otherwise} \end{cases} \tag{4.11}$$

The value of C_j determines the trade-off between the flatness of f_j and the amount up to which deviations larger than ε are tolerated [31].

Here, nonlinear regression employing the kernel approach is described, in which the linear function $\boldsymbol{w}_j^T\boldsymbol{x} + b_j$ is replaced by the nonlinear function $\Sigma_i\gamma_{j\cdot i}K(\boldsymbol{x}^k,\boldsymbol{x})+b_j$, where K is a kernel function. Thus, the SVR model is defined as:

$$\boldsymbol{s}(\boldsymbol{x}) = \begin{bmatrix} \sum_{i=1}^{N}\gamma_{1\cdot i}K(\boldsymbol{x}^i,\boldsymbol{x}) + b_1 \\ \vdots \\ \sum_{i=1}^{N}\gamma_{m\cdot i}K(\boldsymbol{x}^i,\boldsymbol{x}) + b_m \end{bmatrix} \tag{4.12}$$

with parameters $\gamma_{j\cdot i}$ and b_j, $j = 1,\ldots, m$, $i = 1,\ldots, N$ obtained according to a general SVR methodology. In particular, Gaussian kernels of the form $K(\boldsymbol{x},\boldsymbol{y}) = \exp(-0.5\cdot\|\boldsymbol{x}-\boldsymbol{y}\|^2/c^2)$ with $c>0$ can be used, where c is the scaling parameter. Both c as well as parameters C_j and ε can be adjusted to minimize the generalization error calculated using a cross-validation method [1].

4.4.5 Neural networks

Artificial Neural Networks (ANNs) is a large area of research by itself [37]. From the perspective of surrogate modeling, ANNs can be viewed as just another way of approximating sampled high-fidelity model data to create a surrogate model.

The most important component of an NN is the neuron (or single-unit perceptron) [37]. A neuron realizes a nonlinear operation illustrated in Figure 4.4(a), where w_1 through w_n are regression coefficients, β is the bias value of the neuron, and T is a user-defined slope parameter. The most common NN architecture is the multilayer feed-forward network shown in Figure 4.4(b).

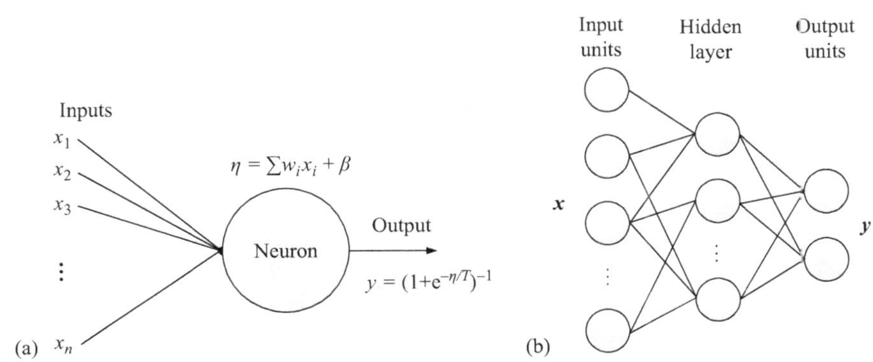

Figure 4.4 Basic concepts of artificial neural networks: (a) structure of a neuron and (b) two-layer feed-forward neural network architecture

Construction of an NN model is a two-step process: (i) selection of the network architecture and (ii) training, i.e., the assignment of the values to the regression parameters. The network training can be formulated as a nonlinear least-squares regression problem. A popular technique for solving this regression problem is the error back-propagation algorithm [3,37].

4.4.6 Other methods

As mentioned before, a large variety of data-driven modeling techniques are available, the ones outlined in this section are the most widely used ones. Another popular approach is moving least squares [38], where the error contribution from each training point $\boldsymbol{x}^{(i)}$ is multiplied by a weight ω_i that depends on the distance between $\boldsymbol{x}$ and $\boldsymbol{x}^{(i)}$. A typical choice for the weights is:

$$\omega_i\left(\|\boldsymbol{x} - \boldsymbol{x}^{(i)}\|\right) = \exp\left(-\|\boldsymbol{x} - \boldsymbol{x}^{(i)}\|^2\right). \tag{4.13}$$

It should be noted that incorporating weights improves the flexibility of the model, however, at the expense of increased computational complexity, since computing the approximation for each point $\boldsymbol{x}$ requires solving a new optimization problem.

Gaussian process regression (GPR) [29] is another data-driven modeling technique that, as kriging, addresses the approximation problem from a stochastic point view. From this perspective, and since Gaussian processes are mathematically tractable, it is relatively easy to compute error estimations for GPR-based surrogates in the form of uncertainty distributions. Under certain conditions, GPR models can be shown to be equivalent to large NNs while requiring significantly less regression parameters than NNs.

The last technique to be mentioned here is cokriging [39,40]. It is an interesting variation of the standard kriging interpolation, which allows for combining information from computational models of various fidelity. The major advantage of cokriging is that—by exploiting knowledge embedded in the low-fidelity model—the surrogate can be created at much lower computational cost than for the models exclusively based on high-fidelity data. Cokriging is a relatively recent method with a few (but growing) applications in engineering [40–43].

4.5 Physics-based models

Data-driven models are suitable for solving a range of practical problems. Their two major advantages are versatility and low evaluation cost. A disadvantage is considerable computational cost of training data acquisition, which grows quickly with dimensionality of the design space and the ranges of parameters. Consequently, in many cases, construction of approximation surrogates becomes impractical.

Physics-based models are quantitatively different because of being constructed by suitable correction or enhancement of an underlying low-fidelity model. The correction process aims at improving alignment between the low-fidelity and the high-fidelity models, either locally or across the entire design space. As the low-fidelity model embeds certain knowledge about the system of interest (e.g., due to evaluating the same system at the level of circuit theory rather than EM analysis as in many microwave engineering problems), the models are normally well correlated. As a result, only a limited amount of high-fidelity data is normally sufficient to ensure reasonable accuracy of the surrogate model. For the same reason, physics-based surrogates typically feature good generalization.

To illustrate the concept of physics-based surrogate modeling, let us consider a simple example. We denote by $c(\boldsymbol{x})$ a low-fidelity model of the device or system of interest. Consider a simple case of multiplicative response correction, considered in the context of the surrogate-based optimization [14]. The optimization algorithm produces a sequence $\{\boldsymbol{x}^{(i)}\}$ of approximate solutions to the original problem $\boldsymbol{x}^* = argmin\{\boldsymbol{x} : f(\boldsymbol{x})\}$. From the algorithm convergence standpoint (particularly if the algorithm is embedded in the trust-region framework [14]), a local alignment between the surrogate and the high-fidelity model is of fundamental importance. The surrogate $\boldsymbol{s}^{(i)}(\boldsymbol{x})$ at iteration i can be constructed as:

$$\boldsymbol{s}^{(i)}(\boldsymbol{x}) = \beta_k(\boldsymbol{x})\boldsymbol{c}(\boldsymbol{x}), \tag{4.14}$$

where $\beta_k(\boldsymbol{x}) = \beta_k(\boldsymbol{x}^{(i)}) + \nabla\beta(\boldsymbol{x}^{(i)})^T(\boldsymbol{x}-\boldsymbol{x}^{(i)})$, where $\beta(\boldsymbol{x}) = \boldsymbol{f}(\boldsymbol{x})/\boldsymbol{c}(\boldsymbol{x})$. This ensures a so-called zero- and first-order consistency between $\boldsymbol{s}$ and $\boldsymbol{f}$, i.e., agreement of function values and their gradients at $\boldsymbol{x}^{(i)}$ [44]. Figure 4.5 illustrates correction (4.14) for the exemplary models based on analytical functions.

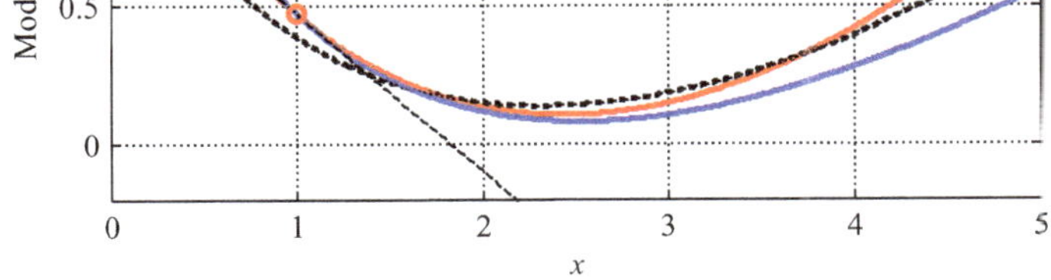

Figure 4.5 Visualization of the response correction (4.14) for the analytical functions **c** *(low-fidelity model) and f (high-fidelity model). The correction established at* $\mathbf{x}_0 = 1$*. It can be observed that corrected model (surrogate* **s***) exhibits good alignment with the high-fidelity model in relatively wide vicinity of* $\mathbf{x}_0$*, especially compared to the first-order Taylor model set up using the same data from f (the value and the gradient at* $\mathbf{x}_0$*)*

4.5.1 Variable fidelity models

Low-fidelity models are fundamental components of physics-based surrogates. They are normally problem dependent, i.e., need to be set up individually for a given high-fidelity model. In electrical engineering, especially high-frequency electronics, full-wave EM analysis is often used as an original (high-fidelity) model. Typical low-fidelity model choices include analytical representations (rather rare due to complexity of contemporary components and devices), equivalent circuit models (commonly used for, e.g., microwave devices, such as filters, couplers, and power dividers), or coarse-mesh full-wave EM simulation models (utilized for, e.g., antenna structures [12]). The last group is the most versatile because it can always be set up for any EM-based high-fidelity model. Here, we use it to discuss certain practical options concerning low-fidelity model selection.

Consider a microstrip antenna shown in Figure 4.6(a) and discretization of its high-fidelity model (here, using a tetrahedral mesh). With the discrete solvers, it is the discretization density that has the strongest impact on the accuracy and computational time of a particular antenna model. At the same time, the discretization density or mesh quality is probably the most efficient way to trade accuracy for speed. Therefore, a straightforward way to create a low-fidelity model of the antenna is through coarser mesh settings compared to those of the high-fidelity antenna model, e.g., as illustrated in Figure 4.6(b). Because of possible simplifications, the low-fidelity model can be faster than the high-fidelity one by a factor 10–50. However, the low-fidelity model is obviously not as accurate as the high-fidelity one.

Figure 4.7 shows the high- and low-fidelity model responses at a specific design for the antenna of Figure 4.6 obtained with different meshes, as well as the relationship between mesh coarseness and simulation time. Clearly, coarser mesh results in shorter simulation time but it is obtained at the expense of compromising

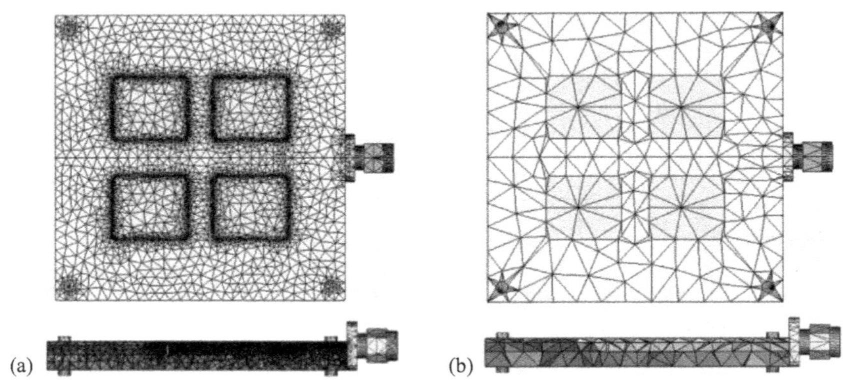

Figure 4.6 Microstrip antenna [12]: (a) high-fidelity model shown with a fine tetrahedral mesh and (b) low-fidelity model shown with a much coarser mesh

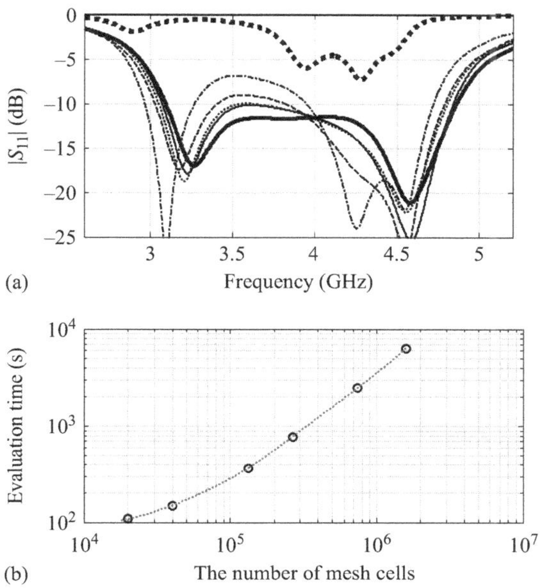

Figure 4.7 Antenna of Figure 4.6 at a selected design simulated with the CST Microwave Studio: (a) reflection response of different discretization densities, 19,866 cells (), 40,068 cells (·—·), 266,396 cells (– –), 413,946 cells (···), 740,740 cells (—), and 1,588,608 cells (▬); and (b) the antenna simulation time versus the number of mesh cells [12]

the model alignment. In practice, the appropriate choice of the low-fidelity model discretization density is not a trivial task and maybe critical for the predictive power of the physics-based surrogate, in which the low-fidelity model is used. More information about possible simplifications that can be made to establish the low-fidelity model and model selection trade-offs can be found in the literature [12,45].

4.5.2 Space mapping

SM [46] is one of the most popular physics-based modeling techniques in high-frequency electronics. Here, we briefly discuss several types of SM assuming single-point correction (multipoint generalization is straightforward [47]). The first case is input SM (ISM), where correction is applied at the level of the model domain. Here, we use vector notation for the models, i.e., $\boldsymbol{c}$ and $\boldsymbol{s}$ for the low-fidelity and the surrogate models, respectively. The surrogate is defined as:

$$\boldsymbol{s}(\boldsymbol{x}) = \boldsymbol{c}(\boldsymbol{x} + \boldsymbol{q}). \tag{4.15}$$

The parameter vector $\boldsymbol{q}$ is obtained by minimizing $||\boldsymbol{f}(\boldsymbol{x}) - \boldsymbol{c}(\boldsymbol{x} + \boldsymbol{q})||$, which represents the misalignment between the surrogate and the high-fidelity model.

Figure 4.8 shows an example of a microwave filter structure [48] evaluated using EM simulation (high-fidelity model), its equivalent circuit (low-fidelity model), and the corresponding responses (here, so-called transmission characteristics $|S_{21}|$ versus frequency) before and after applying the ISM correction. Figure 4.8(d) indicates excellent generalization capability of the surrogate. A multipoint version of IMS may use more involved domain mapping, e.g., $\boldsymbol{s}(\boldsymbol{x}) = \boldsymbol{c}(\boldsymbol{B}\boldsymbol{x} + \boldsymbol{q})$, where $\boldsymbol{B}$ is a square matrix.

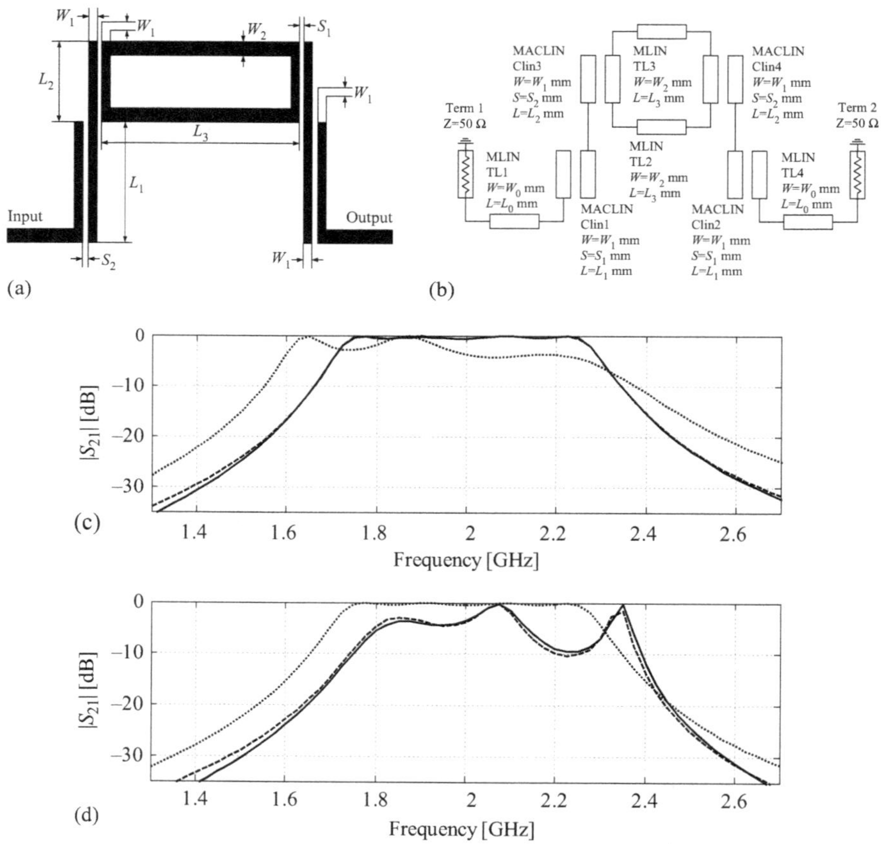

Figure 4.8 Low-fidelity model correction through parameter shift (input space mapping) [14]: (a) microstrip filter geometry (high-fidelity model **f** *evaluated using EM simulation); (b) low-fidelity model* **c** *(equivalent circuit); (c) response of* **f** *(—) and* **c** *(····), as well as response of the surrogate model* **s** *(- - -) created using input space mapping; (d) surrogate model verification at a different design (other than that at which the model was created)* **f** *(—),* **c** *(····), and* **s** *(- - -)*

Another type of model correction may involve additional parameters that are normally fixed in the high-fidelity model but can be adjusted in the low-fidelity one. The latter is just an auxiliary design tool after all: it is not supposed to be built or measured. This concept is utilized, among others, in the so-called implicit SM technique [49], where the surrogate is created as:

$$\boldsymbol{s}(\boldsymbol{x}) = \boldsymbol{c}_I(\boldsymbol{x}, \boldsymbol{p}), \tag{4.16}$$

where $\boldsymbol{c}_I$ is the low-fidelity model with the explicit dependence on additional parameters $\boldsymbol{p}$. The vector $\boldsymbol{p}$ is obtained by minimizing $||\boldsymbol{f}(\boldsymbol{x}) - \boldsymbol{c}_I(\boldsymbol{x},\boldsymbol{p})||$.

For the sake of illustration, consider again the filter of Figure 4.8(a) and (b) with the implicit SM parameters being dielectric permittivities of the microstrip-line component substrates (rectangle elements in Figure 4.8(b)). Normally, the entire circuit is fabricated on a dielectric substrate with a specified (and fixed) characteristics. For the sake of correcting the low-fidelity model, the designer is free to adjust these characteristics (in particular, the value of dielectric permittivity) individually for each component of the equivalent circuit. Figure 4.9 shows the responses before and after applying the implicit SM correction. Again, good generalization of the surrogate model can be observed.

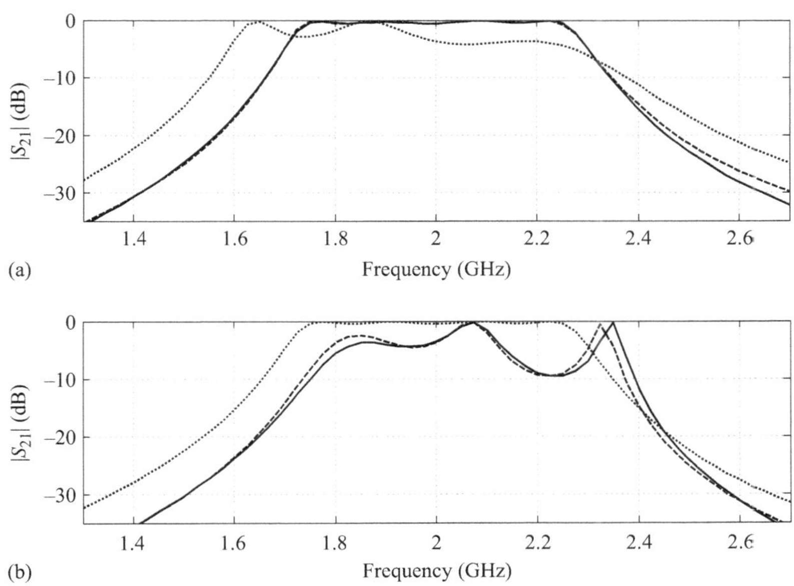

Figure 4.9 Low-fidelity model correction through implicit space mapping applied to a microstrip filter of Figure 4.8 [14]: (a) response of $\boldsymbol{f}$ *(—) and* **c** *(····), as well as response of the surrogate model* **s** *(- - -) created using implicit space mapping; (b) surrogate model verification at a different design (other than that at which the model was created)* $\boldsymbol{f}$ *(—),* **c** *(····), and* **s** *(- - -)*

In many cases, vector-valued responses of the system are actually evaluations of the same design but at different values of certain parameters, such as the time, frequency (e.g., for microwave structures), or a specific geometry parameter (e.g., chord line coordinate for the airfoil profiles). In such situations, it might be convenient to apply a linear or nonlinear scaling to this parameter so as to shape the response accordingly. A good example of such a correction procedure is frequency scaling often utilized in electrical engineering [47]. As a matter of fact, for many components simulated using EM solvers, a frequency shift is the major type of discrepancy between the low- and high-fidelity models. Let us consider a simple frequency scaling procedure, also referred to as frequency SM [47]. We assume that $\boldsymbol{f}(\boldsymbol{x}) = [f(\boldsymbol{x},\omega_1)\ \ f(\boldsymbol{x},\omega_2)\cdots f(\boldsymbol{x},\omega_m)]^T$, where $f(\boldsymbol{x},\omega_k)$ is the evaluation of the high-fidelity model at a frequency ω_k, whereas ω_1 through ω_m represent the entire discrete set of frequencies at which the model is evaluated. A similar convention is used for the low-fidelity model. The frequency-scaled model $\boldsymbol{s}_F(\boldsymbol{x})$ is defined as:

$$\boldsymbol{s}_F(\boldsymbol{x},[F_0\ \ F_1]) = [\boldsymbol{c}(\boldsymbol{x},F_0+F_1\omega_1)\cdots\boldsymbol{c}(\boldsymbol{x},F_0+F_1\omega_m)]^T, \tag{4.17}$$

where F_0 and F_1 are scaling parameters obtained to minimize misalignment between $\boldsymbol{s}$ and $\boldsymbol{f}$ at a certain reference design $\boldsymbol{x}^{(i)}$ as:

$$[F_0,F_1] = \arg\min_{[F_0,F_1]} ||\boldsymbol{f}(\boldsymbol{x}) - \boldsymbol{s}_F(\boldsymbol{x},[F_0\ \ F_1])||. \tag{4.18}$$

An example of frequency scaling applied to the low-fidelity model of a substrate-integrated cavity antenna [12] is shown in Figure 4.10. Here, both the low- and high-fidelity models are evaluated using EM simulation ($\boldsymbol{c}$ with coarse discretization).

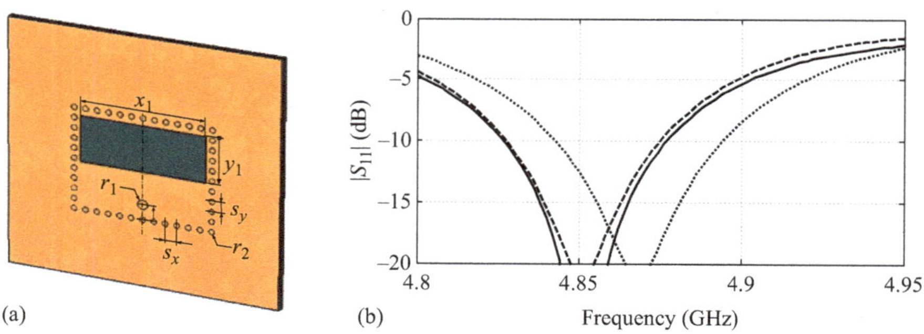

*Figure 4.10 Low-fidelity model correction through frequency scaling [14]: (a) antenna geometry (both **f** and **c** evaluated using EM simulation, coarse discretization used for **c**); (b) response of **f** (—) and **c** (····), as well as response of the surrogate model **s** (- - -) created using frequency scaling as in (4.18) and (4.19)*

4.5.3 Response correction methods

Response correction techniques aim at reducing misalignment between the low- and high-fidelity models at the level of the model output. In the optimization context, it is often performed to improve the model matching at the current design, so that the surrogate model obtained this way maybe used as a reliable prediction tool that allows us to find a better design. There are three main groups of the function correction techniques: compositional, additive, or multiplicative corrections. We will briefly illustrate each of these categories for correcting the low-fidelity model $c(\boldsymbol{x})$, as well as discuss if zero- and first-order consistency conditions with $f(\boldsymbol{x})$ [44] can be satisfied. Here, we assume that the models are scalar functions; however, generalizations for vector-valued functions are straightforward in some cases.

The compositional correction [4]:

$$s^{(i+1)}(\boldsymbol{x}) = g(c(\boldsymbol{x})) \tag{4.19}$$

is a simple scaling of the objective function. Since the mapping g is a real-valued function of a real variable, a compositional correction will not in general yield first-order consistency conditions. By selecting a mapping g that satisfies:

$$g'(c(\boldsymbol{x}^{(i)})) = \frac{\nabla f(\boldsymbol{x}^{(i)})\nabla c(\boldsymbol{x}^{(i)})^T}{\nabla c(\boldsymbol{x}^{(i)})\nabla c(\boldsymbol{x}^{(i)})^T} \tag{4.20}$$

the discrepancy between $\nabla f(\boldsymbol{x}^{(i)})$ and $\nabla s^{(i+1)}(\boldsymbol{x}^{(i)})$ (expressed in Euclidean norm) is minimized. The compositional correction can be also introduced in the parameter space [46]:

$$s^{(i+1)}(\boldsymbol{x}) = c(\boldsymbol{p}(\boldsymbol{x})). \tag{4.21}$$

If the ranges of $f(\boldsymbol{x})$ and $c(\boldsymbol{x})$ are different, then the condition $c(\boldsymbol{p}(\boldsymbol{x}^{(i)})) = f(\boldsymbol{x}^{(i)})$ is not achievable. This difficulty can be alleviated by combining both compositional corrections so that g and $\boldsymbol{p}$ take the following forms:

$$g(t) = t - c(\boldsymbol{x}^{(i)}) + f(\boldsymbol{x}^{(i)}), \tag{4.22}$$

$$\boldsymbol{p}(\boldsymbol{x}) = \boldsymbol{x}^{(i)} + \boldsymbol{J}_p(\boldsymbol{x} - \boldsymbol{x}^{(i)}), \tag{4.23}$$

where $\boldsymbol{J}_p$ is an $n \times n$ matrix for which $\boldsymbol{J}_p^T \nabla c = \nabla f(\boldsymbol{x}^{(i)})$ guarantees consistency.

Additive and multiplicative corrections allow obtaining first-order consistency conditions. For the additive case, we can generally express the correction as:

$$s^{(i+1)}(\boldsymbol{x}) = \lambda(\boldsymbol{x}) + s^{(i)}(\boldsymbol{x}). \tag{4.24}$$

The associated consistency conditions require that $\lambda(\boldsymbol{x})$ satisfies $\lambda(\boldsymbol{x}^{(i)}) = f(\boldsymbol{x}^{(i)}) - c(\boldsymbol{x}^{(i)})$ and $\nabla\lambda(\boldsymbol{x}^{(i)}) = \nabla f(\boldsymbol{x}^{(i)}) - \nabla c(\boldsymbol{x}^{(i)})$. These can be obtained by the following linear additive correction:

$$s^{(i+1)}(\boldsymbol{x}) = f(\boldsymbol{x}^{(i)}) - c(\boldsymbol{x}^{(i)}) + (\nabla f(\boldsymbol{x}^{(i)}) - \nabla c(\boldsymbol{x}^{(i)}))(\boldsymbol{x} - \boldsymbol{x}^{(i)}) + c(\boldsymbol{x}). \tag{4.25}$$

Multiplicative corrections (also known as the β-correlation method [4]) can be represented generically by:

$$s^{(i+1)}(\boldsymbol{x}) = \alpha(\boldsymbol{x})s^{(i)}(\boldsymbol{x}). \tag{4.26}$$

If $c(\boldsymbol{x}^{(i)}) \neq 0$, zero- and first-order consistency can be achieved if $\alpha(\boldsymbol{x}^{(i)}) = f(\boldsymbol{x}^{(i)})/c(\boldsymbol{x}^{(i)})$ and $\nabla\alpha(\boldsymbol{x}^{(i)}) = [\nabla f(\boldsymbol{x}^{(i)}) - f(\boldsymbol{x}^{(i)})/c(\boldsymbol{x}^{(i)})\nabla c(\boldsymbol{x}^{(i)})]/c(\boldsymbol{x}^{(i)})$.

The requirement $c(\boldsymbol{x}^{(i)}) \neq 0$ is not strong in practice, since very often the range of $f(\boldsymbol{x})$ (and thus, of the surrogate $c(\boldsymbol{x})$) is known beforehand, and hence, a bias can be introduced both for $f(\boldsymbol{x})$ and $c(\boldsymbol{x})$ to avoid cost function values equal to zero. In these circumstances, the following multiplicative correction:

$$s^{(i+1)}(\boldsymbol{x}) = \left[\frac{f(\boldsymbol{x}^{(i)})}{c(\boldsymbol{x}^{(i)})} + \frac{\nabla f(\boldsymbol{x}^{(i)})c(\boldsymbol{x}^{(i)}) - f(\boldsymbol{x}^{(i)})\nabla c(\boldsymbol{x}^{(i)})}{c(\boldsymbol{x}^{(i)})^2}(\boldsymbol{x} - \boldsymbol{x}^{(i)})\right]c(\boldsymbol{x}) \tag{4.27}$$

ensures both zero- and first-order consistency between $s^{(i+1)}$ and f.

Response correction maybe realized in a multipoint manner (generalizing a so-called output SM concept [46]), with the low-fidelity model correction determined using the high-fidelity model data at several points (designs). The multipoint response correction surrogate can be formulated as (here, for vector-valued models, and assuming the optimization context with the surrogate model $\boldsymbol{s}^{(i)}$ established at the current iteration point $\boldsymbol{x}^{(i)}$) [14]:

$$\boldsymbol{s}^{(i)}(\boldsymbol{x}) = \boldsymbol{\Lambda}^{(i)} \circ \boldsymbol{c}(\boldsymbol{x}) + \boldsymbol{\Delta}_r^{(i)} + \boldsymbol{\delta}^{(i)}, \tag{4.28}$$

where $\Lambda^{(i)}$, $\Delta_r^{(i)}$, and $\delta^{(i)}$ are column vectors, whereas $\circ$ denotes component-wise multiplication. The global response correction parameters $\Lambda^{(i)}$ and $\Delta_r^{(i)}$ are obtained as:

$$[\boldsymbol{\Lambda}^{(i)}, \boldsymbol{\Delta}_r^{(i)}] = \arg\min_{[\Lambda,\Delta_r]} \sum_{k=0}^{i} \|\boldsymbol{f}(\boldsymbol{x}^{(k)}) - \boldsymbol{\Lambda} \circ \boldsymbol{c}(\boldsymbol{x}^{(k)}) + \boldsymbol{\Delta}_r\|^2 \tag{4.29}$$

i.e., the response scaling is supposed to improve the matching for all previous iteration points. The points $\boldsymbol{x}^{(k)}$, $k = 1, \ldots, i$ may represent the optimization history (previously considered design on the optimization path) or be elements of the training set allocated for the purpose of surrogate model construction. The (local) additive response correction term $\delta^{(i)}$ is then defined as:

$$\boldsymbol{\delta}^{(i)} = \boldsymbol{f}(\boldsymbol{x}^{(i)}) - [\boldsymbol{\Lambda}^{(i)} \circ \boldsymbol{c}(\boldsymbol{x}^{(i)}) + \boldsymbol{\Delta}_r^{(i)}], \tag{4.30}$$

i.e., it ensures a perfect match between the surrogate and the high-fidelity model at the current design $\boldsymbol{x}^{(i)}$, $\boldsymbol{s}^{(i)}(\boldsymbol{x}^{(i)}) = \boldsymbol{f}(\boldsymbol{x}^{(i)})$. Note that the local correction (4.30) is primarily used in the optimization context, when the surrogate has to be at least zero-order consistent with the high-fidelity model at the current iteration point $\boldsymbol{x}^{(i)}$.

It should be noted that all of the correction terms $\Lambda^{(i)}$, $\boldsymbol{\Delta}_r^{(i)}$, and $\delta^{(i)}$ can be obtained analytically. Let $\boldsymbol{\Lambda}^{(i)} = \left[\lambda_1^{(i)}\ \lambda_2^{(i)} \cdots \lambda_m^{(i)}\right]^T$, $\boldsymbol{\Delta}_r^{(i)} = \left[\delta_1^{(i)}\ \delta_2^{(i)} \cdots \delta_m^{(i)}\right]^T$, $\boldsymbol{c}(\boldsymbol{x}) = [c_1(\boldsymbol{x})\ c_2(\boldsymbol{x}) \cdots c_m(\boldsymbol{x})]^T$, and $\boldsymbol{f}(\boldsymbol{x}) = [f_1(\boldsymbol{x})\ f_2(\boldsymbol{x}) \cdots f_m(\boldsymbol{x})]^T$. We have:

$$\begin{bmatrix} \lambda_k^{(i)} \\ \delta_k^{(i)} \end{bmatrix} = (\boldsymbol{C}_k^T \boldsymbol{C}_k)^{-1} \boldsymbol{C}_k^T \boldsymbol{F}_k, \tag{4.31}$$

where:

$$\boldsymbol{C}_k = \begin{bmatrix} c_k(\boldsymbol{x}^{(0)}) & \cdots & c_k(\boldsymbol{x}^{(i)}) \\ 1 & \cdots & 1 \end{bmatrix}^T \tag{4.32}$$

$$\boldsymbol{F}_k = \left[f_k(\boldsymbol{x}^{(0)}) \quad \cdots \quad f_k(\boldsymbol{x}^{(i)})\right]^T \tag{4.33}$$

for $k = 1, \ldots, m$, which is a least-square optimal solution to the linear regression problems $c_k\lambda_k^{(i)} + \delta_k^{(i)} = f_k, \quad k = 1, \ldots, m$, equivalent to (4.29). Note that the matrices $\boldsymbol{C}_k^T\boldsymbol{C}_k$ are non-singular for $i > 1$. For $i = 1$, only the multiplicative correction with $\Lambda^{(i)}$ components is used (calculated in a similar way).

In order to illustrate the multipoint response correction, consider a dielectric resonator antenna (DRA) shown in Figure 4.11 [50]. The DRA is suspended above the ground plane in order to achieve enhanced impedance bandwidth. The design specifications imposed on the reflection coefficient of the antenna are $|S_{11}| \le -15$ dB for 5.1–5.9 GHz. The design variables are $\boldsymbol{x} = [a_x\ a_y\ a_z\ a_c\ u_s\ w_s\ y_s\ g_1\ b_y]^T$. The high-fidelity model $\boldsymbol{f}$ is simulated using the CST MWS transient solver (~800,000 mesh cells, evaluation time 20 min). The low-fidelity model $\boldsymbol{c}$ is also evaluated in CST

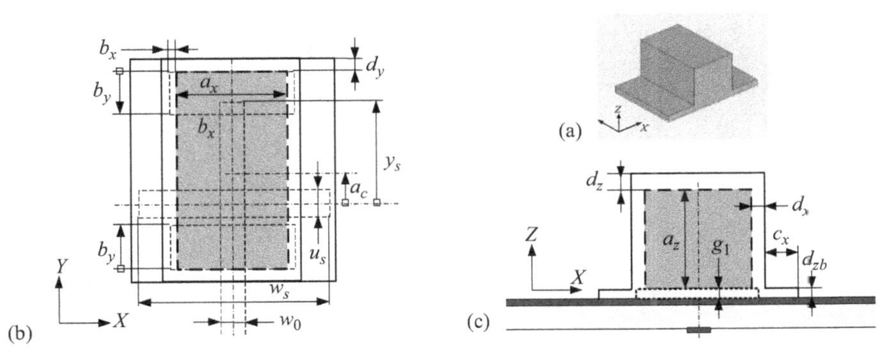

Figure 4.11 Suspended DRA [50]: (a) 3D view of its housing, top (b) and front (c) views

(~30,000 mesh cells, evaluation time 40 s). The initial design is $\boldsymbol{x}^{(0)} = [8.0\ 14.0\ 8.0\ 0.0\ 2.0\ 8.0\ 4.0\ 2.0\ 6.0]^T$ mm. The antenna responses at the initial design and the design found using multipoint response correction surrogate are shown in Figure 4.12 (see [50] for details). Figure 4.13 demonstrates a better generalization capability of the multipoint surrogate as compared to a single-point correction. The misalignment between $\boldsymbol{c}$ and $\boldsymbol{f}$ shown in Figure 4.13(a) is perfectly removed by single-point correction at the design where it is established, but the alignment is not as good for other

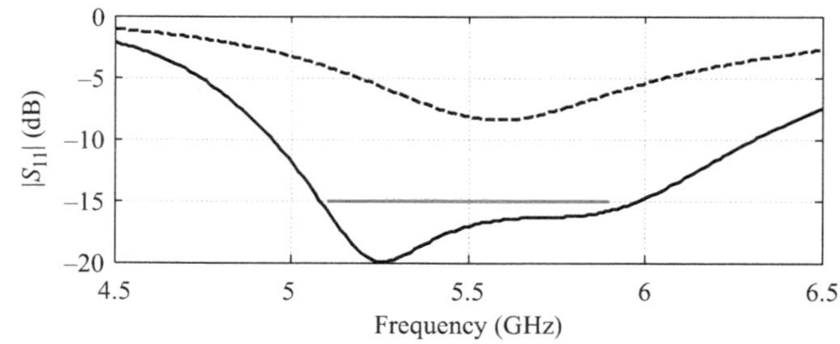

Figure 4.12 Suspended DRA: high-fidelity model response at the initial design (- - -) and at the optimized design obtained using multipoint response correction (—)

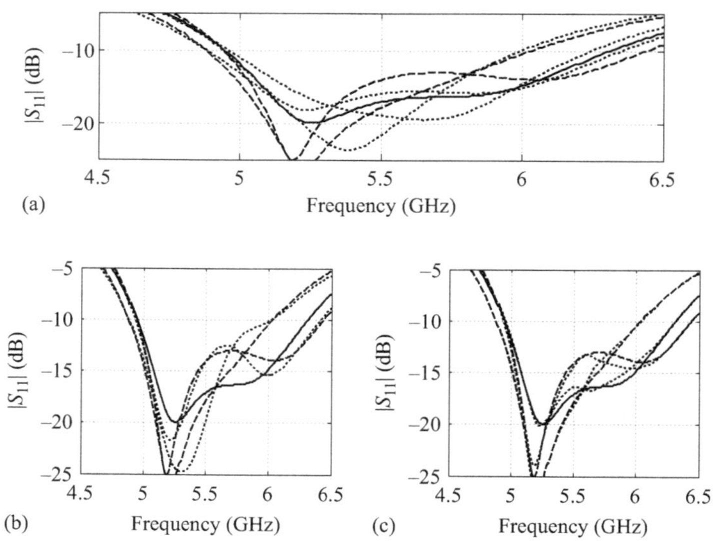

Figure 4.13 Suspended DRA: (a) low- (····) and high- (- - - and —) fidelity model responses at three designs, (b) single-point-corrected low- (····) and high- (- - - and —) fidelity model responses at the same designs (single-point correction at the design marked —), (c) multipoint-corrected low- (····) and high- (- - - and —) fidelity model responses

designs. On the other hand, as shown in Figure 4.13(c), the multipoint response correction improves model alignment at all designs involved in the model construction (Figure 4.13(c) only shows the global part $\Lambda^{(i)\circ}c(x) + \Delta_r^{(i)}$ without $\delta^{(i)}$ which would give a perfect alignment at $x^{(i)}$).

4.5.4 Feature-based modeling

The physics-based modeling techniques presented so far in this section are all parameterized ones, i.e., the low-fidelity model correction is realized using explicit certain transformations, the coefficient of which has to be extracted from the training data. Parameter-less methods are also available, including adaptive response prediction [51] or shape-preserving response prediction [52]. Here, a brief overview of another approach, referred to as feature-based modeling [53], is provided.

Feature-based modeling aims at reducing the number of training data samples necessary to construct an accurate surrogate model by reformulating modeling process and conducting it at the level of appropriately defined characteristic points of the system responses (so-called response features) [53]. The methodology is explained and demonstrated using the case of narrow-band antenna modeling. The high-fidelity model $f(x)$ is obtained from full-wave EM simulation. It represents the reflection coefficient $|S_{11}|$ of the antenna evaluated at m frequencies, ω_1 to ω_m, i.e., we have $f(x) = [f(x, \omega_1) \cdots f(x, \omega_m)]^T$. The objective is to build a replacement model (surrogate) s. The surrogate should represent the EM model in a given region X of the design space.

The set of training samples is denoted as $X_T = \{x^1, x^2, \ldots, x^N\} \subset X$. The corresponding EM model responses $f(x^k)$ are acquired beforehand. Within a conventional data-driven surrogate modeling paradigm, the responses $f(x,\omega_j)$, $j = 1, \ldots, m$, are approximated directly (either separately for each frequency or by treating the frequency as an additional input parameter of the model). The fundamental problem is nonlinearity of the responses as shown in Figure 4.14.

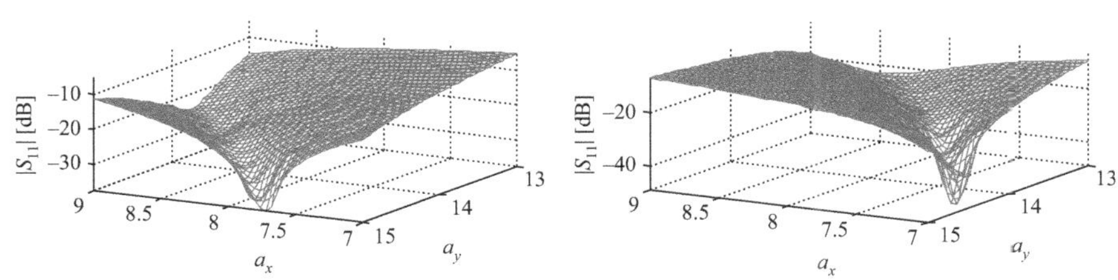

Figure 4.14 Exemplary responses of the dielectric resonator antenna considered in Section 3.1 (here, reflection coefficient $|S_{11}|$). The responses are evaluated in the region $7.0 \leq a_x \leq 9.0$ and $13.0 \leq a_y \leq 15.0$ at the frequencies of (a) 5.3 GHz and (b) 5.5 GHz. Other variables are fixed to the following values: $a_z = 9$ $a_c = 0$ $u_s = 2$ $w_s = 10$ $y_s = 8$ (all in mm). (Note: Rapid changes of the response levels at certain regions of the design space.)

Figure 4.15 clarifies the definition of the feature points in the case of narrow-band antennas. The characteristic point set is constructed sequentially as follows: (i) identification of the primary point which corresponds to the center frequency (antenna resonance) and the response level at that frequency; (ii) allocation of the supplemental points (here, uniformly with respect to the level and separately on the left- and right-hand-side of the primary point); (iii) allocation of the infill points uniformly in frequency in between the supplemental points. Clearly, one needs to ensure that the number of characteristic points is sufficient so as to allow reliable synthesis of the antenna response (here, through interpolation).

On the other hand, although it is important that the major features of the response (e.g., antenna resonance) are accounted for, particular point allocation is not critical. The response features, once defined, can be easily extracted using simple postprocessing of EM simulation results. For subsequent considerations, the jth feature point of $\boldsymbol{f}(\boldsymbol{x}^k)$ ($j = 1,\ldots, K$, $k = 1,\ldots, N$) will be denoted as $\boldsymbol{f}_k^j = \left[\omega_k^j\ l_k^j\right]$. Here, ω_k^j and l_k^j represent the frequency and the magnitude (level) components of $\boldsymbol{f}_k^j$, respectively.

For the sake of illustration, the frequency and level components of the selected feature points have been shown in Figure 4.16. The considered design space region is the same as the one shown in Figure 4.15. It is important that the functional landscapes of the feature points are not as nonlinear as those shown in Figure 4.15. This is particularly the case of the frequency component. Clearly, it is to be expected that the construction of a reliable surrogate model at the feature point level will require a smaller number of training samples than modeling the reflection response in a traditional manner.

Having the response features defined, the surrogate modeling process works as follows: first, the data-driven models $s_{\omega.j}(\boldsymbol{x})$ and $s_{l.j}(\boldsymbol{x})$, $j = 1,\ldots, K$, of the sets of

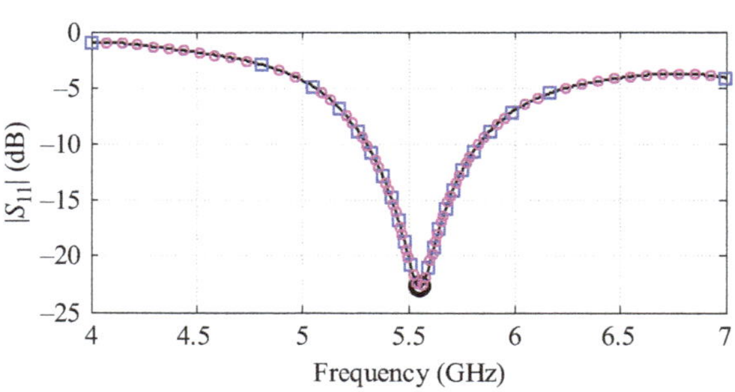

Figure 4.15 Definition of response features in the case of a narrow-band antenna reflection characteristic: the primary point (here, corresponding to the antenna resonance) is represented as ○; □ represents supplemental points distributed equally with respect to response level, (○) denotes infill points distributed equally in frequency between the main and supplemental points (note that the number of points maybe different for various intervals)

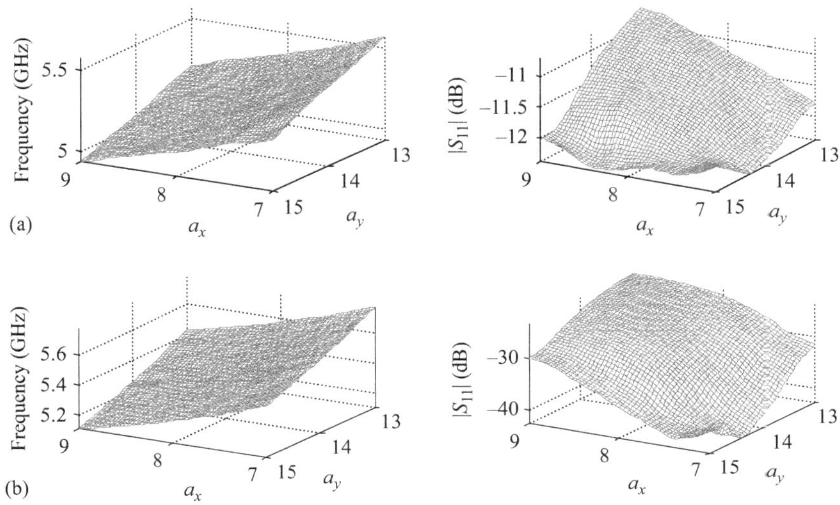

Figure 4.16 Frequency (left panel) and level (right panel) components as functions of geometry parameters a_x and a_y (with other variables fixed) for the three selected feature points. The responses are evaluated for the same design space region as that considered in Figure 4.15. Note that the functional landscapes of the feature point coordinates are considerably less nonlinear than those for original responses (cf. Figure 4.14)

corresponding feature points are constructed using available training designs, $\{f^j_1, f^j_2, \ldots, f^j_N\}$, $j = 1, \ldots, K$ [13]. At this stage, we utilize kriging interpolation. The surrogate model itself is defined as:

$$\boldsymbol{s}(\boldsymbol{x}) = [s(\boldsymbol{x}, \omega_1) \cdots s(\boldsymbol{x}, \omega_m)]^T \tag{4.34}$$

where its jth component is given as:

$$s(\boldsymbol{x}, \omega_j) = I(\boldsymbol{\Omega}(\boldsymbol{x}), \boldsymbol{L}(\boldsymbol{x}), \omega_j) \tag{4.35}$$

In (4.35), $\boldsymbol{L}(\boldsymbol{x}) = [s_{l\cdot 1}(\boldsymbol{x}) \;\; s_{l\cdot 2}(\boldsymbol{x}) \cdots s_{l\cdot K}(\boldsymbol{x})]$ and $\boldsymbol{\Omega}(\boldsymbol{x}) = [s_{\omega\cdot 1}(\boldsymbol{x}) s_{\omega\cdot 2}(\boldsymbol{x}) \cdots s_{\omega\cdot K}(\boldsymbol{x})]$ are the predicted feature point locations corresponding to the evaluation design $\boldsymbol{x}$. As we are interested in evaluating the antenna response at a discrete set of frequencies ω_1 through ω_m, it is necessary to interpolate both the level vector $\boldsymbol{L}$ and frequency vector Ω into the response at the previous set of frequencies. This interpolation is represented as $I(\Omega, \boldsymbol{L}, \omega)$.

Feature-based modeling is illustrated using a rectangular-slot-coupled patch antenna shown in Figure 4.17, implemented on a 0.762-mm thick Taconic RF-35 substrate ($\varepsilon_r = 3.5$, $\tan\delta = 0.0018$). Antenna parameters are represented by vector $\boldsymbol{x} = [d \; d_1 \; l_1 \; s_1 \; s_2 \; w_1]^T$. Variable $d_2 = d_1$, whereas dimensions $w_0 = 3$, $l_0 = 30$, and $s_0 = 0.15$ remain fixed (all parameters are in mm). The high-fidelity

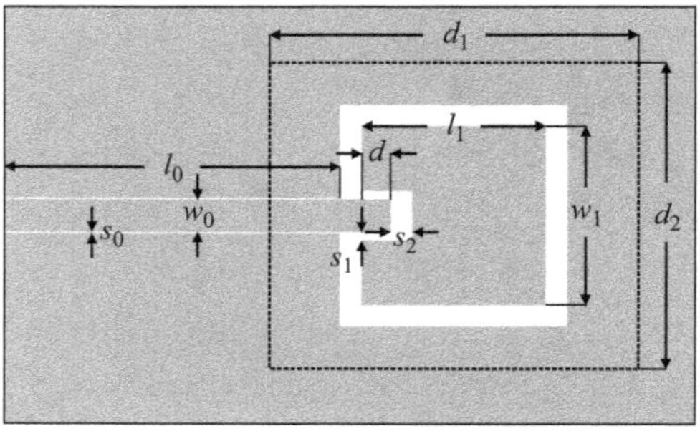

Figure 4.17 Rectangular-slot-coupled patch antenna: geometry with marked design parameters. Gray and white colors represent the ground plane and slots, respectively, whereas the patch radiator (located on the opposite side of the substrate) is marked using a dashed line

Table 4.1 Modeling results of the patch antenna coupled to rectangular slot

Model	**Average error (%)**					
	$N^a = 20$	$N = 50$	$N = 100$	$N = 200$	$N = 400$	$N = 800$
Feature-based surrogate	32.3	25.9	14.4	11.7	9.3	7.5
Kriging interpolation[b]	99.5	99.2	74.2	71.9	53.9	46.4

[a] N stands for the number of training points.
[b] Direct kriging interpolation of high-fidelity model $|S_{11}|$ responses.

model is implemented in CST Microwave Studio. The frequency range considered for modeling is from 1 to 4 GHz. The surrogate model of the reflection coefficient $|S_{11}|$ is constructed in the region $X = [\boldsymbol{x}^{(0)} - \boldsymbol{d}, \boldsymbol{x}^{(0)} + \boldsymbol{d}]$, where $\boldsymbol{x}^{(0)} = [4.5\ 4.0\ 15.0\ 1.2\ 1.2\ 15.0]^T$ mm and $\boldsymbol{d} = [2.5\ 4.0\ 5.0\ 1.0\ 1.0\ 5.0]^T$ mm.

The feature-based surrogate has been constructed using six different training sets. Table 4.1 shows the average error values for particular training sets. It should be noted that because of high nonlinearity of the antenna responses and wide range of geometry parameters, conventional modeling is of poor quality, whereas feature-based modeling ensures practically usable prediction power. The high-fidelity and feature-based surrogate response at the selected test designs are shown in Figure 4.18.

For additional verification, we use the feature-based model for optimization of the antenna structure at two operating frequencies, 1.85 and 3.2 GHz. In all cases, the objective is to minimize reflection level $|S_{11}|$ in the frequency range of ± 50 MHz around a given operating frequency. Figure 4.19 shows the antenna responses at the designs obtained by optimizing the feature-based surrogate. No further design tuning is necessary because of good accuracy of the surrogate.

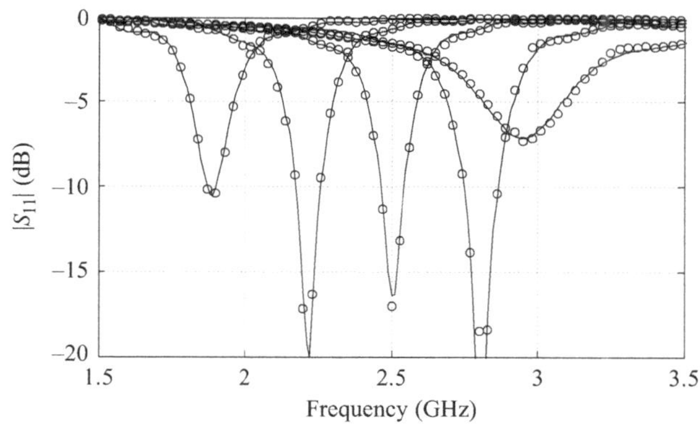

Figure 4.18 Patch antenna: high-fidelity (—) and feature-based model (o) at the selected test designs

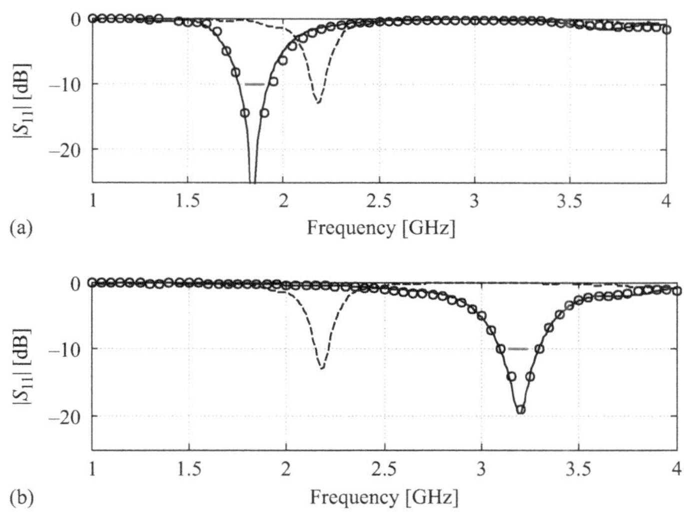

Figure 4.19 Rectangular-slot-coupled patch antenna optimization: the high-fidelity model response at the initial design (- - -) and at the design obtained by optimizing the feature-based surrogate (—). The surrogate model response marked as (o). Results for the operating frequency of (a) 1.83 GHz and (b) 3.2 GHz

4.6 Model selection and validation

Having the model identified, its predictive power has to be estimated in order to find out the actual usefulness of the surrogate. Obviously, it is not a good idea to evaluate the model quality only at the training set locations, because this would usually give an overly optimistic assessment. In particular, interpolative models exhibit zero error by definition. Some of the techniques described previously identify a surrogate model together with some estimation of the attendant approximation error (e.g., kriging or GPR). In general, the model validation can be realized using stand-alone procedures specifically designed for this purpose. The simplest and probably the most popular way for validating a model is the split-sample method [1], where part of the available data set (the training set) is used to construct the surrogate, whereas the second part (the test set) serves purely for model validation. It should be emphasized here that the error estimated by a split-sample method depends strongly on how the set of data samples is partitioned.

A more accurate estimation of the model generalization error can be obtained by cross-validation [1]. In this method, the data set is divided into K subsets, and each of these subsets is sequentially used as testing set for a surrogate constructed on the other $K-1$ subsets. The prediction error can be estimated with all the K error measures obtained in this process (e.g., as an average value). Cross-validation provides an error estimation that is less biased than with the split-sample method. The disadvantage of this method is that the surrogate has to be constructed K times.

As mentioned on several occasions, the entire procedure of allocating samples, acquiring data, model identification, and validation can be iterated until the prescribed surrogate accuracy level has been reached. In each repetition, a new set of training samples is added to the existing ones. The strategies of allocating the new samples (referred to as infill criteria [5]) usually aim at improving the global accuracy of the model, i.e., inserting new samples at the locations where the estimated modeling error is the highest. When the model is constructed in the context of design optimization, other infill criteria can be used. For example, maximization of the probability of improvement, i.e., identifying locations that gives the highest chance of improving the objective function value [5].

Selection of the best type of surrogate model to be used in a particular situation is very much problem dependent. For example, from the point of view of simulation-driven design, the main advantage of data-driven surrogates is that they are very fast. Unfortunately, the high computational cost of setting up such models (related to training data acquisition) is a significant disadvantage. In order to ensure decent accuracy, hundreds or even thousands of data samples are required and the number of training points quickly grows with dimensionality of the design space. Therefore, approximation surrogates are mostly suitable to build multiple-use library models or sub-models for decomposable systems [54]. Their use for ad hoc optimization of expensive computational models is rather limited unless the dimensionality of the parameter space is relatively low.

Availability of fast low-fidelity models is an indication that physics-based surrogates maybe a good choice. For example, in high-frequency electronics, a standard choice of the low-fidelity model is equivalent circuit (versus full-wave EM simulation as the high-fidelity one). In many cases, fast low-fidelity models are not available but coarse-mesh (or simplified physics) simulation models may successfully play such a role, e.g., in antenna design [12] or aerospace engineering [45]. Physics-based surrogates are most typically used for solving simulation-driven optimization tasks.

4.7 Summary

Fast replacement models (or surrogates) are becoming more and more popular in contemporary engineering and science as they are capable of facilitating various simulation-driven design tasks, such as parametric optimization, statistical analysis, or robust design. This chapter gives a brief outline of surrogate modeling techniques. Classification of surrogate models has been followed by a description of the surrogate modeling process, including DoEs, data acquisition, and model identification and validation. Two groups of models have been discussed, including data-driven and physics-based ones. The most popular techniques pertinent to each group were formulated, including polynomial regression, radial basis function interpolation, kriging, SVR, SM, response correction techniques, and feature-based modeling. Finally, qualitative comparison of various modeling techniques as well as their advantages and disadvantages has been carried out. The material in this chapter can be used as the first guide to surrogate modeling and related techniques.

Acknowledgments

The author thanks Dassault Systemes, France, for making CST Microwave Studio available. This work is partially supported by the Icelandic Centre for Research (RANNIS) Grant 174114051.

References

[1] Queipo, N.V., Haftka, R.T., Shyy, W., Goel, T., Vaidynathan, R., and Tucker, P.K. (2005) Surrogate-based analysis and optimization. Progress in Aerospace Sciences, 41, 1–28.
[2] Koziel, S., Yang, X.S., and Zhang, Q.J. (Eds.) (2013) Simulation-Driven Design Optimization and Modeling for Microwave Engineering. London: Imperial College Press.
[3] Simpson, T.W., Peplinski, J., Koch, P.N., and Allen, J.K. (2001) Metamodels for computer-based engineering design: survey and recommendations. Engineering with Computers, 17, 129–150.
[4] Søndergaard, J. (2003) Optimization using surrogate models – by the space mapping technique. *Ph.D. Thesis*, Informatics and Mathematical Modelling, Technical University of Denmark, Lyngby.
[5] Forrester, A.I.J., and Keane, A.J. (2009) Recent advances in surrogate-based optimization. Progress in Aerospace Sciences, 45, 50–79.
[6] Couckuyt, I. (2013) Forward and inverse surrogate modeling of computationally expensive problems. *PhD Thesis*, Ghent University.
[7] Lophaven, S.N., Nielsen, H.B., and Søndergaard, J. (2002) DACE: A Matlab Kriging Toolbox. Technical University of Denmark.

[8] Gorissen, D., Crombecq, K., Couckuyt, I., Dhaene, T., and Demeester, P. (2010) A surrogate modeling and adaptive sampling toolbox for computer based design. Journal of Machine Learning Research, 11, 2051–2055.

[9] Gorissen, D., Couckuyt, I., Laermans, E., and Dhaene, T. (2010b) Multiobjective global surrogate modeling, dealing with the 5-percent problem. Engineering with Computers, 26, 81–98.

[10] Bandler, J.W., Biernacki, R.M., Chen, S.H., Grobelny, P.A., and Hemmers, R.H. (1994) Space mapping technique for electromagnetic optimization. IEEE Transactions on Microwave Theory and Techniques, 42, 2536–2544.

[11] Bandler, J.W., Cheng, Q.S., Gebre-Mariam, D.H., Madsen, K., Pedersen, F., and Søndergaard, J. (2003) EM-based surrogate modeling and design exploiting implicit, frequency and output space mappings. *IEEE Int. Microwave Symp. Digest*, Philadelphia, PA, 1003–1006.

[12] Koziel, S., and Ogurtsov, S. (2014) Antenna Design by Simulation-Driven Optimization. Surrogate-Based Approach. Springer.

[13] Koziel, S., and Bandler, J.W. (2008) Support-vector-regression-based output space-mapping for microwave device modeling. *IEEE MTT-S Int. Microwave Symp. Dig*, Atlanta, GA, 615–618.

[14] Koziel, S., and Leifsson, L. (2016) Simulation-Driven Design by Knowledge-Based Response Correction Techniques. New York, NY: Springer.

[15] Kleijnen, J.P.C. (2009) Kriging metamodeling in simulation: a review. European Journal of Operational Research, 192, 707–716.

[16] Rayas-Sanchez, J.E. (2004) EM-based optimization of microwave circuits using artificial neural networks: the state-of-the-art. IEEE Transactions on Microwave Theory and Techniques, 52, 420–435.

[17] Giunta, A.A., Wojtkiewicz, S.F., and Eldred, M.S. (2003) Overview of modern design of experiments methods for computational simulations. Paper AIAA 2003-0649. American Institute of Aeronautics and Astronautics.

[18] Santner, T.J., Williams, B., and Notz, W. (2003) The Design and Analysis of Computer Experiments. Springer-Verlag.

[19] Koehler, J.R., and Owen, A.B. (1996) Computer experiments. In S. Ghosh and C.R. Rao (Eds.) Handbook of Statistics. Elsevier Science B.V. 13, pp. 261–308.

[20] Devabhaktuni, V.K., Yagoub, M.C.E., and Zhang, Q.J. (2001) A robust algorithm for automatic development of neural-network models for microwave applications. IEEE Transactions on Microwave Theory and Techniques, 49, 2282–2291.

[21] Cheng, Q.S., Koziel, S., and Bandler, J.W. (2006) Simplified space mapping approach to enhancement of microwave device models. International Journal of RF and Microwave Computer-Aided Engineering, 16, 518–535.

[22] McKay, M., Conover, W., and Beckman, R. (1979) A comparison of three methods for selecting values of input variables in the analysis of output from a computer code. Technometrics, 21, 239–245.

[23] Beachkofski, B., and Grandhi, R. (2002) Improved distributed hypercube sampling. Paper AIAA 2002-1274. American Institute of Aeronautics and Astronautics.

[24] Leary, S., Bhaskar, A., and Keane, A. (2003) Optimal orthogonal-array-based Latin hypercubes. Journal of Applied Statistics, 30, 585–598.

[25] Ye, K.Q. (1998) Orthogonal column Latin hypercubes and their application in computer experiments. Journal of the American Statistical Association, 93, 1430–1439.

[26] Palmer, K., and Tsui, K.-L. (2001) A minimum bias Latin hypercube design. IIE Transactions, 33, 793–808.

[27] Golub, G.H., and Van Loan, C.F. (1996) Matrix Computations. 3rd ed. The Johns Hopkins University Press.

[28] Wild, S.M., Regis, R.G., and Shoemaker, C.A. (2008) ORBIT: optimization by radial basis function interpolation in trust-regions. SIAM Journal on Scientific Computing, 30, 3197–3219.

[29] Rasmussen, C.E., and Williams, C.K.I. (2006) Gaussian Processes for Machine Learning. MIT Press, Cambridge, MA.

[30] Journel, A.G., and Huijbregts, C.J. (1981) Mining Geostatistics. Academic Press.

[31] Gunn, S.R. (1998) Support vector machines for classification and regression. Technical Report. School of Electronics and Computer Science, University of Southampton, Southampton.

[32] Angiulli, G., Cacciola, M., and Versaci, M. (2007) Microwave devices and antennas modelling by support vector regression machines. IEEE Transactions on Magnetics, 43, 1589–1592.

[33] Smola, A.J., and Schölkopf, B. (2004) A tutorial on support vector regression. Statistics and Computing, 14, 199–222.

[34] Meng, J., and Xia, L. (2007) Support-vector regression model for millimeter wave transition. Journal of Infrared, Millimeter, and Terahertz Waves, 28, 413–421.

[35] Andrés, E., Salcedo-Sanz, S., Monge, F., and Pérez-Bellido, A.M. (2012) Efficient aerodynamic design through evolutionary programming and support vector regression algorithms. International Journal Expert Systems with Applications, 39, 10700–10708.

[36] Zhang, K., and Han, Z. (2013) Support vector regression-based multidisciplinary design optimization in aircraft conceptual design. *AIAA Aerospace Sciences Meeting*. AIAA paper 2013-1160.

[37] Haykin, S. (1998) Neural Networks: A Comprehensive Foundation. 2nd ed. Upper Saddle River, NJ: Prentice Hall.

[38] Levin, D. (1998) The approximation power of moving least-squares. Mathematics of Computation, 67, 1517–1531.

[39] Forrester, A.I.J., Sóbester, A., and Keane, A.J. (2007) Multi-fidelity optimization via surrogate modeling. Proceedings of the Royal Society A, 463, 3251–3269.

[40] Huang, L., and Gao, Z. (2012) Wing-body optimization based on multi-fidelity surrogate model. *28th Int. Congress of the Aeronautical Sciences*, Brisbane, Australia.

[41] Toal, D.J.J., and Keane, A.J. (2011) Efficient multipoint aerodynamic design optimization via cokriging. Journal of Aircraft, 48, 1685–1695.

[42] Laurenceau, J., and Sagaut, P. (2008) Building efficient response surfaces of aerodynamic functions with kriging and cokriging. AIAA Journal, 46, 498–507.

[43] Koziel, S., Ogurtsov, S., Couckuyt, I., and Dhaene, T. (2013) Variable-fidelity electromagnetic simulations and co-kriging for accurate modeling of antennas. IEEE Transactions on Antennas and Propagation, 61, 1301–1308.

[44] Alexandrov, N.M., and Lewis, R.M. (2001) An overview of first-order model management for engineering optimization. Optimization and Engineering, 2, 413–430.

[45] Leifsson, L., and Koziel, S. (2015) Simulation-Driven Aerodynamic Design Using Variable-Fidelity Models. Imperial College Press.

[46] Bandler, J.W., Cheng, Q.S., Dakroury, S.A., *et al.* (2004) Space mapping: the state of the art. IEEE Transactions on Microwave Theory and Techniques, 52, 337–361.

[47] Koziel, S., Bandler, J.W., and Madsen, K. (2006) A space mapping framework for engineering optimization: theory and implementation. IEEE Transactions on Microwave Theory and Techniques, 54, 3721–3730.

[48] Salleh, M.K.M., Pringent, G., Pigaglio, O., and Crampagne, R. (2008) Quarter-wavelength side-coupled ring resonator for bandpass filters. IEEE Transactions on Microwave Theory and Techniques, vol. 27, no. 4, paper no. e21077.

[49] Koziel, S., Cheng, Q.S., and Bandler, J.W. (2010b) Implicit space mapping with adaptive selection of preassigned parameters. IET Microwaves, Antennas & Propagation, 4, 361–373.

[50] Koziel, S., and Leifsson, L. (2013) Multi-point response correction for reduced-cost EM-simulation-driven design of antenna structures. Microwave and Optical Technology Letters, 55 (9), 2070–2074.

[51] Leifsson, L., and Koziel, S. (2017) Adaptive response prediction for aerodynamic shape optimization. Engineering Computations, 34 (5), 1485–1500.

[52] Koziel, S. (2010) Shape-preserving response prediction for microwave design optimization. IEEE Transactions on Microwave Theory and Techniques, 58 (11), 2829–2837.

[53] Koziel, S., and Bekasiewicz, A. (2017) Computationally feasible narrow-band antenna modeling using response features. International Journal of RF and Microwave Computer, 27.

[54] Koziel, S., and Ogurtsov, S. (2013) Decomposition, response surface approximations and space mapping for EM-driven design of microwave filters. Microwave and Optical Technology Letters, 55 (9), 2137–2141.

Chapter 5

Verification of modeling: metrics and methodologies

Ahmad Tarraf[1] *and Lars Hedrich*[1]

5.1 Overview

Behavioral modeling of analog circuits is still an open problem with a long research history [1]. Nowadays, designers have the choice of either manually write models that roughly described the system behavior with high simulation speed-up or to spend a lot of time generating complex models that are much more accurate with low speed-ups.

As the systems integrable on a chip become more complex and heterogeneous, the use of accurate behavioral models for analog signal processing and interfacing would enhance design, simulation and validation routines. Thus, the goal is to achieve a fully automatic modeling process with high speed-up and best accuracy.

Due to the lack of automatic tools, most design groups manually write Verilog-A, VHDL-AMS or MAST models to perform a module or system simulation. Even if a behavioral model is available, the succeeding problem is to prove the validity of the model. For example, due to increased demand in safety for autonomous driving applications, the need for a verifiable methodology of model generation is obvious [2]. Hence, we can state that there exists a lack of formally verified models today. In this chapter we want to tackle both problems:

- Propose metrics to judge the model-verification process.
- Describe approaches and methodologies going systematically beyond the standard simulation driven model verification.
- Propose approaches to generate models in a correct-by-construction manner.

5.1.1 State space and normal form

Many of the later presented approaches depend on the idea of a continuous state space of the underlying analog circuit. Here we give a short description of this concept. In most simulators and more sophisticated tools, the analog circuit is described by a netlist of electrical compact models. The transistors are often modeled by a

[1]Institute for Computer Science, University of Frankfurt, Frankfurt, Germany

complex compact model like the BSIM [3] model. Following, the circuits are mathematically described by a system of implicit differential algebraic equations (DAEs) by applying the modified nodal approach:

$$\vec{f}(\dot{\vec{x}}(t), \vec{x}(t), \vec{u}(t)) = \vec{0} \tag{5.1}$$

where $\vec{x}$ is a vector containing the system voltages and currents, and $\vec{u}$ is a vector of input variables. This system can be locally linearized to:

$$\underline{C} \cdot \dot{\vec{x}} + \underline{G} \cdot \vec{x} = \vec{b} \cdot u \tag{5.2}$$

$$y = \vec{r}^T \cdot \vec{x} \tag{5.3}$$

With the conduction matrix $\underline{G}$ and the capacitance matrix $\underline{C}$, the input and output vectors are $\vec{b}$ and $\vec{r}^T$, respectively, and the output is y. Here only single-input–single-output systems are considered. However, the methodology can be transferred to multiple inputs and multiple outputs.

By using the transformation matrices $\underline{E}$ and $\underline{F}$, the system can be transformed from the $\vec{x} \in \mathbb{R}^n$ space to the $\vec{x}_s \in \mathbb{R}^n$ space. This can be done with the following transformation:

$$\vec{x} = \underline{F}\, \vec{x}_s \tag{5.4}$$

When additionally (5.2) is multiplied with $\underline{E}$ from the left, (5.2) and (5.3) can be written as:

$$s \cdot \underline{E} \cdot \underline{C} \cdot \underline{F} \cdot \vec{x}_s + \underline{E} \cdot \underline{G} \cdot \underline{F} \cdot \vec{x}_s = \underline{E} \cdot \vec{b} \cdot u \tag{5.5}$$

$$y = \vec{r}^T \cdot \underline{F} \cdot \vec{x}_s \tag{5.6}$$

Equations (5.5) and (5.6) can be rewritten as:

$$s \cdot \underline{\widetilde{C}} \cdot \vec{x}_s + \underline{\widetilde{G}} \cdot \vec{x}_s = \vec{\widetilde{b}} \cdot u \tag{5.7}$$

$$y = \vec{\widetilde{r}}^T \cdot \vec{x}_s \tag{5.8}$$

With the appropriate transformation matrices $\underline{F}$ and $\underline{E}$, (5.7) and (5.8) represent the Kronecker canonical form of the system, with $\underline{\widetilde{C}}$ being a block diagonal unity or zero matrix and $\underline{\widetilde{G}}$ being a block diagonal and unity matrix (see (5.11)). For that purpose, $\underline{F}$ is computed from the right eigenvectors $\underline{V}_R$ of the generalized eigenproblem, while $\underline{E}$ is a proper calculated matrix from the same problem, such that:

$$\underline{F} = \underline{V}_R \tag{5.9}$$

$$\underline{E} = \underline{\widetilde{G}} \cdot \underline{V}_R^{-1} \cdot \underline{G}^{-1} \tag{5.10}$$

Expanding (5.7) and (5.8) results in the following expanded *Kronecker canonical form* representation:

$$s \cdot \begin{bmatrix} \underline{I} & \underline{0} \\ \underline{0} & \underline{0} \end{bmatrix} \begin{bmatrix} \vec{x}_{s,\lambda} \\ \vec{x}_{s,\infty} \end{bmatrix} + \begin{bmatrix} \underline{\Lambda} & \underline{0} \\ \underline{0} & \underline{I} \end{bmatrix} \begin{bmatrix} \vec{x}_{s,\lambda} \\ \vec{x}_{s,\infty} \end{bmatrix} = \begin{bmatrix} \vec{\tilde{b}}_{\lambda} \\ \vec{\tilde{b}}_{\infty} \end{bmatrix} \cdot u \tag{5.11}$$

$$y = \begin{bmatrix} \vec{\tilde{r}}^{T}_{\lambda} \vec{\tilde{r}}^{T}_{\infty} \end{bmatrix} \begin{bmatrix} \vec{x}_{s,\lambda} \\ \vec{x}_{s,\infty} \end{bmatrix} \tag{5.12}$$

where $\underline{I}$ is the identity matrix, while $\underline{\Lambda}$ is a diagonal or block-diagonal matrix with numerically decreasing generalized eigenvalues of the system. As indicated, transformed vectors are marked with a tilde (~). Equations (5.11) and (5.12) can be divided into a differential (subscript λ) and an algebraic part (subscript ∞). The transformation matrices are also split into two parts, which are later used in Section 5.5.1.3.

$$\underline{F} = [\underline{F}_{\lambda} \underline{F}_{\infty}] \tag{5.13}$$

$$\underline{E} = \begin{bmatrix} \underline{E}_{\lambda} \\ \underline{E}_{\infty} \end{bmatrix} \tag{5.14}$$

The number of nonzeros diagonal elements in $\underline{\tilde{C}}$ corresponds to the number of eigenvalues λ_i from the generalized eigenproblem. As known, the number of eigenvalues result from the energy storing elements of the system.

However, some of those states are the results of parasitic components. For example, the ports of a transistor have a parasitic resistance throughout a current path as well as a parasitic capacitance between each other. The eigenvalues of those states are far to the left from zero, resulting in high-frequency values. To neglect them a *dominant pole order reduction* [4] can be carried out by simply reducing the active dynamic part (subscript λ) in the opposite to the algebraic part of (5.11), resulting at the same time in a reduction of (5.12). This can also be seen as converting a finite eigenvalue λ_i into an infinite eigenvalue.

All together, we have now a *reduced system in Kronecker canonical form* (Equations (5.11) and (5.12)) representing a normalized linear state space for a single operating point of a nonlinear analog circuit.

As the systems are mostly nonlinear, lot of operating points $\vec{x}_{sp}$ must be sampled and used to build the reduced Kronecker canonical form, to get an overall impression of the system. However, only the reachable sample points:

$$\vec{x}_{sp} \in \Sigma_R \tag{5.15}$$

being part of the reachable set Σ_R, are important. This is due to the fact that the other points will never be visited by the system. Therefore, a reachability analysis on top of the sampling process and the reduced Kronecker canonical form computation is used [5]. It calculates the reachable set of sample points Σ_R, approximating the true reachable state-space region.

We can now use the set of reachable sample points together with the reduced Kronecker canonical form for coverage measures (see Section 5.2.1.1), equivalence checking (EC) (see Section 5.4.1) and model abstraction and generation (see Section 5.5.1).

5.2 Model validation

The main challenge for the validation of behavioral models is to ensure that all important behaviors have been captured and no unwanted, may be spurious behaviors, have been included in the model.

This is done in general by simulation. The behavioral model can be compared with the original netlist—if it is already existing—by putting both circuits in the same test bench (see Figure 5.1). The input waveforms and the test bench have to be provided by the designer.

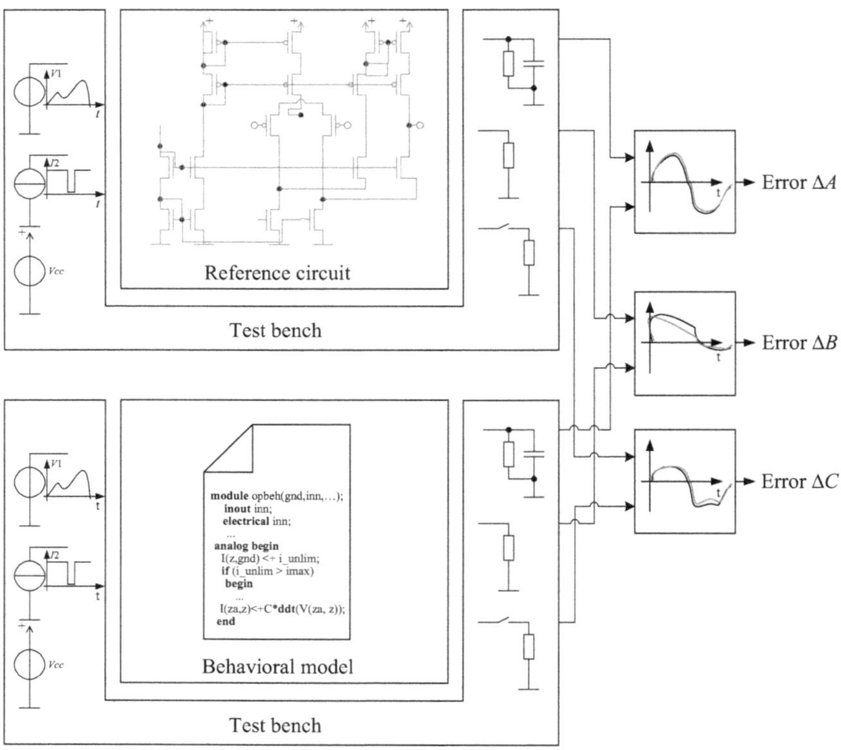

Figure 5.1 Model validation using simulation with the same test bench: the input waveforms V1, I2 and the test bench have to be provided by the designer. The outputs have to be compared using an appropriate error measure

The outputs of the two circuits have to be compared by optical inspection or with an appropriate error measurement. These errors can be pure voltage or current differences or more sophisticated error measures like Hausdorff distances or cross correlation coefficients [6]. A limit on the maximum error: $\max(\Delta A, \Delta B, \Delta C) \leq \Delta_{lim}$ will help to decide if the model is correct or not.

5.2.1 Model validation metrics

The main shortcoming of the above designer-driven simulation approach is that we are not sure that the designer has chosen the right input excitations and test benches. For example, another load resistance or input stimulus may lead to bigger deviation of a model compared to the reference circuit. Furthermore, we are not able to determine if the actual test bench or input stimuli are sufficient. Hence, a development of metrics to judge the quality of the verification of the behavioral model is needed. Our metrics are presented next.

5.2.1.1 Code coverage

We start with code coverage metrics. The simplest form of coverage in the digital domain is the statement or line coverage. This very straightforward method points a designer to parts of the implementation, which were not executed by a test set. Missing lines of code can be based on some misleading input stimuli or a wrong test bench. A direct adaption of this coverage metric fails at a very basic mechanism of analog circuit simulators. These simulators solve complex differential equations by several numerical methods, for example a Newton iteration. It is very important not to count visited code lines at every iteration step. In fact, code lines have to be counted if and only if they were executed in the last, valid evaluation of the behavioral model for each time step.

In this metric, only code lines, being executable are counted skipping comments, parameter definitions, control statements, etc. The line coverage C_{line} is expressed as the following ratio:

$$C_{line} = \frac{|executed\ lines|}{|all\ executable\ lines|}. \tag{5.16}$$

In addition to this simple form, we can define a branch coverage. It does not count simple code lines but has a more abstract view on the code, counting which branches are visited and which not. If an implementation deals with a given range of values in the enabling condition of the *if* statements, this coverage metric forces the designer to think about the complementary values as well. In contrary to C_{line} the nominator of this coverage is not only the sum of all available branches in the implementation. Moreover, missing *else* branches need to be counted as well even if missing in the code. The branch coverage C_{branch} is expressed as the following ratio:

$$C_{branch} = \frac{|executed\ branches|}{|all\ branches|}. \tag{5.17}$$

These both coverages can reach 100% in real examples. Moreover, for the definable path coverage C_{path}, this is in general not possible with practical examples, as the paths to be visited are all paths through the control graph implicitly defined by the behavioral model [7]. For realistic numbers of *if-then-else* clauses, the number of paths grows exponentially. In fact, it would require input stimuli having an exponential number of parts to excite the behavioral model into all paths. Furthermore, there may be some paths not reachable, which cannot easily be detected.

In Figure 5.2 an example with a very primitive behavioral model in Verilog-A is presented. It also visualizes the state-space coverage, which is defined in the next section.

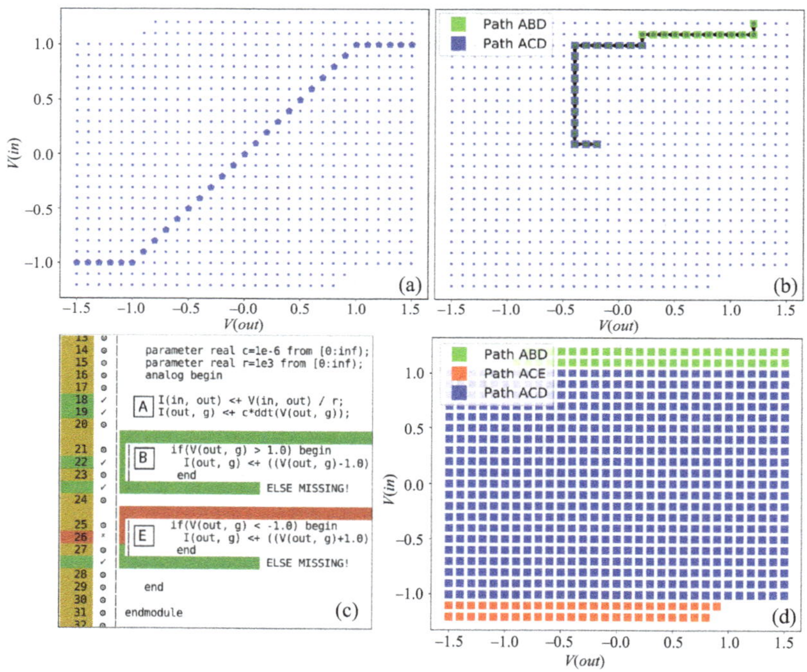

Figure 5.2 Mapping between code, branch and state-space coverage: (a) full discretization of the reachable state space of the implementation; (b) a trajectory through the state space with colorized paths back-annotated from the Verilog-A model ($C_{line} = 75\%$, $C_{branch} = 75\%$ and $C_{state} = 9.6\%$); (c) annotated Verilog-A code corresponding to that trajectory. The path elements A, B and E are marked, while the elements C and D are nonexisting in the code but part of a corresponding control flow graph; (d) situation after a full state-space coverage has been executed with back-annotated colorized paths ($C_{line} = 100\%$, $C_{branch} = 100\%$ and $C_{state} = 100\%$)

5.2.1.2 State-space coverage

In addition to code-based coverage metrics, a state-space coverage metric can be defined and may lead to a different point of view on the quality of a test set. It concentrates on the question if for a given parameter set and test bench the reachable state space of the system of underlying DAEs (see (5.1)) has been visited by the given input stimuli. Hence, it cannot judge the parameter set and the test bench like the code coverage. Only the internal states are investigated for being visited. However, these internal states are investigated in a formal way.

The state-space coverage C_{state} can be defined as the following ratio:

$$C_{state} = \frac{|\Sigma_{Sim}|}{|\Sigma_R|} \tag{5.18}$$

with Σ_{Sim}, being the set of the visited sample points corresponding to a simulation and Σ_R being the reachable set of all states Σ.

The definition enables a coverage computation by marking for a given simulation trajectory the sample points lying in the neighborhood. The resulted marked set is the set Σ_{Sim}. For details, see [8]. In Figure 5.2(b) such a situation is shown. The blue and green path are the marked sample points belonging to a simulation trajectory. Dividing the cardinality of that set by the cardinality of the set of the reachable state space (all blue points) results in 9.6% state-space coverage. Each simulation step can also be directly matched to a path through the implemented behavioral code Figure 5.2(c), which is indicated by the colors of the sampled points. If a lot of stimuli are simulated, the situation for this simple example can look like that in Figure 5.2(d), where all points are visited and the state-space coverage reaches 100%.

In Table 5.1 some results about the achievable coverages for different complexities of behavioral models are shown. All three coverages can reach 100%. However, for the complex EKV transistor model the transistor parameters as well as the voltages sources in the test bench has to be varied. The resulting six different test runs have additionally different piecewise linear input stimuli, some provided manually and some generated with a coverage maximizing method described in [9]. Concluding, it can be noticed that for medium complex behavioral models, a complicated test procedure has to be provided, which varies the test bench, the parameters and the input stimuli. Having a code, a state-space coverage and a coverage measure at hand—here branch coverage is the most pessimistic one—enables a clear judgment if the test setup is sufficient.

Some other measures like fault coverage are only indirectly able to help verifying the behavioral model. If the fault coverage for a given test set is high, one can expect that the corresponding input stimuli and test benches will also drive the behavioral model in the important regions.

All these metrics can also be used to improve the verification process approach (see Figure 5.1) by optimizing the underlying coverage metric. The optimization process can be manual or automatic, e.g., an automatic optimization of the state-space coverage has been proposed in [10].

Table 5.1 Results of the coverage metrics C_{line}, C_{branch}, C_{state}. In order to reach high coverage values the test benches, the transistor parameter sets and the input stimuli has to be varied. vs and vd are the voltages at drain and source node, respectively

Behavioral model	Lines of code	Id	Test bench	Stimuli/ Parameter	C_{line} (%)	C_{branch} (%)	C_{state} (%)
Low pass	31	1	Standard	Full stim.[a]	100	100	100
OP	129	2	Standard	Full stim.[a]	100	100	100
EKV	726	3	$(vs = 0)$	DC[b]	65.7	39.7	0.2
		4	$(vs = 0)$	pwl gen.[c]	85.7	67.2	4.9
		5	$(vs = vd - 0.1)$	pwl man.[c]	83.4	60.3	8.4
		6	$(vs = vd)$	pwl man.[d]	81.7	46.5	7.2
		7	$(vs = vd - 1)$	pwl gen.[c]	87.4	65.5	30.1
		3–7			100	100	40.7
		8	$(vs = 0)$	Full stim.[a]	88.5	67.2	100

[a]Method from [8] used to generate the stimuli.
[b]Parameters AD, PD, CJ, AS and JS of transistor model at slightly changed values.
[c]Additionally parameter CJSW shifted.
[d]Parameters AD, PD, CJ, AS and JS on default values.

5.3 Semiformal model verification

A semiformal approach uses the simulation-based verification method of Section 5.2 and generates input stimuli to drive the circuit into the not-tested regions of the state space. An approach in [11] does exactly this by using a modified version of rapid random trees (RRTs). The method needs a goal defined in the state space or in other variables of the circuit to be reached. For example, a minimal or maximal value of a certain output. It generates an input stimuli using RRTs to reach that goal. A drawback may be the needed knowledge about the goals to be defined and long generation and simulation runtimes. An advantage is that complex circuits, with a lot of states and strong nonlinearities, can be handled, as the state space is not sampled explicitly.

In [12] another approach for checking the correctness of a behavioral model is presented, which linearizes the analog behavior in order to apply regressions methods to generate input stimuli for that behavior. The approach is more focused on complex mixed-signal circuits with a lot of inputs and outputs. Hence, a linear model can help excite at least all inputs to some interesting value and help the modeler notice some missing inputs or essential behaviors. Typical applications are ADCs, DACs or phase interpolators.

An even more digital-oriented approach is given in [13], concentrating on the digital part using a formal SAT (satisfiability) engine to calculate digital input stimuli. The analog input stimuli are indirectly derived from the equations of the behavioral description. A shortcoming is founded in this last concept, as regions in a state space or parameter space emerging from the circuit behavior may not be visited.

5.4 Formal model verification

The coverage definition of Section 5.2.1.2 can be used to formally check a given model against a netlist. The idea is to find an input stimulus, which covers the whole reachable state space and hence reaches 100% state-space coverage. Checking the model versus the netlist with these stimuli with the simulation-based method from Section 5.2 achieves a formal result. The approach is explained in more detail in [8].

However, it is not always possible to reach 100% state-space coverage, and the simulation runtime due to generating a large input stimulus may be too large. Hence, a more direct method may be better, which is explained in the next section.

5.4.1 Equivalence checking

A very conservative and save method to check the correctness of a behavioral model is the use of an EC approach. In Figure 5.3 the principle off an approach explained in detail in [5] is shown. It is strongly based on the Kronecker form explained in Section 5.1.1.

The EC methodology starts with both systems, a netlist and a behavioral description, or any other combinations. These are given in SPICE syntax and are read by a modified version of a SPICE simulator. We support Gnucap [14] and an industrial in-house simulator. Hence, the simulators are based on the state-of-the-art BSIM4 or other transistor models and can handle both Verilog-A. In strong coupling with the verification algorithm, the simulators are forced to sample the

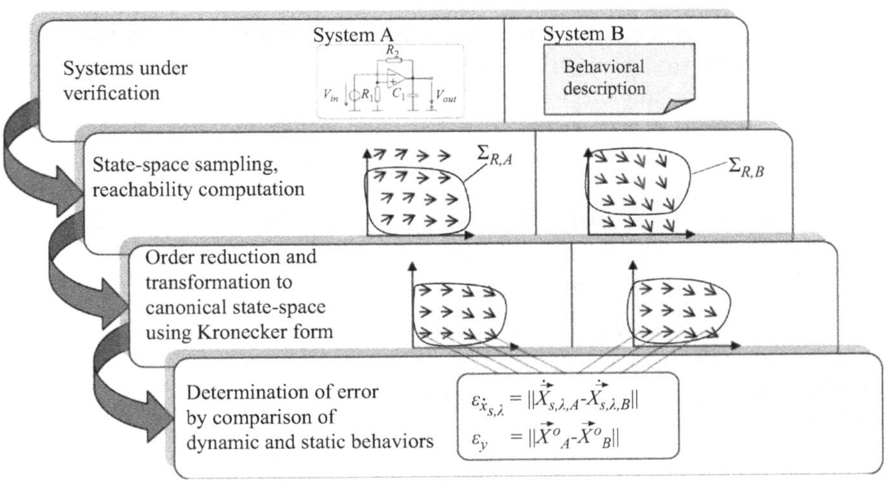

Figure 5.3 Principle of equivalence checking using sampling, reachability analysis and the Kronecker canonical form for transforming both systems into a nonlinear canonical state space. The four steps result in error measures indicating equivalence or difference

Table 5.2 Results of equivalence checking between manual generated behavioral models and netlists

Circuit/ Architecture	Abstraction levels	Transistors	Inputs/ States	Sample points	Run time (s)	Error $\varepsilon_{\dot{x}_{s,\lambda}}$ (%)	Error ε_y (%)
Low pass/ Active-RC	Makro ⇔ Verilog-A	0	1/1	917	0.5	10.7	13.7
OP/ Two stage	Transistor ⇔ Verilog-A	19	1/1	23,754	32	3.8	10.4
Low pass/ Gm/C	Transistor ⇔ Verilog-A	46	1/2	5,359	26	0.4	3.5
Bandpass/ Active-RC	Transistor ⇔ Verilog-A	8	1/2	40,680	160	0.1	0.04

state space (step 2 in Figure 5.3) and calculate the linearized system (see (5.3)) and a reachable region ($\Sigma_{R,A/B}$) of sample points.

In step 3, these regions are reduced in order and transformed to the Kronecker canonical form (Equation (5.11)), as described in Section 5.1.1. As the state spaces are canonical and the investigations are limited to the reachable sample points, we can expect an identical phase portrait in the state space (see the figures in step 3). This can be checked by calculating a dynamical error $\varepsilon_{\dot{x}_{s,\lambda}}$ between the derivatives of the transformed state variables $\dot{\vec{x}}_{s,\lambda,A/B}$ and a static error ε_y between the output variables $\vec{x}^o_{A/B}$ in the original state space as shown in step 4.

In Table 5.2 some results for the circuits from Sections 5.2.1.2 and 5.5.1.4 are given. The runtimes are in minutes for transistor netlist with full BSIM accuracy and up to 46 transistors. The errors are not going down to 0% like in EC for digital circuits due to the continuous variables and nonlinear characteristics of the circuits. Hence, values below 10% are a good indicator for having equivalent circuits. Additionally, a behavioral model has to be very complex to reach smaller error levels and will also never reach 0%. Overall for these examples, the models have a good quality and an equivalence could be stated.

5.4.2 Other formal techniques

In [15] a technique for calculating the reachable state space by a search technique using forward simulations is presented. On top of this, a discrete automaton is generated from the state-space dynamics. This discrete automaton can be used for model checking, which is done in the approach with a Computation Tree Logic (CTL)-like language. However, this can also be used for checking a behavioral model's properties. We can define a bound on the errors Δ_{lim} between the outputs of the behavioral model and the circuit (see Figure 5.1). Finally, we prove formally that the error bound is never crossed. The use of model checking can also help prove the correctness of the model in the absence of a golden circuit netlist, e.g., in early stages of a top-down design. The behavior of the model can formally be defined and proved using these approaches [15,16].

5.5 Formal modeling

An alternative to manual model generation and proving afterward that the model is correct and covering the behavior of the underlying electrical circuit is to generate the model directly from the circuit. This problem has been tackled for a long time. For linear systems, there are many successful efforts [17]. For strongly nonlinear circuits, there are only a few approaches. In terms of speed up with sufficient accuracy, they are struggling to reach the speed-up factors of manually written behavioral models. A direct method is nonlinear symbolic analysis, which can generate behavioral models [18] by simplifying the nonlinear DAE system (see (5.1)). Another method is to follow the signal path in the circuit and model the behavior of all the corresponding poles with small nonlinear static functions and small linear dynamic circuits, modeling the poles at that position in the signal path [19]. Though this becomes tricky if more than one path or feedback loops exist.

In summary, this motivates to look for a direct bottom-up approach for an automatic model generation using a formal methodology, which is presented in the next section.

5.5.1 Correct by construction: (automatic) abstract model generation via hybrid automata

We aim to construct a methodology that tries to extract the partly wanted linear behavior, together with the may-be-wanted or parasitic nonlinear behavior, from an transistor-level circuit. As it is the nonlinear behavior of the system that causes failure, we are in need of an approach that models this behavior but still reaches a very high abstraction level in the linear parts, to reach a high speed-up factor. Additionally, the models should be formally correct by construction. Hence, we present approaches to do the following:

- First, automatically generate a behavioral model in Verilog-A from a transistor-level netlist on SPICE BSIM accuracy based on the complete reachable state-space description and thus being a formal modeling approach.
- Second, compare the model against the netlist with a formal method (EC) to show the correctness.

To tackle the first problem, we abstract the linear and nonlinear behavior of the transistor-level netlist to a hybrid automaton (HA), where each of the reachable states of the automaton models the behavior of the system piecewise linear. The generated Verilog-A models are more precise than the standard models generated through assumptions. Thus, these models accurately describe the circuit behavior.

After the circuit has been modeled as HA, the specifications can be easily verified. This is due to the fact that the behavioral models are abstracted, resulting in a less complex description. Thus, this approach aims to reduce the complexity of circuits, decrease simulation time and allows grouping of systems by modeling the blocks of an analog signal processing chain in a compositional manner, as will be shown later.

As the verification of the specifications has been introduced in the previous sections, this section deals with the automatic generation of abstracted behavioral models in the form as HA from a SPICE netlist with full BSIM accuracy. This is covered in the next three Sections 5.5.1.1–5.5.1.3. In order to compare the generated models against the original netlist, Section 5.5.1.4 shows the experimental results of some simulations on different netlists. Moreover, an example of a model generated from second-order low-pass filter is compared against its original netlist using a formal method (EC) in Section 5.5.1.5.

5.5.1.1 Abstraction by sampling the state space

An overview of our methodology is illustrated in Figure 5.4. The first step in model generation is to sample the state space of the original netlist at transistor level. In order to keep the overall behavior, it is not sufficient to sample along some transient trajectories. Rather, the circuit has to be sampled in the reachable state space in an adequate manner. For that purpose, we use an algorithm to step through the state space [5], borrowed from the EC algorithm in Section 5.4.1. This gives us a set of sample points, along with their nonlinear operating (sample) points $\vec{x}_P$ and their local linearized system description $\underline{G}$, $\underline{C}$, etc. Using the Kronecker canonical form technique of Section 5.1.1 we can try to abstract this linear system for a given sample point by reducing the number of considered eigenvalues with respect to the operating frequency, thus neglecting some parasitic ones. This leads to a reduction of the system order to the first n_r finite eigenvalues of the differential part in (5.11), such that the eigenvalues in Λ are sorted from the ones closest to zero to the ones at infinity. In fact, this realizes the nonlinear order reduction [4]. Note that by

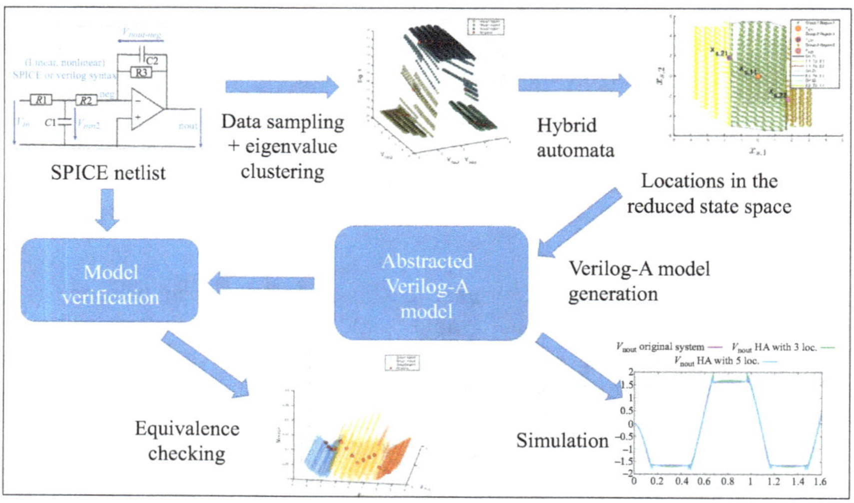

Figure 5.4 An overview of our approach. After the model generation, the model can be used for simulation or for verification against the original netlist

neglecting a parasitic eigenvalue, $\underline{F}$ and $\underline{E}$ lose a column, $\underline{\widetilde{G}}$, $\underline{\widetilde{C}}$ and $\vec{\widetilde{b}}$ a row and $\vec{x}_{s,\lambda}$ an element. The reduced state space is then in the form:

$$s \cdot \underline{I} \cdot \vec{x}_{s,\lambda_r} + \underline{\Lambda}_r \cdot \vec{x}_{s,\lambda_r} = \vec{\widetilde{b}}_{\lambda_r} \cdot u \tag{5.19}$$

To sample the reachable state space, both, the initial values of the energy states of the system and the value of the input voltages, must be considered. For a value in the range of the input voltages, the DC operating point of the system is calculated using the Newton–Raphson iteration. The system in the form shown in (5.1) can be locally linearized around this operating point to establish (5.2). With the transformation indicated in (5.4) and the nonlinear order reduction, the system can be transformed to the reduced state space shown in (5.19). At this point, the energy states of the circuit are charged, which is equivalent to changing the initial conditions of the system. The charged system is transformed back into the original $\vec{x}$ space, and the consistent solution is then calculated by solving an equation similar to (5.1) with constrains. For the solution found, (5.1) is linearized again, and the same procedure is repeated. Thus by changing both, the initial values of the energy states and the values of the input voltages, the reachable space can be sampled as a data set of points connected by a directed graph [20]. In fact, since for each point in the state space the neighbors are considered, we are able to distinguish between the reachable and nonreachable points. The set also contains the nonlinear consistent operating points $\vec{x}_P \in \mathbb{R}^n$, the eigenvalues λ_i and the matrices $\underline{F}$ and $\underline{E}$ at each point in the reduced state space $\vec{x}_{s,\lambda_r}$. To avoid confusion from now on, $\vec{x}_s$ stands for the reduced state space $\vec{x}_{s,\lambda} \in \mathbb{R}^{n_r}$.

5.5.1.2 Linear location identification

The n_r eigenvalues from each of the sampled points are used to identify the regions of the HAs. This can be achieved in two steps. First, the eigenvalues are clustered into groups. Second, a correction is done on this clustering in order to separate the regions that have the same eigenvalue, but belong to different spatial areas in the state space. Consider the example of a second-order low-pass filter circuit shown in Figure 5.5. The filter has a nonlinear limitation at ±1.5 V and is described in

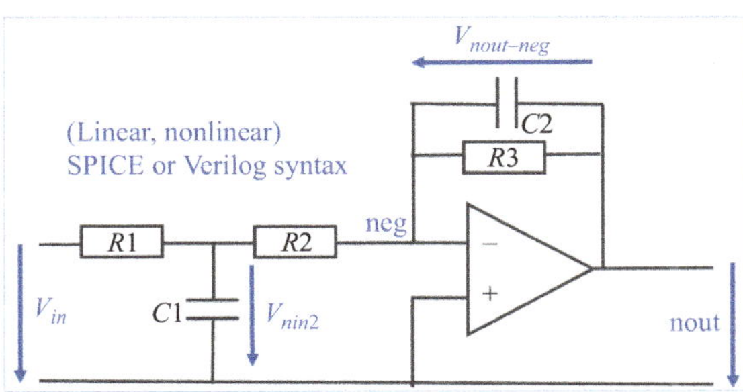

Figure 5.5 A second-order low-pass filter with a nonlinear limitation at ±1.5 V

SPICE syntax. It consists of 11 transistors in a 350-nm CMOS technology [21]. The result of the first step of this clustering for the circuit shown in Figure 5.5 is illustrated in Figure 5.6. As shown, two groups can be identified. Note that this clustering is done with an extended *k*-means algorithm, and that the number of clusters either is determined by our algorithm or can be specified by the user for higher precision.

In the clustering correction step, group 2 is separated into two distinct regions, since the limiting behavior happens in both directions of the input voltage. This can be achieved by examining the connection graph of the state space. The result of this clustering is shown in Figure 5.6. As illustrated, group 2 has been separated into two distinct regions. Thus, the data set consists of two groups of different eigenvalues, with one group having two distinct regions in the state space. The red points shown in Figure 5.6 represent the DC operating points of the system. These points were calculated for an input voltage ranging in the interval [−4,4] with a step size of 0.5 V. Each of the colored data set in Figure 5.6 represents a *location* of the generated HA. For each location of the HA, a representing DC point is chosen. For the example at hand, with G representing the group and R the region, $G1R1$ representing DC point is the one located at $V_{nin2} = 0$ and $V_{nout} - V_{neg} = 0$. $G1R1$ is referred to as the center region. The representing DC point for each of the other locations $G2R1$ and $G2R2$ can be chosen as the closest DC point with respect to location $G1R1$.

5.5.1.3 State-space representation of the HA locations

A location of the HAs consists of an *invariant*, *guards*, *reset* and a *state equation*. The invariants are only used to calculate the guards as seen later. A transition

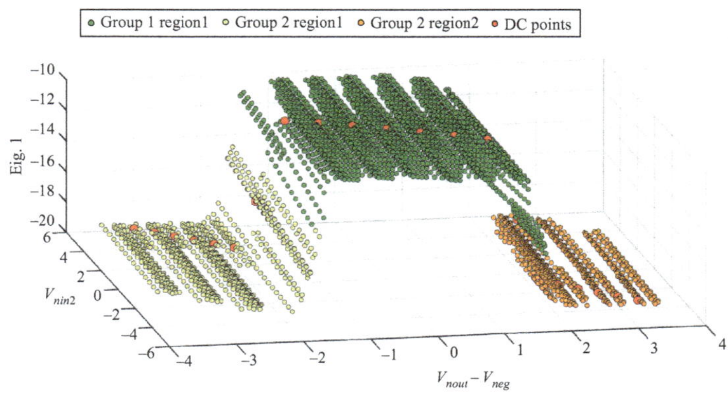

Figure 5.6 State clustering into two groups. Group 2 has two regions (yellow and orange). Group 1 has only one region (dark green). As illustrated, the yellow and orange have nearly the same eigenvalues but still represent two distinct regions. The red points represent the DC operating points. Note the missing dimension of the input voltage

occurs as soon as a guard becomes valid. All of these equations and variables should be calculated in the $\vec{x}_s$ domain. Thus, for further analysis the transformed state space $(\vec{x}_s)$ for each of the locations of the HA has to be calculated. If we define the first linear system representation around the origin, the movement from one location to another will result in a jump of the state variables $\vec{x}_s$. This is due to the new coordinate system in the new location.

To prevent this jump, we propose to shift the state variables by a constant at each location, in order to simulate the new system behavior in the new coordinate system. For this purpose, an operating point in the $\vec{x}_s$ domain has to be calculated. With the DC points in the $\vec{x}$ domain at hand, the DC shift $\vec{x}_{S,DC,k}$ for each location k in the $\vec{x}_s$ domain can be calculated by solving the following equation for $\vec{x}_{S,DC,k+1}$:

$$
\begin{aligned}
\underline{F}_{\lambda,k}\vec{x}_{s,DC,k+1} = \vec{x}_{DC,k+1} - \vec{x}_{DC,k} + \underline{F}_{\lambda,k}\vec{x}_{s,DC,k} \\
+ \underline{F}_{\infty,k}\underline{E}_{\infty,k}\vec{b}(u_{DC,k} - u_{DC,k+1})
\end{aligned}
\tag{5.20}
$$

where $u_{DC,k}$ is the input voltage corresponding to the DC point at location k. The calculation for each location is done in a recursive manner stating from the most inner one, the center location, outward. From this location ($k = 0$), the DC shift $(\vec{x}_{S,DC,1})$ of the next closest location ($k = 1$) is then calculated and so forth. The DC points are labeled by the locations they belong to $\vec{x}_{s,DC,GR}$ or simply $\vec{x}_{s,GR}$ in the figures.

In order to determine the guards and invariants of the HA, all points of the original $\vec{x}$ domain have to be transformed into the new $\vec{x}_S$ domain. This is done for each point i by solving the following equation for $\vec{x}_{s,i}$:

$$
\underline{F}_{\lambda,k}\vec{x}_{s,i} = \vec{x}_i - \vec{x}_{DC,k} + \underline{F}_{\lambda,k}\vec{x}_{s,DC,k} + \underline{F}_{\infty,k}\underline{E}_{\infty,k}\vec{b}(u_{DC,k} - u_i) \tag{5.21}
$$

where i stands for an index of a point in the location k.

Invariants

Having transformed all points into x_s domain, the invariants of the locations of the HA can be identified. This is done by enclosing the points of a given location by a convex hull (CH). Our methodology allows the user to choose the type of the CH. Thereby, three different types of representations for the CHs exist: Polytope, interval and zonotope representation. The polytope representation is the most accurate one, while the remaining two overapproximate. Depending on the type of the CH chosen, the invariants overapproximate differently the data set.

Guards

From the edges of the CHs, the guards to the neighbor locations can be identified. This can be done by searching for the presence of points from the neighbor locations to the edges of the CHs. For the guards, three different types of representation exist: halfspace, interval and zonotope representation. While guards of types zonotope and interval overapproximate in most cases the transitions, guards of type halfspace make clear cuts. For reachability analysis, the first might be more suitable. However, since we are interested in simulating the HA in Verilog-A, we will focus on halfspaces. The type of invariant used affects the guard calculations.

Besides the type of guards, one can also choose the number of guards between the locations. Either all guards can be selected, or the guards are reduced to one dominant one. The dominant guard encloses the highest number of points to the surrounding location. Additionally, the number of guards depends on the number of edges of the CH.

In Figure 5.7(a) our methodology with an invariant of type polytope and guards of type halfspace is illustrated for the circuit shown in Figure 5.5. Only the dominant guards are considered. This setup is used throughout this contribution.

To compare this classification to a different region identification, Figure 5.7(b) has been created using intervals as invariants and guards. As illustrated, the invariants of the locations are overapproximated, thus decreasing the number of available guards compared to the previous region identification. Note that the guards have a negligible thickness (0.01 for the example at hand).

State equations and reset

Along with the guards, the operating points in both domains and the eigenvalues and eigenvectors, we are now able to specify the dynamics of the HA of our running example from Figure 5.5 as follows:

$$\dot{\vec{x}}_{s,LOC} = \underline{\Lambda} \cdot \Delta\vec{x}_{s,LOC} + \underline{E}_{\lambda} \cdot \vec{b} \cdot \Delta u_{LOC} \tag{5.22}$$

The index "LOC" in (5.22) represents a location of the HA in the $\vec{x}_s$ domain. The system matrix $\underline{\Lambda}$ contains the n_r eigenvalues (see (5.19)) and is in block diagonal form. It is important to notice that the equations of the HA at each location are linearized around the DC points in the $\vec{x}_s$ domain, as indicated by the term Δ:

$$\Delta\vec{x}_{s,LOC} = \vec{x}_{s,LOC} - \vec{x}_{s,DC,LOC}$$

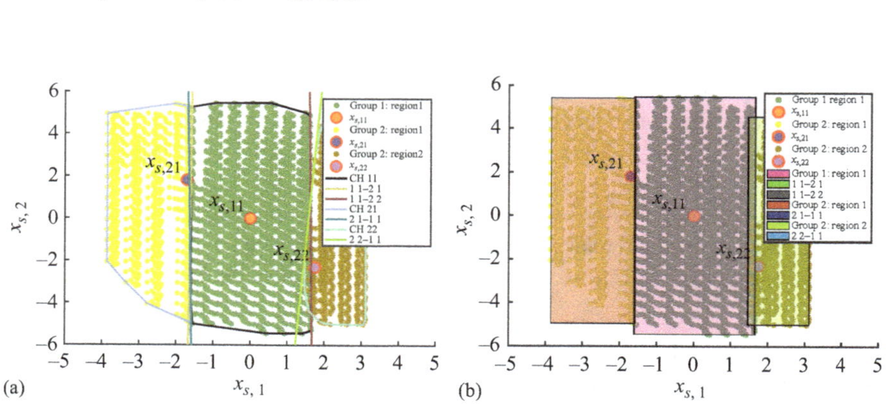

Figure 5.7 *(a) State space of the example illustrated in Figure 5.5 using polytopes as invariants and halfspaces as guards. The number of edges of the convex hulls at G1R1, G2R1 and G2R2 are 29, 10 and 19, respectively. (b) State space of the same example using intervals as invariants and guards*

This means that after taking a guard from location $G1R1$ to location $G2R1$, for example, the DC points in the $\vec{x}_s$ domain change from $\vec{x}_{s,DC,G1R1}$ to $\vec{x}_{s,DC,G2R1}$ as illustrated in Figure 5.7(a).

For the example at hand, the locations of the HAs at $G1R1$ and $G2R1$ are represented in (5.23) and (5.24), respectively. For $G2R2$, the state equation would be similar to (5.24) with the difference that $u_{DC,LOC}$ and $\vec{x}_{s,DC,LOC}$ vary. These equations are later transformed into separated row equations for the Verilog-A model:

$$\dot{\vec{x}}_{s,G1R1} = \begin{bmatrix} -10.4 & 0 \\ 0 & -1,099.7 \end{bmatrix} \Delta\vec{x}_{s,G1R1} + \begin{bmatrix} -8.89 \\ 999.9 \end{bmatrix} \Delta u \tag{5.23}$$

$$\dot{\vec{x}}_{s,G2R1} = \begin{bmatrix} -18.6 & 0 \\ 0 & -1,100.3 \end{bmatrix} \Delta\vec{x}_{s,G2R1} + \begin{bmatrix} -9.51 \\ 999.1 \end{bmatrix} \Delta u \tag{5.24}$$

with:

$$\begin{aligned} \vec{x}_{s,DC,G1R1} &= \begin{bmatrix} 0 & 0 \end{bmatrix}^T & u_{DC,G1R1} &= 0 \\ \vec{x}_{s,DC,G2R1} &= \begin{bmatrix} -1.71 & 1.81 \end{bmatrix}^T & u_{DC,G2R1} &= 2 \end{aligned}$$

As stated earlier, there are different approaches to represent the guards. Regardless of the representation, the guards are transformed into inequalities at the given locations. For example, the guards at location $G1R1$ to $G2R1$ and $G2R2$ are represented in the following two inequalities:

$$\begin{aligned} 0.42 \cdot (x_{s,1} - x_{s,1,DC,G1R1}) + 3.13 \cdot 10^{-4} \cdot (x_{s,2} - x_{s,2,DC,G1R1}) &< -0.71 \\ -0.92 \cdot (x_{s,1} - x_{s,1,DC,G1R1}) + 2.78 \cdot 10^{-3} \cdot (x_{s,2} - x_{s,2,DC,G1R1}) &< -1.81 \end{aligned}$$

Having computed the HA, the system can now be reformulated in Verilog-A. As indicated earlier, the invariants are neglected as the guards are sufficient to use. Therefore, the guards are wrapped in *if* statements and used for the identification of the current location. Furthermore, each location of the HA contains a set of variables specific to it. These variables include each element of the eigenvectors, the operating points in both domains, the input vector $\vec{b}$ and the elements of the system matrix indicated in (5.22). Additionally, in order to obtain the n node voltages and currents in the original $\vec{x}$ domain, a back transformation from the $\vec{x}_s$ space must be performed. This can be done with (5.25), where k indicates the location of the system:

$$\vec{x} = \underline{F}_{\lambda,k} \vec{V}(x_s) + \vec{x}_{DC,k} + \underline{F}_{\infty,k} \underline{E}_{\infty,k} \vec{b}(u - u_{DC,k}) \tag{5.25}$$

All these equations are automatically generated into a Verilog-A behavioral model with n_r ddt operators and n_{LOC} locations. The inputs and output variables are electrical nodes.

5.5.1.4 Experimental results

The experimental results are performed on an Intel Core I5-7300HQ CPU @ 2.50 GHz with 16 GB RAM using the SPICE simulator Gnucap [14] together with the Verilog-A compiler ADMSXml [22]. The circuit from Figure 5.5 is compared against the behavioral Verilog-A model with three locations as previously elaborated. The results of a transient simulation, with a sin wave with an amplitude of 5 V and a frequency of 1 Hz at the input, of this comparison is shown in Figure 5.8. As illustrated, there are some deviations during the switching from the linear to the limiting mode. However, by using HA with five locations this error can be significantly reduced. In both cases, due to the forward linear description of the HA in Verilog-A, the simulator does not react on the edges with strongly reduced time steps. Moreover, these edges can be reduced or even eliminated by several techniques: increasing the number of locations as mentioned, increasing the order of the system by one (to include the input), shifting the guards to the operating points (used in the next example) or by expanding the CH. To understand the result, Figure 5.9 shows the phase portrait along with a reachability analysis of the same HA performed in Cora [23]. The HA for Cora is automatically created in the same manner as the HA in Verilog-A.

In Table 5.3, the errors between the generated HA and the original netlist have been calculated directly during the modeling process for some of the examples. This is done by examining at each point in the sampled state space the difference between the output of the HA and the original Netlist. It can be observed that the maximum error is low and decreasing with more locations. Furthermore, the maximum error occurs at the borders of the locations, which is in congruence with the result in Figure 5.10.

A summary of all computed example circuits together with some properties and performances is illustrated in Table 5.4. The number of time steps during

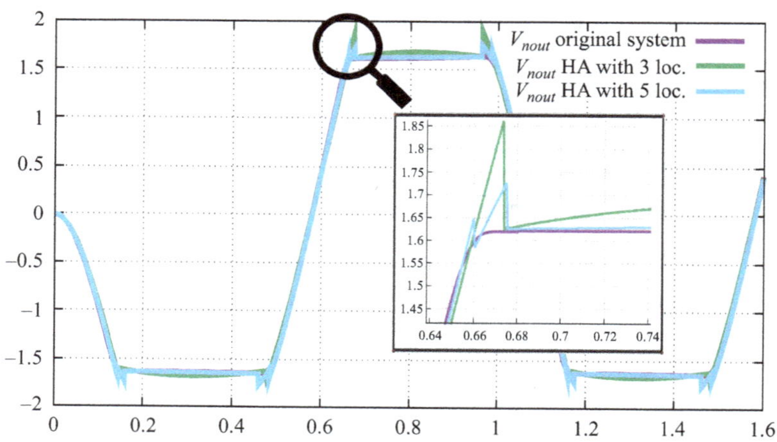

Figure 5.8 SPICE simulation results of the original netlist versus a hybrid automaton (Verilog-A model) with three and five locations, respectively. The output voltage differs depending on the modeling effort

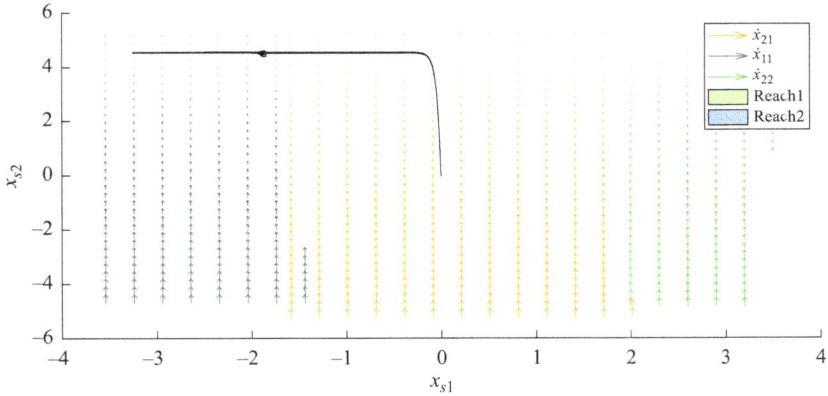

Figure 5.9 Phase diagram of the hybrid automaton for Cora. A reachability analysis is performed via Cora with an input of 5 V. Our tool can generate models in either MATLAB® or Verilog-AMS syntax

Table 5.3 Direct calculated error measure of the abstracted models versus the original Netlists

Circuit	Locations	MSE* (%)	$\mathbf{Error_{mean}}$ (%)	$\mathbf{Error_{max}}$ (%)	$\mathbf{Error_{min}}$ (%)
Differential mode Gm/C filter	3	6.4	5.5	15	0.0
Second-order low-pass filter	3	3.8	2.7	24	0.0
	5	3.2	2.6	11	0.0

*Mean square error.

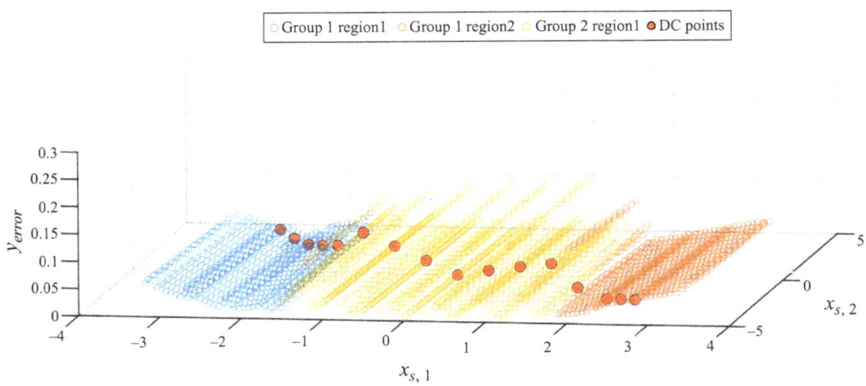

Figure 5.10 Output deviation (y_{error}) between the generated model (with three locations) versus the original netlist shown in Figure 5.5. The used sampled points for linearization (3 DC points) for modeling have the lowest errors

Table 5.4 Comparison of the transistor-level circuits with the presented behavioral models

Circuit	Model	Complexity	Sample points	Sampling + data read time (s)	Modeling time (s)	Locations	Accuracy* (%)	Newton iterations	Time steps	Run time (s)	Speed up
Differential mode Gm/C filter	Netlist	11 OTAs, 46 trans., 38 nodes	–	–	–	–	–	50,204	10,007	17.41	–
	HA	916 lines 5 nodes	17,997	372.2+127.4	6.05	3	0.81	39,221	10,006	0.14	124.3
Low-pass-based on 2-stage OTA	Netlist	1 OTA, 11 trans., 19 nodes	–	–	–	–	–	100,187	20,009	13.31	–
	HA	403 lines 3 nodes	312	0.34+1.68	1.51	3	3.6	61,496	20,006	0.24	55.45
Second-order low-pass filter	Netlist	1 OTA, 11 trans., 19 nodes	–	–	–	–	–	62,572	12,012	6.06	–
	HA	217 lines 4 nodes	4,017	43.12+10.97	4.46	3	15.2	58,202	12,057	0.20	30.29
	HA	765 lines 4 nodes	4,017	43.12+10.97	6.04	5	6.63	58,673	12,121	0.21	28.85

*Maximum relative dynamic output error. Same as y_{error} from Figure 5.10.

transient simulation is comparable between the circuit and the model indicating no problem with location switching in the model. Furthermore, the lower number of Newton iterations indicate that the convergence is faster, which is expected from a piecewise linear behavior. The speed-up is significant and varies between 30 for smaller transistor netlist and up to 130 for bigger netlists. Furthermore, we expect even bigger speed-ups for more sophisticated netlists on modern process nodes. The modeling time, which is the time that the algorithm spends in doing the algorithms of Section 5.5.1.3, is reasonable for mid-size circuits. As indicated in Table 5.4, the models show a good accuracy.

5.5.1.5 Correction and formal verification of the constructed model

To close this tool chain, we can formally verify the generated models. This is done by using the EC tool from Section 5.4.1. In Figure 5.10, the deviations in output voltage in the reachable state space between the original transistor netlist and the generated model for our running low-pass filter example is presented. Amongst others, the figure shows that the closer the system behavior is to the selected sample points for linearization, the more accurate the results are. At the far borders of the state space, the behavior deviates from the original model. Several approaches exist to optimize this modeling. The most obvious one is increasing the number of locations. As indicated in Table 5.4, this leads to more accurate results. Another approach would be to optimize the guards and invariants. This can be achieved either by shifting and rotating them to minimize the error shown in Figure 5.10 or by optimizing the shape and type of the guards. A different approach would also be to choose different linearization points. By applying this, the invariants, guards and state equations of the HA have to be recalculated. Another optimization step that can be done is to choose different eigenvalues for the locations of the HA. As shown in (5.23) and (5.24), each eigenvalue has been chosen as the mean of the eigenvalues in a given location (see Figure 5.6).

Since both, the model and the original netlist, are available, the abstract model could be continuously optimized until it fits the desired behavior or the errors from Table 5.3 are minimized. After that, the abstract model can now be used instead of the original netlist. Since the model is pin compatible and conservative, a compositional approach could be used which can abstract very large circuits. Moreover, by handling a large circuit partwise, the approach can be used to abstract the parts and bind them back together.

5.6 Conclusion

In this chapter we concentrated on checking the correctness or accuracy of analog behavioral models. We presented some approaches to judge the verification process using metrics like code or state-space coverage. These coverage methods are new for the analog domain and can be used to substantially increase the confidence in the verification setup and following in the model.

Additionally, we presented a methodology to generate a model in a complete formal way due to the consideration of the whole reachable state space of the original circuit. This approach reaches a big abstraction in combination with a high speed-up comparable with manually written behavioral models.

Acknowledgements

This work presents result of the project faveAC funded by the DFG under the project number 286525601.

References

[1] Gielen GG, and Phillips JR. Simulation and modeling for analog and mixed-signal integrated circuits. In: Electronic Design Automation for IC Implementation, Circuit Design, and Process Technology. 2016. p. 455–477.

[2] Mariani R. An overview of autonomous vehicles safety. In: Reliability Physics Symposium (IRPS), 2018 IEEE International. IEEE; 2018. p. 6A.1-1–6A.1-6.

[3] Cheng Y, and Hu C. Mosfet Modeling & BSIM3 User's Guide. Kluwer Academic Publishers, Boston/Dordrecht/London. 1999.

[4] Phillips J, Afonso J, Oliveira A, *et al.* Analog macromodeling using kernel methods. In: Proceedings of the 2003 IEEE/ACM International Conference on Computer-Aided Design. IEEE Computer Society; 2003. p. 446.

[5] Steinhorst S, and Hedrich L. Advanced methods for equivalence checking of analog circuits with strong nonlinearities. Formal Methods in System Design. 2010;36(2):131–147.

[6] Ohlendorf O, Steinhorst S, Hartong W, *et al.* Comparing two analog waveforms—a trivial task. In: Zuverlässigkeit und Entwurf (ZuD 2008), Ingolstadt; 2008. p. 153–154.

[7] Zaki M, Mokhtari Y, and Tahar S. A path dependency graph for Verilog program analysis. In: Proceedings of the IEEE Northeast Workshop on Circuits and Systems (NEWCAS'03), Montreal; 2003. p. 109–112.

[8] Fürtig A, Steinhorst S, and Hedrich L. Feature based state space coverage of analog circuits. In: Proceedings of the Forum on Specification and Design Languages (FDL 2016); 2016.

[9] Fürtig A, Paschke M, and Hedrich L. Comparing code coverage metrics for analog behavioral models. In: Synthesis, Modeling, Analysis and Simulation Methods and Applications to Circuit Design (SMACD), 2017 14th International Conference on. IEEE; 2017. p. 1–4.

[10] Fürtig A, and Hedrich L. Formal techniques for verification and coverage analysis of analog systems. In: Formal System Verification. Springer, Cham; 2018. p. 1–35.

[11] Ahmadyan SNA, and Vasudevan S. Automated transient input stimuli generation for analog circuits. IEEE Trans on CAD of Integrated Circuits and Systems. 2016; 35(5):858–871. Available from: http://dx.doi.org/10.1109/TCAD.2015.2488482.

[12] Horowitz M, Jeeradit M, Lau F, *et al.* Fortifying analog models with equivalence checking and coverage analysis. In: Proceedings of the 47th Design Automation Conference. DAC '10. New York, NY, USA; 2010. p. 425–430. Available from: http://doi.acm.org/10.1145/1837274.1837381.

[13] Salem A. Semi-formal verification of VHDL-AMS descriptions. In: IEEE International Symposium on Circuits and Systems (ISCAS '02). vol. 5; 2002. p. 333–336.

[14] Davis A. An overview of algorithms in Gnucap. In: University/Government/IndustryMicroelectronicsSymp.; 2003. p. 360–361.

[15] Dastidar TR, and Chakrabarti PP. A verification system for transient response of analog circuits. ACM Transactions on Design Automation of Electronic Systems. 2008;12(3):31:1–31:39. Available from: http://doi.acm.org/10.1145/1255456.1255468.

[16] Steinhorst S, and Hedrich L. Model checking of analog systems using an analog specification language. In: Proc. Design, Automation and Test in Europe (DATE '08); 2008. p. 324–329.

[17] Freund RW, and Feldmann P. Reduced-order modeling of large linear passive multi-terminal circuits using matrix-Padé approximation. In: Proceedings Design, Automation and Test in Europe, 1998. IEEE; 1998. p. 530–537.

[18] Näthke L, Burkhay V, Hedrich L, *et al.* Hierarchical automatic behavioral model generation of nonlinear analog circuits based on nonlinear symbolic techniques. In: Proc. Design, Automation and Test in Europe Conference and Exhibition. vol. 1; 2004. p. 442–447.

[19] Zheng W, Feng Y, Huang X, *et al.* Ascend: automatic bottom-up behavioral modeling tool for analog circuits. In: IEEE International Symposium on Circuits and Systems, 2005 (ISCAS 2005). vol. 5; 2005. p. 5186–5189.

[20] Steinhorst S, and Hedrich L. Trajectory-directed discrete state space modeling for formal verification of nonlinear analog circuits. In: Proceedings of the International Conference on Computer-Aided Design. ACM; 2012. p. 202–209.

[21] AMS HitKit. Process Design Kit (PDK) for Cadence Tools, Circuit Simulation, Device Models. 2019. Available from: http://asic.ams.com/hitkit/index.html.

[22] Lemaitre L, McAndrew C, and Hamm S. ADMS-automatic device model synthesizer. In: Proceedings of the IEEE 2002 Custom Integrated Circuits Conference, 2002. IEEE; 2002. p. 27–30.

[23] Althoff M. An introduction to CORA 2015. In: Proc of the Workshop on Applied Verification for Continuous and Hybrid Systems. 2015.

Part II

Applications in analogue integrated circuit design

Chapter 6

An overview of modern, automated analog circuit modeling methods: similarities, strengths, and limitations

Alex Doboli[1] and Ciro D'Agostino[1]

Analog circuit modeling is the procedure through which analog circuit operation and performance are described by a succinct, comprehensive yet accurate representation. Circuit modeling has application in numerous design-related activities, like getting insight into circuit operation, circuit topology synthesis and improvement, circuit parameter optimization, fast simulation, design centering and yield improvement, test and validation, and design knowledge representation. This chapter offers an overview of some of the main methodologies and algorithms for analog circuit modeling without claiming that it exhaustively summarizes the entire research domain. It would be very hard considering the large number of publications on analog circuit modeling. This chapter mainly focuses on three main categories of automated circuit modeling techniques: symbolic methods, neural network (NN)-based approaches, and macromodeling. The related techniques are grouped depending on the similarity of their pursued goals and concepts. The chapter refers to close to 100 papers to present the capabilities, strengths, and limitations of the different methods.

6.1 Introduction

Analog circuit modeling is the activity through which the behavior of an analog circuit is characterized by a more succinct yet accurate representation [1,2]. Circuit behavior can refer to a large set of characteristics related to circuit operation, function, performance, and testing. Analog circuit modeling has been an important research topic for long time considering its application to a large number of circuit design activities, such as getting insight into circuit operation [3], circuit structure (topology) synthesis and improvement [4,5], circuit parameter optimization [4,6], fast simulation, design centering and yield improvement [7], and test and validation. Very recently, circuit modeling has been suggested as a solution to creating

[1]Department of Electrical and Computer Engineering, Stony Brook University, Stony Brook, NY, USA

circuit design knowledge representations that cluster circuits into categories based on their structural and performance similarity [8–10] as well as systematically comparing two different circuit designs [11], and extracting the causal relations between the circuit parameters of a design. Such a "data-driven" approach to circuit modeling might offer new opportunities for circuit topology selection, incremental topology modification [12], circuit reuse during technology migration, new topology synthesis [12,13], circuit verification, new building block (BB) identification [14], synthesis for creative design [15,16], and capturing the design style pursued by an individual or team [10,16]. Other possible novel applications include the description of the design community's knowledge evolution to identify new design opportunities or those have not been sufficiently considered before [17].

It can be argued that circuit modeling is an abstraction and/or approximation process through which a detailed circuit is represented by an abstract description (often a certain type of functions) with a similar behavior of interest. The new behavior should be within certain error from the simulated or measured behavior of the electronic design. The acceptable error sets the *modeling accuracy*, which ultimately decides the usefulness of a model for real-world design. A second model characteristic is its *development efforts*, such as the complexity of model generation. It refers to the required memory and execution time to produce models as well as the amount of data needed to train and validate the models, such as in the case of regression models and models based on NNs. The third attribute represents the *evaluation effort*, such as the required computations to obtain the described behavior using the model. Another characteristic refers to the capability of a model of *being understood* by the user (transparency), such as models that offer insight about a circuit's behavior and what produces that behavior versus black box models which only predict performance without giving an explanation of the result. Fifth, modeling algorithms also differ by their capability of *incorporating new knowledge* (i.e., simulation data) (generalizability) as it becomes available. Other model characteristics might include *domain sense*, which means offering the right insight to the user, *consistency* between different models, *parsimony* which indicates that the simplest model should be produced to express a certain behavior, and *fidelity* that describes the soundness and completeness of the modeled behavior. In general, the model characteristics that are deemed to be relevant are decided by the nature of the application in which the model is used.

This chapter presents an overview of some of the main methodologies and algorithms for analog circuit modeling. We do not claim to offer a comprehensive description of all methods as the related research literature is extremely broad. We have mainly focused on automated modeling methods that have been published in *IEEE Transactions on Computer-Aided Design of Integrated Circuits and Systems* journal over the last 30 years as well as in some of the main electronic design automation conferences, such as Design Automation Conference (DAC) and Design, Automation and Test in Europe Conference (DATE). The reviewed papers were grouped into three analog circuit modeling approaches: symbolic analysis, macromodeling, and NN-based methods. Each approach was further divided depending on the main goal pursued by the related work. For example, symbolic analysis was distinguished in four subcategories: fundamental symbolic methods, simplification and hierarchical methods, and advanced representations for symbolic analysis. Macromodeling methods were divided into basic macromodeling techniques, structural (white box) modeling, modeling of weakly nonlinear systems, and modeling of oscillator circuits. Note that this division into subcategories of papers addressing the same goal corresponds to a fairly well-defined progression (evolution) of the modeling research domain over time, even though some of the goals recurred at a later time too. The main advantages and limitations of the methods were also summarized as they are stated in the papers.

This chapter argues that most of the circuit modeling methods are abstraction procedures that attempt to find the parameters of a generic model (implicitly or explicitly expressed), so that it closely matches a representative set of numerical data (usually expressed as minimizing the square error). Semantically, such models express a circuit by the collection of its relevant features. Symbolic methods, NN-based models, and some macromodeling methods (e.g., posynomial models) pertain to this class. Another class of modeling methods represents a circuit by the set of steps (activities) that produce the model. These models correspond to an operational semantics. In this case, only after "executing" the operational semantics will the model features be revealed. Certain macromodeling methods, such as those for symbolic analysis and oscillator modeling, are in this class. Within their corresponding class, each modeling method executes the same steps. For example, methods pursuing feature-based modeling must (i) select the generic model and identify its (ii) structure and (iii) parameters to obtain the needed model quality. Activities, like (i) data sampling, (ii) parameter optimization, and (iii) simplification, are critical. Macromodeling algorithms in the second class (operational semantics) start from the analytical equations of a circuit (e.g., Kirchhoff's current law (KCL) and Kirchhoff's voltage law (KVL), state equations expressed as linear or nonlinear differential equations), which are then solved through a number of steps involving (iv) linearization, (v) separation of slow and fast components, and (vi) approximation to tractable forms that can be solved. Often, such modeling methods use minimal circuit design knowledge for model construction, especially in the case of NN-based models. Another distinction between modeling methods depends on how they describe model parameters: through (i) a comprehensive set of relevant values, (ii) a set of disjoint subranges that possibly satisfy certain constraints, and (iii) symbolic parameters in a domain. We argue that analysis is important for comparing the similarities of the existing modeling methods, including situations when methods belong to different types, as well as their unique strengths and limitations. Moreover, the analysis might suggest modeling opportunities that have not been previously considered by the community or identify new directions worthwhile to be pursued.

This chapter also discusses the evolution of different circuit modeling methods over time in an early attempt to understand some of the invariant characteristics of this process. Moreover, it is important to get insight on how specific application needs to shape the progress of circuit modeling domain, the nature of the studied methodologies and algorithms, their incremental development over time, as well as the efficiencies and limitations of the group-level process through which the research community identifies and tackles problems of interest. The process sheds interesting light on how certain design theories and procedures emerge and get adopted by the broad community. This process is similar to belief formation in cognitive psychology. It refers to the conditions under which a community will converge toward a set of accepted concepts or diverge with a high likelihood without reaching sufficient common ground. Another interesting question of interest is whether the main conceptual ideas related to modeling emerged during the early stages of the domain (followed by an incremental development of the initial concepts), or if new conceptual ideas, fundamentally different from the existing concepts, are likely to emerge at certain points of the evolution. Another interesting issue refers to the nature of the research work that is likely to have a higher contribution at a certain point. Would it be more beneficial to mainly explore incremental improvements of existing solutions, or attempt to combine concepts developed by different subareas, or search for approaches that diverge from the existing methods, or consider devising generalizing, encompassing theories for the existing techniques? Moreover, what would be an "ideal" ration of mixing the four different research strategies, so that the community benefits as a whole while still maintaining sufficient freedom and flexibility to produce healthy domain knowledge progress? Such questions are very important to understand and devise strategies to optimize a research community's effectiveness in uncovering new and robust knowledge while having enough flexibility to adapt to new challenges beyond the current one. Specifically, related issues include required funding levels, effective funding policies, the critical mass of needed research personnel, the need for alternative knowledge dissemination mechanisms, the likelihood of pursuing "equivalent" work, the importance and impact of domain-related prestige as well as more effective ways of capturing it, and the existence of inconsistencies (e.g., contradictions) within the current corpus of knowledge. This chapter offers an early discussion on some of these issues and suggests some intriguing hypotheses that are important to consider in future work.

This chapter has the following structure. Section 6.2 presents algorithms for symbolic analysis. Section 6.3 summarizes work in analog circuit macromodeling. Section 6.4 discusses modeling techniques based on NNs. A discussion of the work in circuit modeling is offered in Section 6.5. Section 6.6 concludes the chapter.

6.2 Symbolic analysis

The goal of symbolic analysis has been to automatically compute accurate mathematical relationships (called symbolic expressions) that relate circuit outputs, transfer functions (TFs), and performance to circuit parameters, like transistor sizes [1,2]. The computed symbolic expressions represent mostly linear behavior in the frequency domain, but also the effect of linear noise, weakly nonlinear behavior, process parameter variations, periodical systems, and layout-induced parasitic.

The initial application of symbolic analysis included finding filter TF used then in implementation (filter synthesis) [18–21], obtaining insight about circuit operation and performance [3,22–24] to select or improve the circuit topology used in an application [3,12,13,25], and circuit modeling and fast simulation for analog electronic design at high levels (e.g., mixed-signal frontends, high performance filters, analog-to-digital converters (ADCs), phase-locked loop (PLL) circuits, and so on) [25–29]. The specific nature of the applications requires computing symbolic expressions that are understandable by a human designer, so that the obtained insight (like information about performance trade-offs and bottlenecks) can be used to improve a circuit design. More recently, a new application of symbolic analysis addressed the automated generation of specifications that are used then as inputs by certain transistor sizing tools to complete a design. For example, symbolic circuit models described as posynomials have been automatically produced to be used as inputs for geometric-programming-based sizing tools [30]. Finally, very recent work proposed methods to compute symbolic expressions that compare a circuit against another circuit, so that the specific (unique) performance capabilities and bottlenecks of a circuit versus other circuits can be understood [11]. The authors explain that the method is useful for selecting a topology from a library of existing topologies, incrementally improving a given circuit by changing the build blocks causing the performance bottlenecks, and by organizing (i.e., hierarchically clustering) circuits based on their structural and performance similarities and differences [9,10]. Also, circuit testing through formal verification has emerged as a recent yet important application domain for symbolic analysis [31].

While obtaining insight on circuit operation and performance has remained arguably the main objective of symbolic analysis, research has pointed out some major difficulties challenging this approach: (i) the computational intractability of finding closed-form mathematical expressions for certain cases, e.g., strongly coupled circuits analyzed in the time domain or if performance parameters are not computed by explicit formulas and/or if calculating derivatives is needed, (ii) the difficulty of creating white-box models that present (some) causal dependencies between circuit parameters and performance, and (iii) the high complexity of any symbolic expression, including those of linear circuits (i.e., the huge size of symbolic expressions). These challenges influenced the evolution of research on symbolic analysis, including the devising of methods that attempt to conduct symbolic analysis beyond linear circuits (e.g., weakly nonlinear circuits), to offer insight about how circuit parameters decide performance trade-offs and bottlenecks, and to reduce the size of the computed symbolic expressions by developing more compact representations.

6.2.1 Fundamental symbolic methods

In a survey of applications of symbolic analysis, Lin in [22] mentions six types of algorithms to compute circuit TFs as relations of single variable s (tree enumeration, signal-flow graph (SFG) method, state-variable-eigenvalue method, iterative method, nodal-analysis-eigenvalue method, and parameter extraction method) and three types of algorithms to calculate general symbolic TFs (STFs) of multiple parameters (SFG method, tree enumeration, and parameter extraction method).

According to Lin, the algorithms for general STF computation use Cramer's rule to find the STF. SFG method computes the related determinants by using loop enumeration and path identification for SFG method and directed two trees in the tree enumeration method. A main limitation of SFG method is the generation by Mason's formula [32] for determinant computation of terms that cancel out each other [33]. This is one of the reasons that increase the size of the computed expressions [33,34]. To address this limitation, Mielke [33] discusses necessary and sufficient conditions for SFG topology, so that canceling terms are not produced by Mason's formula. An SFG method without term canceling has been also proposed. Sannuti and Puri [34] describe an improvement to SFG method based on the diagonal expansion of determinants. Invalid symbol combinations are avoided through a new method that uses each symbolic parameter only once, along the principal diagonal. The method has been proposed for calculating pure symbolic expressions and mixtures of numerical-symbolic expressions.

Tree enumeration methods were proposed as an alternative to SFG algorithms [32,35–39]. Early work by Kurth and Moschytz [19] discusses z-domain analysis of linear switched capacitor (SC) networks, e.g., SC filters. The analysis method is based on setting up the nodal charge equations for the network, transforming the nodal charge equations in the z-domain using a table of predefined cases, and setting up the current and voltage equations of the four-port models of an SC network (2 m port networks in the general case). From an algorithmic point of view, the method performs a sequence of port composition steps starting from a set of z-domain expressions describing the network components. Another approach proposes a computational procedure based on the concept of k-tree terms describing the partial sum of k-trees [40]. A k-tree has all the original vertices but k-less edges than the original tree. The algorithm decomposes the circuit graph into its passive and active subgraphs containing all circuit vertices. Then, all i-tree terms for different partitions are calculated, followed by computing all directed j-tree terms for different partitions. Then, the list of all possible combinations of i- and j-tree terms is created, so that j-tree with two active edges for the same voltage-controlled current source is not considered. Finally, the required sums of directed k-tree terms are produced. Tsai and Shenoi discuss a number of software optimizations to reduce the memory needs of topological methods: using unused memory word bits to store symbols used during the computing stage to represent groups of symbols, rules to make all exponentials positive, avoidance of common factors, and avoiding round-off errors through fixed-point coefficients operation [41]. Konczykowska and Bon [18] propose an alternative method for the same task. Networks are described as two time-invariant networks describing the odd and even time slots of the network. Each network has conductances corresponding to capacitors in parallel voltage-controlled charge sources. Then, TFs for the network are computed by applying a set of equations and rules to computing the equations depending on the network topology. Another approach is discussed in [42] to transform a network incorporating capacitors, ideal switches, independent voltage and current sources, voltage-controlled voltage sources, and current-controlled current sources into an equivalent, resistive network. This method is further improved in [21] by suggesting a new variable transformation. In summary, these classes of symbolic analysis methods propose a certain kind of network representation, so that tractable rules are available to compute network TFs.

Parameter extraction methods are a third alternative to finding symbolic expressions, especially mixed numerical symbolic functions of variable s. The method in [43] computes STFs starting from nodal admittance matrices. The coefficients in matrix A are found using the expression $[A] = [K]^{-1}[D]$, where K is the Vandermonde matrix of the sampled points and D is the matrix of functional values. Matrix inversion uses Traub algorithm. TFs are found through Lagrange interpolation. The formula might experience stability problems if the determinant values are of the same order of magnitude like the round-off values. Another parameter extraction method is presented in [24]. It computes the polynomials of TFs in the frequency domain starting from the tableau representation of a circuit, in which rows describe KVL and KCL laws (specific to the circuit topology) as well as the constitutive equations. Circuit parameters appear in a single position of the tableau formulation. In general, parameter extraction method finds a polynomial of degree n by considering $n + 1$ arbitrary points and then interpolating the corresponding values. The authors explain that the points should be selected on a unit circle as it improves the stability of interpolation. An intermediate step of the method finds all the derivatives of the numerators and denominators of the TFs using the tableau matrix. The derivative of a variable is found through triangular decomposition by setting the variable to one and the other entries of the column to zero. The rule is extended for partial derivatives too. The algebra of matrices and partial derivatives are then used to compute the coefficients of the polynomials by solving the related equations through back substitution and fast Fourier transformation. The method also includes a set of rules that identifies the nonzero coefficients before the method attempts to generate these coefficients.

6.2.2 Beyond linear filters: simplification and hierarchical symbolic expressions

The next generation of symbolic analysis tools introduced simplification and hierarchical computation of symbolic expressions as means of tackling complexity [3,44,45]. Tools included ISAAC symbolic simulator [3] as one of its most well-known tools. ISAAC computes STFs in the s or z domain for linear or linearized circuits. Its main goals are getting insight into circuit operation, interactive circuit improvement, repeated circuit simulations, and circuit model generation for transistor sizing. We think that gaining insight into circuit operation is arguably the more interesting application. Along this opinion, Gielen *et al.* mention that numerical simulations give little information other than that, that a circuit meets or not the performance requirements. There is no information about any causal aspect, like the reasons for performance degradations and what possible solutions might address it. The authors explain that repeated numerical simulations cannot produce the missing insight, as certain parameters, like power supply rejection ratio (PSRR), depend mainly on design constraints, like matching and layout-related constraints. ISAAC uses compacted modified nodal analysis (MNA), in which branch currents are considered only for current meters and dependent current sources. Symbolic expressions are computed using the well-known determinant expansion formula $|D|_{n\times n} = \sum_{d_{ij}\neq 0}^{n} (-1)^{i+j} d_{ij}|M_{ij}|$, which avoids the nonzero entries of a determinant and the repeated computing of the same terms (stored in a hash table) [3]. Determinant expansion is applied recursively to the minors $|M_{ij}|$. ISAAC also includes approximation of symbolic formulas. Approximation conducts two steps, first the terms below the average accepted error are eliminated, and the second step removes terms in increasing order of their magnitude. The suggested optimizations significantly reduce the size of the symbolic expression, e.g., 982 terms are left from an initial 7,533 terms or 495 symbolic terms from the starting 3,469 terms. Still, the size of the symbolic expressions remains high, i.e., 2,866 terms for an OpAmp circuit. Such a high number of terms are likely to diminish the kind of insight a designer might get from the calculated symbolic expressions.

Vanassche *et al.* discuss calculating symbolic expressions for the harmonic TFs of linear periodic time-varying systems (LPTVS), e.g., mixers and PLLs [28]. Models describe up and down conversion of wanted and unwanted inputs. LPTVS are represented as multiple paths (for different side-band signals) that include a linear time invariant (LTI) filter and a frequency shift. LTI filter models are integrals of the Laplace transformation of harmonic TFs to which harmonic impulses are applied. The TF matrices have as elements the Laplace transformations of the harmonic impulse responses for different side-band signals [28]. The output in each signal band is the summation of all frequency band spectra. Noise analysis for LPTVS calculates the product of the harmonic power spectral density matrices. The analysis process considers the structure of systems, such as the fact that blocks are in series, parallel, or feedback.

Another interesting SFG method was proposed a few years earlier by Starzyk and Konczykowska [20]. The method uses Coates flow graph representation to compute symbolic immitances and TFs. Circuits are described using nodal admittance matrices. The main innovation of the algorithm is using hierarchical decomposition of a circuit's SFG, so that terminal nodes represent circuit BBs, and middle nodes show how BBs are composed with each other. Then, BBs can be analyzed through direct topological analysis by enumerating multi-connections with paths linking different block nodes. Middle node analysis combines the multi-connections without duplicating similar connections and by considering large groups, if possible. The method includes two ways of traversing the graph decomposition tree: the downward hierarchical analysis traverses the tree top-down, so that only the necessary functions at the middle nodes are further evaluated using the next hierarchical level and unnecessary functions are avoided. Upward hierarchical analysis is considered to be more efficient. It starts from the terminal nodes (BBs) and moves up through the middle nodes while considering only the combinations that meet a set of constraints. The method produces symbolic expressions that can be still fairly large for some circuit types, which arguably reduces their usefulness in gaining insight into circuit operation and incremental improvement of designs. However, the method is capable of analyzing circuits with more than 10 nodes, which has been previously reported as being a limitation of topological algorithms [46] and symbolic analysis, in general [47].

An alternative hierarchical symbolic analysis technique is presented in [45]. The main novelty of the approach is in that it produces the sequence of symbolic expressions (e.g., a sequence of assignment operations, where each right side is a symbolic expression) instead of a single expression like in the other hierarchical methods. Circuits are described as Coates flow-graphs, which are hierarchically partitioned. Partitioning creates a binary partitioning tree, in which the leaf nodes represent the terminal blocks of the circuit (sub-circuits) and the middle blocks show how their leaf nodes are connected with each other. Ideally, the partitioning tree should be a well-balanced tree (which reduces the complexity of the analysis), even though there is no guarantee that unbalanced trees are created. STFs are computed for terminal blocks (sub-circuits) through reduced MNA followed by elimination of the internal variables, one variable at a time, through Gaussian elimination and Schur complement. The matrices used in the process include the modified nodal admittance matrix and those corresponding to the branch currents equations. Middle block analysis is performed bottom-up considering connections between two *n*-terminal blocks. Analysis groups the bock nodes in three categories, local tearing nodes, which are eliminated by the current middle block analysis step, nodes shared between the two blocks with other blocks, thus, have to be further considered by the analysis process, and external nodes of a circuit. Experiments show that the proposed hierarchical analysis method is very efficient. For 100 resistor ladder networks, it generates only 248 terms instead of an estimated 3.510^{20} terms if flat symbolic analysis algorithms are used. Also, the number of operations to be performed for expression evaluation is reduced from 87 additions and 302 multiplications to only 8 additions and 8 multiplications for a certain symbolic expression. Results suggest that hierarchical methods can be a viable option for linear symbolic analysis of large analog circuits.

Li and Gu propose a symbolic method based on parameter extraction [27] that is conceptually similar to the algorithm in [24]. The method computes circuit TFs, TF sensitivities, and poles and pole sensitivities using a modified nodal two-graph representation. Matrix inversion is performed using the Kader method. Large-scale sensitivities are calculated using perturbation theory, such as the nominal values are changed by a certain amount. Numerical values are calculated based on 2^n uniformly distributed samples, where 2^n is the lowest value larger than the degree of the interpolated polynomial. Similar to [24], polynomials are found using fast Fourier transformation over the calculated values. Large-scale sensitivities are used to find the STFs, parameter sensitivities, pole expressions, and pole sensitivities.

Fernandez *et al.* [48] focus on methods to simplify symbolic expressions before and during generation. The authors explain that the simplification is required as complete symbolic expressions are huge for typical analog circuits, i.e., more than 8,600 symbolic terms for a Miller OpAmp and about 10^{16} for a μA 741 circuit. The authors argue that simplification after producing a complete symbolic expression is ineffective, as it still requires huge memory and computing time to calculate the complete expressions. Hence, they conclude that only simplification before and during generation can be useful in a broader situation. Simplification before generation (SBG) removes graph branches after comparing its influence on the circuit (e.g., by comparing the response of the complete and reduced circuits). Simplification during generation (SDG) produces symbolic terms in decreasing order of their magnitude, so that the needed accuracy is achieved. The method assumes that the maximum error specifications are given. A challenge is that as the complete symbolic expressions are not known, finding the maximum phase and magnitude errors is not trivial. The authors suggest that interval analysis is used to compute the ranges where the maximum errors are located, followed by finding the errors by using Newton–Raphson algorithm. Both SBG and SDG compute the circuit TFs using MNA and polynomial interpolation based on digital Fourier transformation, similar to the method presented in [24]. The numerator and denominator polynomials are found through LU decomposition and back substitution. Experiments show that SBG and SDG reduce significantly the size of the symbolic expressions for a μA 741 circuit. Simplification for hierarchical symbolic analysis has been discussed in [49].

Hsu and Sechen [23] propose another SBG method based on a sifting concept that includes six algorithms. Each algorithm implements another way of simplifying determinant expansion. First, device parameter elimination attempts to remove entire devices, pairs of device parameters, and single parameters by estimating their error starting from the determinant derivatives. Determinant dimension reduction and factorization expands a determinant along the rows or columns with a single nonzero entry. Heuristic row and column operation increases the number of zero entries or a row or a column by summing up two rows or columns or by multiplying them with a scalar. Determinant expansion computes the determinants of sparse matrices as products of sums as the authors argue that this form offers superior insight into circuit operation compared to other forms, e.g., the sum of products. Factorization of expanded expressions identifies common factors among expressions. This algorithm is suggested for expressions with less than 100 terms. Finally, cancelation of common factors removes factors that are common for the numerator and denominator of a TF, e.g., the factors due to bias sub-circuits, which have a small influence on the circuit TFs in the frequency domain. The methods produce significantly smaller symbolic expressions, while error is below 12%. Yu and Sechen [29] suggest two SDG methods that eliminate terms of small impact (e.g., magnitude) computed using rules stated for two-graph enumeration method [50]. The first simplification algorithm identifies the color-constrained spanning trees in increasing order of their weight. The idea is to generate first the spanning trees in graph G_V in decreasing order of the magnitude and then verify if they are spanning trees in graph G_I too. The authors suggest for using Gabow's algorithm for this task, in which the $k + 1$ lowest weight spanning tree is found from one of the k lowest spanning trees with one of their edges replaced by an edge previously not included in the tree. To approximate the numerator and denominator of a TF, the authors generate spanning trees in G_V with exactly i capacitor edges. The constraint is modeled as red edges in a red-green spanning tree. The k lowest weight colored spanning tree is found from the lowest colored spanning tree found using Gabow and Tarjan's procedure. The number of common graphs to be analyzed might be still very high. The second simplification algorithm performs a sensitivity-based simplification of the two graphs. First, the numerator and denominators are simplified, followed by a simplification of their ratio. The algorithm computes the contribution of every parameter of the two symbolic polynomials and removes those that have a small contribution to the polynomials and do not significantly modify the contribution of another parameter. Entire groups of symbols can be also removed. Finally, edges are candidates for being removed if they have or have not (simultaneously) a parameter. The authors report that the proposed simplification methods allow symbolic analysis of circuits with up to 39 nodes as compared to previous techniques that can tackle only circuits with up to 15 nodes.

An alternative simplification algorithm has been described in [51]. It considers SFG symbolic method. SFGs are built starting from a library of SFGs for the basic circuit elements. SFGs are transformed through merging of vertices and edge lumping, and rearrangement of the voltage sources that share the common voltage nodes of current dependent sources [51]. The authors suggest that frequency sub-ranges are calculated using numerical computing of a circuit's poles and zeros and then estimating the observability of poles and zeros. Roots can be clustered based on their frequency. The related roots are of two kinds, open-loop roots and roots originating from summing points. Network simplification is controlled by the complexity of estimating the symbolic expressions. The mentioned simplifications include the removal of signal paths, open-loop root removal, vertex-pair contraction, and subgraph substitution. Other important work on pole and zero modeling has been presented in [52–55].

6.2.3 Tackling complexity: advanced symbolic representations and beyond linear analysis

The following generation of symbolic analysis methods has focused on using advanced representations as means to reduce the complexity of symbolic expressions. Arguably, these methods employ previously developed solutions for calculating and simplifying symbolic expressions. Hierarchical representations are also being employed. In [25], Shi and Tan suggest a very efficient representation, called determinant decision diagrams (DDDs) for representation of symbolic expressions. The authors explain that previous symbolic methods are limited to circuits of small size. Circuits are described using MNA, and similar to other work such as [3], Cramer's rule is used to compute the determinants of matrices and cofactors. Parameter sensitivities are calculated by adding small perturbations δ to the representations. The produced symbolic expressions are the sum of products in nested form. DDDs are inspired after zero-suppressed decision diagrams representation [56] to exploit the sharing of terms and sub-expressions. The symbolic products corresponding to determinant computation in Cramer's rule are paths in DDDs. The authors indicate that the size of the symbolic representation depends on vertex ordering. A heuristic has been proposed based on Laplace expansion of sparse matrices in which dense columns and rows are analyzed first. The method is extended for hierarchical symbolic analysis in [57]. The work has been subsequently extended for multi-root DDDs [58], in which each root is a symbolic expression of the coefficients of a separate power of variable s (called coefficient DDD). Common sub-expressions are shared between different root trees. The same vertex ordering is used to construct the DDDs of the trees that share common sub-expressions. Each term is assigned a unique label that makes vertex sharing possible. The authors introduce two operations on multi-root DDDs: s-expand extends a vertex with its s-expanded DDD vertices, and s-extract only keeps a label coefficient.

Another work focusing on devising efficient representations for linear symbolic analysis has been discussed in [26]. The symbolic methods use two kinds of regularities to dramatically reduce the size of symbolic expressions: (i) the structural regularity of analog networks, e.g., analog filters but also circuits and (ii) the regularity of symbolic expressions. For example, analog networks, i.e., filters, ADCs, and PLLs, include often multiple stages in which BBs, like adders, integrators, and amplifiers, are connected with each other according to similar templates. The work calls such templates *generic interconnection templates* (GITs). The symbolic expressions describing GITs are reused (as part of a middle block analysis [[20,45]]) independently of whether the terminal BBs are transistors or OpAmps. The authors show that GITs can handle not only modular networks, e.g., formed out of stages, but also the networks that are less structured, e.g., various OpAmps. Second, the regularity of the symbolic expressions refers to that similar operations are performed for various kinds of GITs and BBs but obviously applied to the different block parameters. Then, the uniform set of operations can be encapsulated in a *generic symbolic function*, so that it can be shared for all parameters that use the function. Conceptually, this step is similar to extracting a function for a set of computations and then sharing the functions for repeated yet similar calculations for different actual parameter sets. The proposed representation for symbolic analysis is called *analog performance trees* (APT). The internal nodes of APTs describe the connections between the BBs of the node's children (not limited to only two children as connections between multiple blocks can be handled). The leaf nodes describe the BBs of the network. Each internal call represents a set of calls to the shared symbolic functions to indicate how the parameters of the internal node are computed using the parameters of the children. APTs are constructed top-down through a process of partitioning (tearing) the network. Partitioning uses any hierarchical structure or connection regularity of the analyzed network. A partitioning-based technique has been also proposed to automatically identify a network's regularity. Partitioning uses a cost function that considers three aspects: (i) minimizes the number of distinct GITs in the partitions as structurally, more similar clusters are likely to use the same GITs, (ii) the GIT of the current partitioning step is already present among the GITs that have been already selected, and (iii) selecting GITs that form similar sequences of partitions. Then, such sequences correspond to similar connection templates for more than two blocks. A theoretical analysis shows that the size of the produced symbolic expressions remains always polynomial [26]. This property has been used to speed up the execution time of network parameter optimization [4]. Another interesting application would be to explore simplification after generation, an approach previously discarded as being infeasible, to create approximate symbolic expressions for gaining insight into circuit operation.

In subsequent work [59], the concepts proposed in [26] were extended to devise a compiled-code simulator to time-effectively simulate analog systems with nonlinear parameters. The idea behind the simulator is that optimized and customized simulation code can be generated for a given analog system or circuit, following techniques from compiler design, so that the code is much faster than the simulation time using a general-purpose simulator, like Spice. In addition to producing sequences of symbolic expressions, similar to [45], the code also includes functions to describe the symbolic expressions of GITs (called *instances of structure* (IST) in the paper), and function calls on how the GITs are used to connect specific blocks. The simulator handles time and frequency domains. Each IST is coded as a C^{++} class C_{IST}. Every IST instance in a system topology becomes an object of class C_{IST} with the identities of the involved blocks being set depending on the topological connection. The produced code represents the BBs of the system, e.g., OpAmps, operational transconductance amplifiers (OTAs), and comparators of a $\Delta\Sigma$ ADC, as macromodels with a rich set on nonidealities and nonlinearities, like finite gain, poles, zeros, common-mode rejection ratio (CMRR), phase margin, and so on. Nonlinear parameters are modeled as piecewise linear (PWL) models. PWL models are extracted from trained NNs. NNs are trained through back-propagation for a set of sampled values describing the design space hypercube of interest. The trained NNs are pruned to remove weights and neurons with small contributions to the modeled parameter. The sigmoidal functions of an NN are replaced by PWL approximations, in which the number of linear segments, their limits, and linear approximation on each segment are automatically found through clustering. The PWL approximations of the NN output are produced by composing the PWL approximations of the hidden neurons. The activation regions for each PWL segment are found by solving a linear system. Experiments show that when simulating $\Delta\Sigma$ ADC, the proposed compiled-code simulation methods have fewer stability problems and offer speed-ups of about 100 times as compared to numerical simulation with Spectre simulator. Such a speed-up is important for synthesis tools meant to optimize at the high level of the design parameters of $\Delta\Sigma$ ADC, but also other analog systems and frontends.

Daems *et al.* suggest a method to produce posynomial response surface models for linear and nonlinear performance attributes [30]. Automatically creating posynomial models is important because geometric programing, a powerful optimization method, uses posynomial circuit representations as inputs [60]. The method pertains to the category of parameter extraction methods, similar to [24,43], in which sampled values are fitted to predefined symbolic models, usually in polynomial formats. The main contribution being in the way the values used in modeling are being sampled. Numerical performance samples are obtained through Spice simulations, in which sampling is performed based on design of experiments (DOE) theory. DOE includes the steps of variable screening, sample generation, parametric or nonparametric model regression, and model validation. Sampling uses traditional methods like Latin cube or orthogonal array-based sampling. Posynomial models are first expressed using small-signal parameters, but these models are transformed by substituting the small-signal parameters with design parameters like transistor sizes, and bias voltages and currents. The work proposes three performance parameter fitting approaches: indirect fitting, direct fitting with template estimation, and direct fitting without template estimation. Indirect fitting first creates polynomial models, which are then converted into posynomial models. Polynomial models are found by solving a convex minimization problem, which is to minimize the least square error between the models and sampled values. Polynomial conversion tackles three cases: negative coefficients, negative interaction terms, and negative quadratic terms. Direct fitting with template estimation fits the performance models to a certain kind of posynomial formats. Estimation is based on an a priori procedure. Finally, direct fitting without template estimation minimizes the sum of the square errors of the posynomial and sampled points. Models are up to second-order polynomials and their coefficients are found through conjugate-gradient minimization, claimed to be more efficient than other optimization algorithms. This fitting method supports more general posynomials than fitting with template estimation; however, the authors mention that not all performance attributes can be accurately modeled as posynomials, and that the method requires effective transistor models in posynomial form, which might not be trivial to produce.

6.3 Circuit macromodeling

Wei calls a circuit model structural (white box) if (i) all circuit nodes are present in the model, (ii) KCL and KVL for all nodes are preserved in the model, and every circuit component corresponds to a set of model devices. Two distinct components have disjoint devices in the model [61]. A structural model is uncoupled, if there are no cross-dependencies between model parameters, otherwise, the model is coupled.

A number of circuit macromodels exist in the literature produced on the basis of manual design analysis and insight. For example, Gomez *et al.* describe an OTA circuit model that includes offset voltage, common-mode capacitor, first and second pole, and first and second zero [62]. The model was extended by Zhang *et al.* [59] for fully differential mode in which nonlinear parameters are expressed by PWL models extracted from trained NNs [63]. The structural macromodel of a fully differential OpAmp circuit is presented in [59]. It includes three stages: an input stage, similar to the OTA model, the intermediate stage that represents two dominant poles in differential mode, and the output stage that includes two dominant poles in differential mode and a DC bias voltage for transient analysis. Among others, a comparator macromodel has been proposed by Moscovici [64]. This macromodel was extended for $\Delta\Sigma$ modulator modeling in [59]. The comparator macromodel has the same input stage as the OTA circuit, the intermediate stage to express the self-limiting behavior of the comparator, the slew limited mode, two dominant poles, delay time for large input overdrove voltages, and an output stage that describes the minimum and maximum output voltages. The remaining part of the subsection discusses automated techniques for circuit macromodel creation.

Huang *et al.* propose a circuit modeling method based on the topological location of the circuit poles and zeros through a novel root localization algorithm [65]. It models the localized effects that decide pole and zero location, and which change with the operating conditions of the circuits. The technique creates white box models, which can offer insight on the frequency performance of a circuit. The method includes six steps: first, signal-tracing algorithm identifies the way in which the signals flow through a circuit. The analysis is important because it establishes a facet of a circuit's design "causality," e.g., the way in which the signals produce circuit performance. The authors propose a set of rules to decide the signal flow of a circuit. The second step uses another set of rules that separate the circuit nodes into two sets: nodes that are part of the signal path and nodes that fix the circuit's operating point (i.e., biasing circuits). Obviously, using static rules for steps one and two limits the method only to circuits with topologies corresponding to the existing rules. Next, the π models are used for bipolar transistors and small-signal models for MOSFET devices. This step replaces active devices with their models, hence, enabling small-signal analysis. Step four uses MNA to compute the nodal voltages and currents and then calculates a circuit's poles, zeros, and sensitivities. Step five uses root localization method to identify the local and global roots, which represent poles and zeros. The final structures of the circuit macromodel is built in step six by introducing signal mixing nodes and nodes for large signal modeling.

Wei and Doboli describe a structural linear macromodeling technique to produce uncoupled circuit models [66]. The method has five steps. First, it grounds all biasing voltages. Biasing transistors can influence the TF, introduce additional noise into the circuit, or impact the PSRR of the circuit. The second step partitions the original circuit into its BBs. Designer knowledge is needed for this step to identify BBs that have a meaning in design. All BBs are separately modeled, and their models are connected together to form the circuit model. The third step replaces devices with their macromodels, e.g., the hybrid-π model for bipolar junction transistors (BJTs) and small-signal model for MOSFET devices. The used device models are uncoupled; hence, there are no cross-coupled dependencies between a device's parameters. Step four replaces the transistors of each BB with their uncoupled models. Independent variables are the BB's terminal voltages. The currents flowing into a terminal is set by the circuit parameters and terminal voltages. All nodes of a BB are also visible outside the block, so that it can be coupled to the nodes from other BBs. Step five connects all BBs of the circuit. This generates the coupled building block behavioral (CBBB) model of the circuit. Each node has a voltage characterized by the following expression: $v_i = (R_i + sC_i)\sum_j(sC_{mj} \pm G_{mj})v_j$, where R_i and C_i are the total resistance and capacitance associated to the node, and $(sC_{mj} \pm G_{mj})v_j$ is the voltage-controlled current source associated to the inspected node. Next, CBBBs are transformed into uncoupled building block behavioral models. This activity includes four steps: first, it identifies the decoupling sequence by performing signal path tracing algorithm [65]. It identifies the order in which nodal voltages are decoupled. Next, the CBBBs of a circuit model are rearranged according to the decoupling sequence. All feedback-dependent voltages v_i are replaced by their equivalent voltages $v_{i,eq}$. Next, the symbolic expressions of all rational functions of voltages $v_{i,eq}$ are computed. The work proves that such a coefficient $a_{i,j}$ has the generic form $a_{i,j} = \sum_t (\pm \Pi_{k_i,k_j} G^{\alpha_k}_{m,k_ik_j} \Pi_{l_il_j} C^{\beta_l}_{m,l_il_j})$, where $\alpha_k, \beta_l = 0$ or 1. Parameters G_{m,k_ik_j} are the transconductances between modes k_i and k_j, and C_{m,l_il_j} are the transcapacitances between nodes l_i and l_j. Then, computing the symbolic coefficients $a_{i,j}$ implies finding the values of parameters α_k and β_l. It is achieved by using an optimization algorithm with a cost function that minimizes the modeling error. Note that the generated structural model includes symbolic expressions for their parameters. Step four replaces the feedback dependencies with their equivalent symbolic expressions. The macromodeling method also captures white noise, $1/f$ noise, generation recombination noise, and shot noise. Clipping is also described by the macromodels. The work demonstrates the method for a number of OpAmp circuits, such as single-stage OpAmp, three-stage OpAmp, two-stage folded cascode Miller OpAmp, and resistor-based OpAmp [61].

Zhang and Doboli propose a structural macromodeling method to include MOSFET process parameter variations [67,68]. It considers global and local variations as shown in the formula: $P_i(total) = P_0 + P_G + P_{iL}$, where P_0 is the mean value, P_G is the global variation, and P_{iL} is the local variations [69]. Local variations depend on the circuit layout, e.g., device orientation and adjacencies. Process parameter variations are usually simulated using two nested loops: the outer loop considers global variations approximated as Gaussian distributions with zero expectation and the inner loop describes local mismatches. Local mismatches have two parts, a deterministic component and a Gaussian stochastic part depending on STI and layout elements. Correlations between mismatch variances are described as in σ-space representation [70]. The mismatch $P_i - P_j$ between the process parameters P of MOSFETs i and j is expressed by the following equation: $\sigma^2(P_i - P_j) = \sigma^2_{ij} = \Sigma_k(a_{ik} - a_{jk})^2$ [70]. Parameters a_{ik}, a_{jk} are found by solving a set of equations, including variances of each process parameter of the layout and covariance values between any two layout devices. Variances and covariances are extracted from circuit layouts. Additional equations are considered to describe layout design-specific aspects, e.g., larger MOSFET devices have less variance than smaller MOSFETs as well as the variance relationships between MOSFETs close to each other in the layout. The number of equations is reduced using layout regularity to identify similar equations and principal equation analysis to eliminate less significant parameters. Compared to other similar methods, the technique in [67,68] expresses correctly both process parameter mismatch and process parameter variations of individual devices. Hence, experiments show that this technique is more accurate than Monte Carlo (MC) or σ SA methods.

Li and Pileggi suggest a linear centric approach for modeling periodically time-varying weakly nonlinear circuits [7]. It uses the successive chord method and linearization of nonlinear elements. The proposed solution is arguably straightforward compared to the more complex, Volterra-based solutions and also can perform higher order analysis than Volterra methods. A certain challenge might be due to the design insight required by the method.

Demir *et al.* have introduced phase noise analysis in oscillator circuits by characterizing the per cycle jitter [71,72]. The steady state of an oscillator is computed by solving the mathematical model of the circuit. However, the authors show that computing the eigenvalues of the monodromy matrix (state-transition matrix) is hard. Therefore, subsequent work by Demir and Roychowdhury [73] tackles this limitation by suggesting reverse time-domain integration realized through frequency domain simulation using harmonic balance (HB). A central concept of the method is perturbation projection vector (PPV) that describes the circuit sensitivity to perturbations (trajectory). Finding PPV requires solving equation $(\partial q(x)/\partial t) + f(x) = 0$. After linearizing around the solution point, a solution has the following form, $\Phi(t,s) = U(t)D(t-s)V^t(s)C(s)$ with $V^tCU = I$. Matrix V^t is computed from the augmented HB matrix and solving the corresponding augmented Jacobian. Moreover, $D(t) = (e^{\mu_i t}, \ldots, e^{\mu_m t}, 0, \ldots, 0)$, where values μ_j are the Floquet eigenvalues. U is found from large-signal periodic solutions.

Vanassche *et al.* present a macromodeling method for the steady-state and transient behavior of coupled harmonic oscillators [74]. The method conducts perturbation analysis and computes the amplitude and phase differences as well as the departure from the trajectory in the phase space. The proposed technique uses averaging to separate the slow and the fast changing components, such as steady state and transients with widely separated time constants. The transient behavior is represented as a sequence of transformations from one behavior to another. The technique does arguably a hierarchical, two-level analysis as an oscillator is partitioned into its resonator tank and feedback transistor. The authors show that the method requires low computational cost and extends previous oscillator modeling work to include common-mode phase. An important observation is that the macromodeling method requires design-related knowledge about the energy injected by the feedback into the resonant tank. Also, a number of design-based, simplifying assumptions have been made to help model equation solving. Automating the finding of the required design knowledge is hard.

6.4 Neural networks for circuit modeling

NNs have been used over a long period of time to produce black-box models for the performance of analog circuits [75–79] or circuit faults [80,81]. A tutorial on NN modeling of microwave circuits is offered in [79]. NNs are a strong modeling candidate considering their property being universal approximators [82]. Hence, at least in theory, they can express any real function provided that NN is properly trained. NNs have been used to model performance in the time and frequency domains. The user has to select the structure of the NN, e.g., the number of neurons and the number of layers. NNs are trained using a set of numerical performance data obtained through simulation, i.e., using Spice simulator, and then validated using a second set of data.

In earlier work, Zaabab *et al.* use NNs for performance estimation for optimization of microwave circuits [83]. They use a three layer (with one hidden layer), which is trained online backpropagation algorithm; hence an NN is retrained every time a new training data is available. Wolfe and Vemuri explain that the validation set is maximally distant from the training set [78]. They model seven performance attributes, such as open-loop gain, unity gain frequency, phase margin, CMRR, PSRR, rising slew rate, and falling slew rate. After reducing the parameter space through constraints due to the problem specification and biasing needs, the work suggests that training is performed using a training set of data sampled equally spaced points at a granularity decided by the user. Training uses the traditional Levenberg–Marquardt backpropagation algorithm. The produced NN models are being used for circuit sizing using a genetic algorithm (GA). Suissa *et al.* discuss the modeling with NNs of the power consumption of analog components [77]. NNs describe the nonlinear dependency between power consumption and circuit inputs and parameters. The training data includes real, instantaneous power-consumption measurements and continuous interpolations between the measured points. Similar to [78], Levenberg–Marquardt backpropagation algorithm is trained by sweeping the circuit parameters over their entire ranges. Three-level NNs are suggested for modeling. A similar method is used in [84] to synthesize RF planar on-chip spiral inductors. Two performance attributes of inductors, quality factor (Q) and self-resonance frequency, are modeled using NNs based on layout-related parameters, like spiral outer diameter, number of turns, metal trace widths, and metal spacing. Feedforward NNs with four layers are used in modeling with one input layer, two intermediate layers, and one output layer. The neurons of the hidden layer use hyperbolic-tangent activation function and linear activation for the output layer. NN training used the Levenberg–Marquardt method using 500 spiral inductors that were characterized through detailed electromagnetic simulation (IE3D solver). Planar octagonal spiral inductors were used with dimensions suitable for wireless communication circuits. Dimension values were sampled from a uniform grid, at equal intervals. The NN models are used for synthesis using particle-swarm optimization.

Cao *et al.* propose state-space dynamic NNs (DNNs) (SSDNNs) as a solution to represent the transient behavior of high-speed nonlinear circuits [85]. The work suggests two innovations: the concept of SSDNNs as an extension of DNNs and new stability constraints that are embedded into the training step. Global and local stability requirements are derived. The authors indicate that DNNs have been used to describe the steady-state behavior of nonlinear circuits, like mixers and amplifiers. SSDNNs combine the first-order differential equations describing a steady-space model with multilayer perceptron (MLP) model that describes the functions that are part of the differential equations. Improved stability of SSDNNs compared to DNNs results from the fact that the earlier ones consider first-order differential equations, while the latter higher order differential equations. To derive new Lyapunov stability criteria, the work uses three-layer MLPs, in which weights are separated into three categories: weights between inputs and hidden neurons, weights between state variables and hidden layer, and weights between hidden layer and output neurons. Neuron activation is achieved through bounded, smooth, and monotonically increasing functions. The derived local stability criterion requires that the real parts of the eigenvalues are negative, while the derived global stability constraint states that the real parts of the testing matrix must be negative too. A new gradient-based training algorithm has been proposed to incorporate the two criteria. The derivatives used by training are approximated by differences.

A very interesting NN-based modeling technique has been presented in [86] for high-speed input/output (I/O) buffers. The method is a subsequent development of the concepts in [85]. The method combines design knowledge about I/O buffers with NN modeling, specifically it uses NNs to describe some parameters of the macromodel of the buffer. Traditionally, I/O buffers are modeled using the I/O Buffer Information Specification method. Its main limitation is its inability to describe well transient performance as the input signals must be perfect step functions. The algorithm proposed in the paper addresses this limitation by tackling transient performance parameters, e.g., ringing and attenuation. The work considers that the buffer output is the sum of the nonlinear $I-V$ characteristics of the transistors that form the buffer output. The output current is then as follows: $i_o(k) = w_1(k)f_{NN1}(v_o(k)) + w_2(k)f_{NN2}(v_o(k))$ [86], where functions $f_{NN1}(v_o(k))$ and $f_{NN2}(v_o(k))$ model the static behavior of the buffer and coefficients $w_1(k)$ and $w_2(k)$ the dynamic behavior. Functions $f_{NNi}(v_o(k))$ are described using static feedforward NNs, which are trained using data measured for buffers with a DC voltage source as a load. Load values are swept beyond the operation range in order to capture overshoot and undershoot conditions. Training uses gradient-based search. Coefficients $w_i(k)$ are represented using recurrent NNs (RNNs) to express the delay between inputs and outputs. Delay is learned by selecting the number of delay steps. Training uses transient pulses with different rise and fall times for low and high values. Learning stability is verified using Lyapunov criterion.

Even though the work in [87] discusses NNs for thermal modeling of graphics processing units (GPUs), we consider that it is useful to summarize it here too. The goal of the work is to predict the temperature of GPUs over time for all possible power values and initial temperatures. Thermal models are for 2D and 3D ICs. Training uses a detailed temperature modeling tool called ICE. NN training uses the observation that only active layers have a high impact on integrated circuit temperature, while the ambient temperature is considered through the superposition principle. This observation simplifies the NN complexity. Also, heat flows mainly vertically along the lowest thermal resistance paths, and less between components positioned far apart. This property further simplifies the NN complexity by removing the connections between the neurons that correspond to components positioned at larger distances between each other. Training limits inputs to the maximum power density of the corresponding floorplan elements. Floorplans are also randomized so that the trained NN can predict the thermal dynamics of a GPU independent of its floorplan. Experiments show a speed-up of 35 $\times$ compared to ICE simulation while temperature prediction error is less than 1°C.

Janakiraman *et al.* model through NNs the influence of process parameters on leakage currents [75]. Analytical models are created for mean values and semi-analytical formulations for distributions of leakage currents as functions of voltage and temperature. This work is an alternative to analytical statistical methods, such as linear and quadratic back propagation of variance [88]. Process parameters variations consider die-to-die and within-die variations. Variations have two components, spatially correlated-components and random components. The authors suggest that spatially correlated variations are described by partitioning the geometrical regions into smaller regions within which device parameters are affected in the same way. Random local variations can increase the mean leakage currents. Non-uniform temperature variations across ICs are also modeled by NNs. The authors argue that these models are an effective alternative to traditional leakage currents models, which are exponentials of linear or quadratic polynomials. Mean and variance distributions are computed using the law of iterated expectations and the law of conditional variances. Similar to other work, Levenberg–Marquardt training is used, in which temperature and voltage values are uniformly sampled form the ranges of interest. Pairs (V, T) are randomly permutated, so that as many corner points as possible are sampled from the statistical parameter space. A feedforward NN with a single intermediate layer is suggested, which is trained using 1,500 training data and validated through 500 samples. The authors derive the analytical equations for means and variance by suggesting a new neuron activation function for which the following expectations are computed in the paper: the expectation for the activation output if its inputs are normally distributed, the expectation for the activation functions if its input squares are normally distributed, and the expectation for the activation functions if input pairs are joint normal distributions.

Using similar concepts as [75], Viraraghavan *et al.* present a method for extracting statistical circuit models from trained NNs [89]. The goal is to identify models that relate statistical circuit parameter models to circuit performance. The argument for this work is that traditional MC methods might be cumbersome as they require selection of alternative circuit parameter distributions and MC simulations until the identified parameter simulations produce the performance distributions that were observed in experimental measurements. In the proposed method, the unknown, means and standard deviations of the circuit parameters are found through backpropagation from the NN outputs, a process that is called backpropagation of variance (BPV). BPV is addressed by minimizing the nonlinear error between trained NN outputs and performance statistical parameters expressed through analytical formulas. Process parameter variations are either Gaussian distributions or nonlinear power method based on moment matching. Neuron activation functions are represented using mathematical functions that support the mathematical derivation of expectation formulas, so that means and variances can be backpropagated through an NN. The work considers the following circuit parameters: the statistical distributions of linear region on current, saturation region on current, subthreshold leakage for zero gate-to-source voltage, linear region threshold voltage, and saturation region threshold voltage.

Pratap *et al.* present a design centering method to improve yield, in which NNs are used instead of MC simulation and a GA for design centering [76]. Traditional backpropagation algorithm is used for NN training, in which the synapse weights are optimized with a gradient-descending algorithm that changes the weights depending on the partial derivative of the NN error with respect to the weight. Sigmoidal activation functions are used by the neurons. GA computes the yield of a population and propagates a percentage of the individuals of high fitness to the next generation. The rest of the population is formed through classical GA operators, e.g., mutation and crossover. Successive generations are created by the GA until the required yield is achieved by the circuit population.

A special kind of RNNs called echo state networks (ESNs) have recently been used to model the nonlinear time-response of analog circuits, like DC–DC driver ICs, RF mixer, and local interconnect network interface ICs [90]. Only the output layer of ESNs is trained instead of training all connections of an RNN. The reason is that the impact of input signals on the hidden neurons decreases exponentially over time; therefore, it has been argued that inputs create an "echo"-like effect on these neurons. ESN training is achieved by solving the equation $Y^{train} = W^{out}Z$, where Y^{train} are the training outputs, Z is the matrix of state vectors collected over a sequence of discrete time moments, and W^{out} is the unknown matrix. The equation is solved using ridge regression. The internal weight matrices are generated so that they are diagonally Schur stable. Internal unit activations are set to random variables. Experiments show that ESNs offer good modeling accuracy, while the training time of ESNs is much shorter than the training time of RNNs. Also, it has been argued that the internal units of ESNs have little influence on the outputs. The topology of ESNs is easier to decide than that of RNNs, and ESNs are more stable too. Performance estimations with ESNs are about 77 $\times$ faster than numerical simulation as it has been noted that parasitic modeling does not increase the complexity of ESNs while it produces significant numerical simulation time increase.

Yang *et al.* propose robust heteroscedastic probabilistic NNs (PNNs) (RHPNN) for accurate fault detection and classification through clustering [81]. The fault type is distinguished by its probability. According to the authors, the work attempts to address limitation of previous work on fault modeling with NNs [91–93], e.g., small circuit sizes, low robustness, and non-qualitative prediction. Other limitations include the need for large number of training patterns, the fact that the population size of faulty circuits is much larger than that of fault-free circuits, and that the trained NN might be over-fitted if fault-free circuits are reused during NN training. PNNs, previously used in fault modeling, estimate the class conditional probability density functions. Training uses maximum likelihood, training patterns that are the centers of the Gaussian kernel functions, and common variances or covariances. RHPNNs are a variant of PNNs with four layers: the input layer is the first layer, the second layer has a number of groups depending on the number of fault categories, the third layer estimates the mixture of Gaussian kernels, and the fourth layer produces the fault classification output. RHPNN is trained using estimation maximization. The authors report accuracy close to 100% for OpAmps, even though choosing measurement parameters might be not simple for all circuits.

6.5 Discussion and metatheory on analog circuit modeling

This section summarizes the presented work on analog circuit modeling by analyzing the evolution of modeling methods over time in an early attempt to understand invariant features of this evolutionary process. We think that the analysis is useful to understand how specific application needs shaped the progress of circuit modeling domain, the nature of the studied methodologies and algorithms, their incremental development over time, and the efficiencies and limitations of the group-level processes through which the research community identified and tackled problems of interest. The discussion explains conditions under which the community converged toward a set of accepted concepts (beliefs). A set of interesting questions, as enumerated in the introductory part, can be posed starting from this discussion, i.e., effective management of the community, the likely impact of certain research, and possible funding strategies. This section offers an early discussion of some of these issues and suggests some intriguing hypotheses to be considered by future work.

From a cognitive point of view, it can be argued that any solution (not only circuit modeling solution) includes (i) *an associative part* through which concepts are selected similar to previous work, and (ii) *a causal part* through which the concepts are logically related to each other in a feasible, working solution [10,12]. From this perspective, early solutions must identify a complete set of concepts as well as an entire causal path of a solution. For example, the early work by Fidler and Sewell [43] proposed a complete symbolic analysis method that was capable of automatically creating models starting from circuit schematic descriptions. In spite of its limitations (e.g., linear analysis and complexity constraints), the method suggested early on the potential of symbolic analysis in tackling an important practical design problem, which is to produce effective and reliable circuit models. Hence, together with other similar early work, such as the work by Sannuti and Puri [34], it served as a strong motivator for pursuing similar research. (In fact, papers [43] and [34] are often cited together.) Subsequent work kept improving the elements and sequence of the analysis as new insight about the limitations of the approach became available. Note that this insight can arguably not be inferred at the method level unless the actual modeling algorithms are implemented and experimented. Moreover, the new insight that becomes available might restructure the importance of the various needs that are tackled by research.

From this perspective, the evolution of circuit modeling through symbolic analysis can be arguably separated into five phases, as shown in Figure 6.1. The *first phase*, while noticing the difficulties due to the large size of symbolic expressions, assumed that the challenges can be effectively tackled by devising rules that will produce expressions of minimum size, such as by canceling out opposite symbolic terms. The work pertaining to this phase kept reinforcing this belief as more sophisticated rules were proposed to reduce symbolic expressions. While improvements were reported, reduction as a method did not support tackling

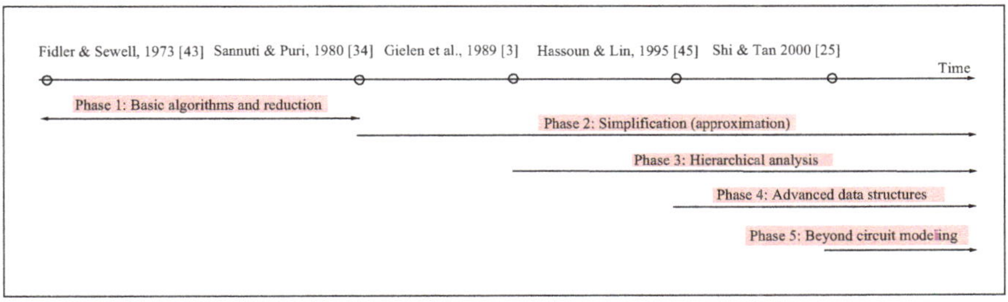

Figure 6.1 *The five phases of the evolution of ideas on symbolic analysis*

of significantly larger circuits. This observation suggests that early rules already captured to some degree the capability of reduction as a method to reduce the size of symbolic expressions; thus, subsequent rules incrementally improved the method, but without providing a major advantage in the kind of circuits that could be analyzed. Moreover, the mathematical theory behind uncovering new rules became increasingly more complex, hence also the time and resource effort required to devise the rules. The insight that reduction is insufficient became available only in small steps.

The *second phase* focused on exploring approximation (simplification) as a method to address the large complexity of symbolic expressions. Approximation, carried on either on the input description or final expressions, showed early on a significant improvement in the size of the circuits that could be tackled (e.g., one order of magnitude or more) without increase in the theoretical difficulty of the methods. Approximation was able to push the limitation borders of symbolic analysis without asking for extremely large efforts, hence, being considered attractive and then adopted by a large number of researchers. However, approximation has been shown not to be the final solution for symbolic analysis. The main insight that emerged was that it is difficult to limit the behavior of a circuit to a small (tractable) number of symbolic terms as the importance of various terms changes for different conditions (e.g., frequency).

The third *phase* has been interleaved with the second phase and considered that a circuit can be hierarchically decomposed into a number of BBs, which are separately analyzed and then joined together. Hierarchical methods are able to tackle much larger circuits; however, difficulties still exist for strongly coupled circuits as well as producing tractable models that offer insight into circuit operation.

The *fourth phase* focused on devising efficient data structures as means of tackling complexity. This step did not significantly change the causal flow of the solutions of the previous steps, but rather the combination of a concept (e.g., binary decision diagrams (BDDs)) developed outside the community with symbolic analysis methods. While larger circuits can be modeled, the limitations of the previous phases exist as improved data structures do not necessarily improve modeling capabilities.

Finally, *the fifth phase* proposed symbolic analysis for applications beyond circuit performance modeling, such as automatically comparing two circuits [8] or constructing knowledge structures [10].

The previous discussion suggests that the evolution of ideas on symbolic analysis (but also other kind of work) pertained to orthogonal categories, where each category corresponds to a certain interpretation (belief) of the way in which the stated goals and needs should be addressed. Multiple interpretations can coexist, like approximation and hierarchical methods. The evolved ideas reflect the acquired insight, and which cannot be directly inferred from the stated goals, needs, or initial ideas. This observation suggests that getting new insight on needs and solutions is not accidental (or serendipity), but the resultant of detailed experimentation and characterization to expose the advantages and limitations of a method. Specifically, careful description of limitations not only exposes the conceptual limitations of a certain class of methods (i.e., reduction simplification, etc.) but also hints toward possible new ways of addressing the limitations. Any new insight and related solution defines a new interpretation (idea category, belief) on how the problem should be solved. Moreover, for other researchers to adopt the new interpretation, should be flexible enough for further improvement, it should offer advantages beyond those of the current interpretation (e.g., separated by a magnitude improvement), and should be cost-effective in terms of the resources required for being adopted and used. If the new interpretation is adopted by other researchers, it becomes a common belief that introduces certain constraints and priorities among the concepts being used in the current solution. Also, a certain bottleneck is implicitly introduced due to the assumed constraints and priorities, but as already explained, the bottleneck cannot be completely understood from the new interpretation and solution. Understanding the bottleneck is part of gaining new insight, which then becomes the starting point of producing a new phase of the idea evolution process.

In summary, the idea evolution process operates in phases, each phase being characterized by (i) accepted, common beliefs, which are (ii) good predictors of the potential benefits offered by the beliefs (hence, motivates others to adopt them) and are (iii) flexible enough, so that their evolution within the phase exposes the insight needed to move beyond the bottlenecks of the beliefs (incremental refinement).

The utility of a certain work (e.g., algorithm and paper) can be quantized as the number of subsequent works that employs it, such as the number of received citations (in a relatively simplistic consideration). In addition, the degree to which work is utilized by subsequent methods depends on elements, like the visibility and prestige (credibility) of the authors (within and beyond the niche of similar work), the possibility of the work to be further developed and improved (flexibility) as well as utilized for achieving other goals (i.e., applications of the method), and the available resources for pursuing this effort. The analysis of the received citations by different modeling papers indicates the following sources: (i) self-citations by the authors, (ii) citations by others who further developed the method, (iii) citations by others who devised alternative methods for the same goals, and (iv) citations by others who utilized the methods for different goals, e.g., used the modeling methods in circuit design and optimization.

Furthermore, the utility of work can be characterized depending on the degree to which authors tend to repeatedly reuse the work in their subsequent efforts, e.g., repeated citations of the same work by a certain author. This aspect depends to a certain degree on the resource availability of the citing authors (i.e., authors with sufficient resources are more likely to pursue a longer "sequence" of related work). Different citation patterns can emerge: for example, the paper presenting ISAAC symbolic analysis tool [3] received a total of 147 citations, out of which about 25% (38 citations) were self-citations, 30% (43 citations) citations of papers that discussed applications of symbolic analysis (like design, synthesis, and verification), 10% (15 citations) were overview papers, 19% (28 citations) from related yet alternative approaches (e.g., work on approximation and hierarchical methods), and 31% (47 citations) from authors that cited it a single time. Similarly, the citations of [28] included 19% self-citations, 34% repeated citations by the same authors (two to four citations in this case), and 45% single citations. A second pattern corresponds to methods that are flexible enough to be combined with new concepts, like [51] 39% of the citations are given by two groups that repeatedly extended the method (i.e., work on error control and heuristics), 30% from related work that gives two to three citations, 9% are self-citations, and 18% correspond to single citations. A third citation pattern emerged for the work in [94]: it received a total of 58 citations, out of which 17% (10 citations) came from similar work, 25% (15 citations) from applications of the method in design, and 34% (20 citations) from authors that cited it a single time. The pattern suggests that the work benefited from high visibility and applicability (as shown by the number of applications and single citations) but might be difficult to expand (given the low number of similar methods). A fourth, more constrained pattern exists for the work presenting the using of DDDs for symbolic analysis [25]: 49% of citations are self-citations, 18% from repeated citations from similar work, and 8% from authors that cite it a single time. A similar citation pattern also exists for [58]: 55% represent self-citations, 24% citations from alternative approaches, and 19% single citations. Also, [95] received 11 citations, out of which 55% were self-citations and 45% represented single citations. The citation patterns suggest less flexibility of the method in being improved by others and also a lower capability in being used in applications of the method. The four citation patterns represent different cases of the flexibility of the new solutions (e.g., capacity to be evolved by others) and the available resources for their expansion (including their perceived utility for the community).

Note that the idea evaluation process did not always pursue all interesting ideas, even though ideas promised significant improvements compared to current approaches, were flexible, and easy to adopt and use. For example, the method of Katzenelson and Unikovski [96] proposes the idea of combining interval calculus with symbolic analysis. However, the idea is not further pursued by the community in spite of its intriguing potential. Moreover, interval calculus can be further generalized into the broader concept of producing sets of nonoverlapping constraints under which symbolic approximation can be significantly more effective. Another unexplored idea is the combination of PWL approximation and symbolic analysis, even though individually both approaches have produced promising results.

Figure 6.2 summarizes the previous discussion as a metamodel arguably presenting the idea evolution process inside the circuit modeling community, and possibly other communities too. The metamodel includes two components: (i) the agent model and (ii) the community model:

- *Agent model* describes the behavior of the participants to the evolution process, also called agents in the model. Every agent has local beliefs representing design knowledge, and local resources available to conduct work. In addition, an agent pursues its long-term goals (which are part of the local belief system) to select new subgoals and specific problems to be addressed. Decision-making identifies a strategy of solving the identified problem, possibly selecting new knowledge from the domain knowledge available to the community and refining the expected outcomes and benefits if the envisioned work is completed. An agent's work produces outcomes and new insight, which become part of the local beliefs and also impact the future selected subgoals and problems and the new knowledge from the community.
- *Community model* includes the domain beliefs, which are the design knowledge available to all members (agents) of the community. The visibility management module refers to the mechanisms used to propagate at the community level the outcomes and insight obtained by the individual agents. The identified new resources module presents the attraction of new resources by the community in response to its progress in satisfying goals and needs. Finally, the resource allocation scheme shows how the resources of the community are assigned to the individual agents.

The trace of the metamodel over time represents the idea evolution process of the community. Figure 6.3 presents the mechanism through which individual outcomes

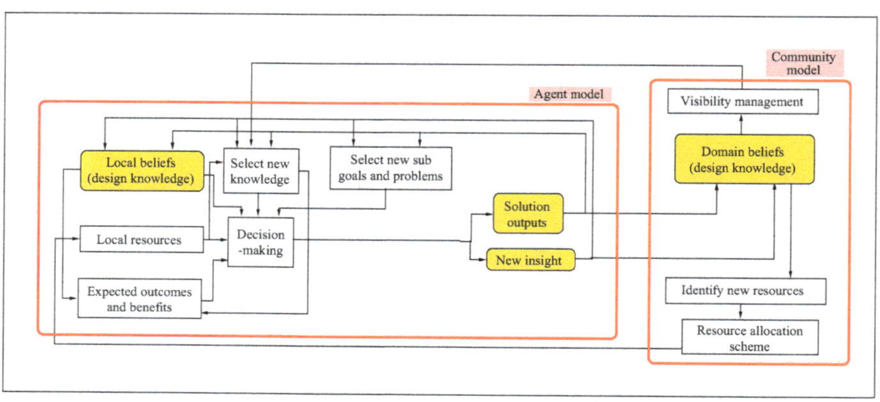

Figure 6.2 Metamodel for idea evolution

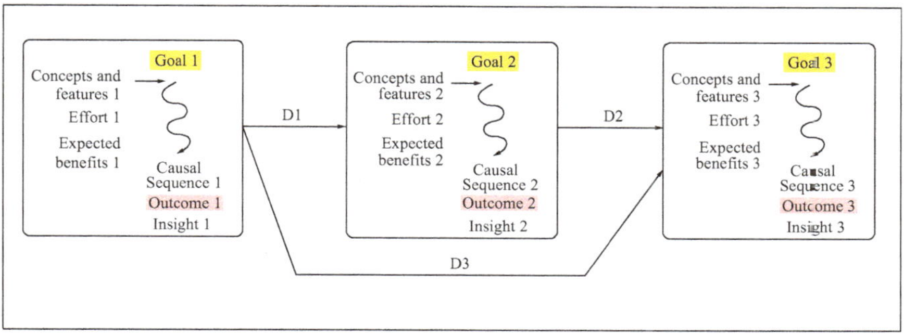

Figure 6.3 Evolution mechanism of the individual outcomes

are evolved by subsequent work. Each individual outcome relates to a certain subgoal and problem selected by the agent, used concepts and features, a causal sequence that creates the outcome, new insight that emerged while solving the problem, the effort that was spent, and expected benefits of the outcomes. The following situations can exist for the evolution mechanism:

- *Selection of goals and subproblems* (see Figure 6.2): There are four cases depending on how $goal_1$, $goal_2$, and $goal_3$ related to each other: (i) $goal_3 \subseteq goal_1, goal_2$; (ii) $goal_3 \bot goal_1, goal_2$; (iii) $goal_3 \bot goal_2$ but not $goal_1$; and (iv) $goal_3 \subseteq goal_2$ but not $goal_1$.
  - In case (i), called *incremental convergence*, $goal_3$ is an instance of $goal_1$ and $goal_2$ meaning that it considers a subset (e.g., subproblem, constraint, feature) of the more general subspace addressed by $goal_1$, $goal_1$, $goal_2$, and $goal_3$ are logically consistent with other, suggesting that the third step represents a *convergent* activity with respect to the first and second steps. Hence, $goal_3$ is logically consistent with $goal_1 \cup insight_1$ and with $goal_2 \cup insight_2$.
  - In case (ii), called *orthogonal divergence*, $goal_3$ is orthogonal to $goal_1$ and $goal_2$, meaning that its requirements cannot be derived from the requirements of $goal_1$ and $goal_2$. Hence, $goal_3$ is not a resultant of the insight from the previous two steps, but the effect of another influence, such as new problems and needs posed to the community (e.g., new applications).
  - In case (iii), called *local alternative*, $goal_3$ is orthogonal to $goal_2$ but not to $goal_1$. This case corresponds to a situation in which the third step pursues an alternative to the second step, both being refinements of the first step. Hence, $goal_3$ is logically consistent with $goal_1 \cup insight_1$ but does not logically relate to $goal_2 \cup insight_2$.
  - In case (iv), called *incremental divergence*, $goal_3$ expands on the specific aspects introduced by $goal_2$ as opposed to $goal_1$ as well as the insight of the second step. Hence, $goal_3$ is logically consistent with $(goal_2 - goal_1) \cup insight_2$.

- *Select new knowledge* (see Figure 6.2): Within, a newly identified goal, related and useful knowledge is selected from the domain knowledge depending on the infrastructure of the domain visibility management (see Figure 6.2). The selection process is a function of the effort required to adopt and use the new knowledge and the expected benefits of the outcome produced for the identified goal.

$$\begin{aligned}\textit{Knowledge selection} \sim \Phi(&\textit{visibility}(\textit{Domain knowledge}),\\ &\textit{expected benefits}, \textit{required effort})\end{aligned} \tag{6.1}$$

  A diversification of the evolution process occurs, if the selected new knowledge is dissimilar to the previously used concepts, possibly from a different domain. Diversification also includes the case in which an agent discovers a new concept, usually as a result of the gained insight on the output limitations and how they can be addressed.
- *Decision-making* (see Figure 6.2): Five decision-making strategies are available to restructure the causal sequence that links together the selected concepts into an outcome for the defined goal. The strategies are similar to those presented in [10,12]. Incremental changes of the outcomes leave most of the causal sequence unchanged with some local extensions to incorporate the additional details of the new outcome. A more significant change of the causal sequence results, if new concepts are incorporated into the causal sequence and/or if orthogonal goals are being tackled. Finally, strong divergent evolution (in which the current evolutionary trend is rejected) imposes a new causal sequence.
- *Local priorities*: They refer to the individual specifics through which an agent frames new subgoals and problems, decides to allocate its local resources as well as the way in which previous experiences are used for assessing expected outcomes and benefits.

  In Figure 6.3, arcs are labeled with quadruples that indicate the differences in the goals, selected new knowledge, decision-making, and local priorities of the two connected outcomes. The nature of the evolution process depends on the relations between the community level elements (identification of new resources, resource allocation schemes, and visibility management) and the local priorities of the agents.

6.6 Conclusions

This chapter gives an overview of some of the main methodologies and algorithms for analog circuit modeling without attempting to exhaustively summarize the entire research domain. It would be very hard considering the large number of publications on analog circuit modeling. Analog circuit modeling is the procedure through which analog circuit operation and performance are described by a succinct, comprehensive yet accurate representation. The chapter mainly focuses on three main categories of automated circuit modeling techniques, such as symbolic methods, NN-based approaches, and macromodeling. The related techniques are grouped depending on the similarity of their pursued goals and concepts. The chapter refers to close to 100 papers to present the capabilities, strengths, and limitations of the different methods. Circuit modeling has application in numerous design-related activities, like getting insight into circuit operation, circuit topology synthesis and improvement, circuit parameter optimization, fast simulation, design centering and yield improvement, test and validation, and design knowledge representation.

References

[1] Gielen G, Wambacq P, and Sansen W. Symbolic analysis methods and applications for analog circuits: a tutorial overview. Proceedings of the IEEE. 1994;82(2):287–304.

[2] Gielen G, and Rutenbar R. Computer aided design of analog and mixed-signal integrated circuits. Proceedings of IEEE. 2000;88:1825–1852.

[3] Gielen G, Walscharts H, and Sansen S. ISAAC: a symbolic simulator for analog integrated circuits. IEEE Journal of Solid-State Circuits. 1989;24(6): 1587–1597.

[4] Doboli A, and Vemuri R. Exploration-based high-level synthesis of linear analog systems operating at low/medium frequencies. IEEE Transactions on Computer-Aided Design of Integrated Circuits and Systems. 2003;22(11): 1556–1568.

[5] McConaghy T, Palmers P, Gao P, *et al.* Variation-Aware Analog Structural Synthesis. London and New York: Springer; 2009.

[6] Tang H, Zhang H, and Doboli A. Refinement based synthesis of continuous-time analog filters through successive domain pruning, plateau search and adaptive sampling. IEEE Transactions on Computer-Aided Design of Integrated Circuits and Systems. 2006;25(8):1421–1440.

[7] Li P, and Pileggi L. Efficient per-nonlinearity distortion analysis for analog and RF circuits. IEEE Transactions on Computer-Aided Design of Integrated Circuits and Systems. 2003;22(10):1297–1309.

[8] Ferent C, and Doboli A. An axiomatic model for concept structure description and its application to circuit design. Knowledge-Based Systems. 2013;45:114–133.

[9] Ferent C, and Doboli A. Analog circuit design space description based on ordered clustering of feature uniqueness and similarity. Integration, the VLSI Journal. 2014;47(2):213–231.

[10] Jiao F, Montano S, Ferent C, *et al.* Analog circuit design knowledge mining: discovering topological similarities and uncovering design reasoning strategies. IEEE Transactions on CADICS. 2015;34(7):1045–1059.

[11] Ferent C, and Doboli A. Symbolic matching and constraint generation for systematic comparison of analog circuits. IEEE Transactions on Computer-Aided Design of Integrated Circuits and Systems. 2013;32(4):616–629.

[12] Jiao F, and Doboli A. Knowledge-intensive, causal reasoning for analog circuit topology synthesis in emergent and innovative applications. In: Proc. Design, Automation and Test in Europe Conference (DATE); 2015. p. 1144–1149.

[13] Ferent C, and Doboli A. Novel circuit topology synthesis method using circuit feature mining and symbolic comparison. In: Proc. Design, Automation and Test in Europe Conference (DATE); 2014. p. 1–4.

[14] Li H, and Doboli A. Analog circuit topological feature extraction with unsupervised learning of new substructures. In: Proc. Design, Automation and Test in Europe Conference (DATE); 2016. p. 1509–1512.

[15] Doboli A, Umbarkar A, Doboli S, *et al.* Modeling semantic knowledge structures for creative problem solving: studies on expressing concepts, categories, associations, goals and context. Knowledge-Based Systems. 2015;34(C):34–50.

[16] Ferent C, and Doboli A. Measuring the uniqueness and variety of analog circuit design features. Integration, the VLSI Journal. 2011;44(1):39–50.

[17] Liu X, and Doboli A. Moving beyond traditional electronic design automation: data-driven design of analog circuits. In: Proc. International Conference on Synthesis, Modeling, Analysis and Simulation Methods and Applications to Circuit Design (SMACD); 2016. p. 1–4.

[18] Konczykowska A, and Bon M. Topological analysis of switched-capacitor networks. Electronics Letters. 1980;16(3):89–90.

[19] Kurth C, and Moschytz G. Nodal analysis of switched-capacitor networks. IEEE Transactions on Circuits and Systems. 1979;26(2):93–105.

[20] Starzyk J, and Konczykowska A. Flowgraph analysis of large electronic networks. IEEE Transactions on Circuits and Systems. 1986;33(3):302–315.

[21] Wehrhahn E. Evaluation of transfer functions of ideal s.c. networks in z-domain using standard linear symbolic or semisymbolic network analysis programs. Electronics Letters. 1980;16(21):801–802.

[22] Lin P. A survey of applications of symbolic network functions. IEEE Transactions on Circuit Theory. 1973;20(6):732–737.

[23] Hsu JJ, and Sechen C. DC small signal symbolic analysis of large analog integrated circuits. IEEE Transactions on Circuits and Systems I: Fundamental Theory and Applications. 1994;41(12):817–828.

[24] Singhal K, and Vlach J. Symbolic analysis of analog and digital circuits. IEEE Transactions on Circuits and Systems. 1977;24(11):598–609.

[25] Shi CR, and Tan XD. Canonical symbolic analysis of large analog circuits with determinant decision diagrams. IEEE Transactions on Computer-Aided Design of Integrated Circuits and Systems. 2000;19(1):1–18.

[26] Doboli A, and Vemuri R. A regularity-based hierarchical symbolic analysis method for large-scale analog networks. IEEE Transactions on Circuits and Systems II: Analog and Digital Signal Processing. 2001;48(11): 1054–1068.

[27] Li B, and Gu D. SSCNAP: a program for symbolic analysis of switched capacitor circuits. IEEE Transactions on Computer-Aided Design of Integrated Circuits and Systems. 1992;11(3):334–340.

[28] Vanassche P, Gielen G, and Sansen W. Symbolic modeling of periodically time-varying systems using harmonic transfer matrices. IEEE Transactions on Computer-Aided Design of Integrated Circuits and Systems. 2002;21(9): 1011–1024.

[29] Yu Q, and Sechen C. A unified approach to the approximate symbolic analysis of large analog integrated circuits. IEEE Transactions on Circuits and Systems I: Fundamental Theory and Applications. 1996;43(8):656–669.

[30] Daems W, Gielen G, and Sansen W. Simulation-based generation of posynomial performance models for the sizing of analog integrated circuits. IEEE Transactions on Computer-Aided Design of Integrated Circuits and Systems. 2003;22(5):517–534.

[31] Walter D, Little S, Myers C, *et al.* Verification of analog/mixed-signal circuits using symbolic methods. IEEE Transactions on Computer-Aided Design of Integrated Circuits and Systems. 2008;27(12):2223–2235.

[32] Mason S. Topological analysis of linear nonreciprocal networks. Proceedings of the IRE. 1957;45:829–838.

[33] Mielke R. A new signal flowgraph formulation of symbolic network functions. IEEE Transactions on Circuits and Systems. 1978;25(6):334–340.

[34] Sannuti P, and Puri N. Symbolic network analysis—an algebraic formulation. IEEE Transactions on Circuits and Systems. 1980;27(8):679–687.

[35] Brown D. New topological formulas for linear networks. IEEE Transactions on Circuit Theory. 1965;CT-12:358–365.

[36] Chen W. Topological analysis for active networks. IEEE Transactions on Circuit Theory. 1965;CT-12:85–91.

[37] Jong M, and Zobrist G. Topological formulas for general linear networks. IEEE Transactions on Circuit Theory. 1968;CT-15:251–259.

[38] Nathan A. Topological rules for linear networks. IEEE Transactions on Circuit Theory. 1965;CT-12:344–358.

[39] Talbot A. Topological analysis of general linear networks. IEEE Transactions on Circuit Theory. 1965;CT-12:170–180.

[40] Shieu SD, and Chan SP. Topological formulation of symbolic network functions and sensitivity analysis of active networks. IEEE Transactions on Circuits and Systems. 1974;21(1):39–45.

[41] Tsai M, and Shenoi B. Generation of symbolic network functions using computer software techniques. IEEE Transactions on Circuits and Systems. 1975;24:344–346.

[42] Kuo YL, Liou M, and Kasinskas J. An equivalent circuit approach to the computer-aided analysis of switched capacitor circuits. IEEE Transactions on Circuits and Systems. 1979;26(9):708–714.

[43] Fidler J, and Sewell J. Symbolic analysis for computer-aided circuit design—the interpolative approach. IEEE Transaction on Circuit Theory. 1973;CT–20(6):738–741.

[44] Wambacq P, Fernandez F, Gielen G, *et al.* Symbolic computation of approximated small-signal characteristics of analog integrated circuits. IEEE Journal on Solid-state Circuits. 1995;30(3):327–330.

[45] Hassoun M, and Lin P. A hierarchical network approach to symbolic analysis of large-scale networks. IEEE Transactions on Circuits and Systems I: Fundamental Theory and Applications. 1995;42(4):201–211.

[46] Chua L, and Lin P. Computer Aided Analysis of Electronic Circuits – Algorithms and Computational Techniques. Englewood Cliffs, NJ: Prentice-Hall; 1975.

[47] Chen W. Applied Graph Theory-Graphs and Electrical Networks. North-Holland; 1976.

[48] Fernandez F, Guerra O, Rodriguez-Garcia J, *et al.* Symbolic analysis of large analog integrated circuits: the numerical reference generation problem. IEEE Transactions on Circuits and Systems II: Analog and Digital Signal Processing. 1998;45(10):1351–1361.

[49] Guerra O, Roca E, Fernandez F, *et al.* Approximate symbolic analysis of hierarchically decomposed analog circuits. Analog Integrated Circuits and Signal Processing. 2002;31(2):131–145.

[50] Chang M, MacKay J, and Wierzba G. Matrix reduction and numerical approximation during computation techniques for symbolic analog circuit analysis. In: Proc. IEEE International Symposium on Circuits and Systems (ISCAS); 1992. p. 1153–1156.

[51] Daems W, Gielen G, and Sansen W. Circuit simplification for the symbolic analysis of analog integrated circuits. IEEE Transactions on Computer-Aided Design of Integrated Circuits and Systems. 2002;21(4):395–407.

[52] Fernandez F, Rodiguez-Vazquez A, and Huertas J. A tool for symbolic analysis of analog integrated circuits including pole/zero extraction. In: Proc. 10th European Conference on Circuit Theory and Design; 1991. p. 752–761.

[53] Henning E. Matrix approximation techniques for symbolic extraction of poles and zeros. Analog Integrated Circuits and Signal Processing. 2002;31:81–100.

[54] Hsu JJ, and Sechen C. Accurate extraction of simplified symbolic pole/zero expressions for large analog IC's. In: Proc. of IEEE International Symposium on Circuits and Systems (ISCAS); 1995. p. 2083–2087.

[55] Nebel G, Kleine U, and Pfleiderer H. Symbolic pole/zero calculation using SANTAFE. IEEE Journal of Solid-State Circuits. 1995;30(7):752–761.

[56] Minato S. Zero-suppressed BDD's for set manipulation in combinatorial problems. In: Proc. IEEE/ACM Design Automation Conference (DAC); 1993. p. 272–277.

[57] Tan XD, and Shi CR. Hierarchical symbolic analysis of analog integrated circuits via determinant decision diagrams. IEEE Transactions on Computer-Aided Design of Integrated Circuits and Systems. 2000;19(4):401–412.

[58] Shi CR, and Tan XD. Compact representation and efficient generation of s-expanded symbolic network functions for computer-aided analog circuit design. IEEE Transactions on Computer-Aided Design of Integrated Circuits and Systems. 2001;20(7):813–827.

[59] Zhang H, Doboli S, Tang H, *et al.* Compiled code simulation of analog and mixed-signal systems using piecewise linear modeling of nonlinear parameters: a case study for $\Delta\Sigma$ modulator simulation. Integration, the VLSI Journal. 2007;40:193–208.

[60] Hershenson M, Boyd S, and Lee T. Optimal design of a CMOS op-amp via geometric programming. IEEE Transactions on Computer-Aided Design. 2001;20:1–21.

[61] Wei Y. Development of Structural Circuit Macromodels and Systematic Design of Reconfigurable $\triangle\Sigma$ Modulator. PhD Thesis, Stony Brook University; 2007.

[62] Gomez G, Embabi S, Sanchez-Sinencio E, *et al.* A nonlinear macromodel for CMOS OTAs. In: Proc. of International Symposium on Circuits and Systems (ISCAS); 1995. p. 920–923.

[63] Doboli S, and Doboli A. Piecewise-linear modeling of analog circuits using trained feed-forward neural networks and adaptive clustering of hidden neurons. In: Proc. International Joint Conference on Neural Networks (IJCNN); 2003. p. 1126–1131, vol. 2.

[64] Moscovici A. High Speed A/D Converters – Understanding Data Converters Through Spice. Boston and Dordrecht: Kluwer; 1999.

[65] Huang X, Gathercole C, and Mantooth A. Modeling nonlinear dynamics in analog circuits via root localization. IEEE Transactions on CADICS. 2003; 2(7):895–907.

[66] Wei Y, and Doboli A. Systematic development of analog circuit structural macromodels through behavioral model decoupling. In: Proc. Design Automation Conference (DAC); 2005. p. 57–62.

[67] Zhang H, and Doboli A. ALAMO: an improved sigma-space based methodology for modeling process parameter variations in analog circuits. In: Proc. Design Automation and Test in Europe Conference (DATE); 2006. p. 156–161.

[68] Zhang H, and Doboli A. A scalable sigma-space based methodology for modeling process parameter variations in analog circuits. Microelectronics Journal. 2008;39(12):1785–1796.

[69] McAndrew C, and Drennan P. Unified statistical modeling for circuit simulation. In: Proc. of Modeling and Simulation of Microsystems; 2002.

[70] Michael C, and Ismail M. Statistical Modeling for Computer-Aided Design of MOS VLSI Circuits. New York, NY: Kluwer; 1993.

[71] Demir A. Phase noise in oscillators: DAEs and colored noise sources. In: Proc. ICCAD; 1998. p. 170–177.

[72] Demir A, Mehrotra A, and Roychowdhury J. Phase noise in oscillators: a unifying theory and numerical methods for characterization. IEEE Transactions on Circuits and Systems I: Fundamental Theory and Applications. 2000; 47(5):655–674.

[73] Demir A, and Roychowdhury J. A reliable and efficient procedure for oscillator PPV computation, with phase noise macromodeling applications. IEEE Transactions on Computer-Aided Design of Integrated Circuits and Systems. 2003;22(2):188–197.

[74] Vanassche P, Gielen G, and Sansen W. Behavioral modeling of (coupled) harmonic oscillators. IEEE Transactions on Computer-Aided Design of Integrated Circuits and Systems. 2003;22(8):1017–1026.

[75] Janakiraman V, Bharadwaj A, and Visvanathan V. Voltage and temperature aware statistical leakage analysis framework using artificial neural networks. IEEE Transactions on Computer-Aided Design of Integrated Circuits and Systems. 2010;29(7):1056–1069.

[76] Pratap R, Sen P, Davis C, *et al.* Neurogenetic design centering. IEEE Transactions on Semiconductor Manufacturing. 2006;19(2):173–182.

[77] Suissa A, Romain O, Denoulet J, *et al.* Empirical method based on neural networks for analog power modeling. IEEE Transactions on Computer-Aided Design of Integrated Circuits and Systems. 2010;29(5):839–844.

[78] Wolfe G, and Vemuri R. Extraction and use of neural network models in automated synthesis of operational amplifiers. IEEE Transactions on Computer-Aided Design of Integrated Circuits and Systems. 2003;22(2):198–212.

[79] Zhang QZ, Gupta K, and Devabhaktuni V. Artificial neural networks for RF and microwave design – from theory to practice. IEEE Transactions on Microwave Theory and Techniques. 2003;51(4):1339–1350.

[80] Stratigopoulos HD, and Makris Y. Nonlinear decision boundaries for testing analog circuits. IEEE Transactions on Computer-Aided Design of Integrated Circuits and Systems. 2005;24(11):1760–1773.

[81] Yang ZR, Zwolinski M, Chalk C, *et al.* Applying a robust heteroscedastic probabilistic neural network to analog fault detection and classification. IEEE Transactions on Computer-Aided Design of Integrated Circuits and Systems. 2000;19(1):142–151.

[82] Scarselli F, and Tsoi A. Universal approximation using feedforward neural networks: a survey of some existing methods, and some new results. Neural Networks. 1998;11(1):15–37.

[83] Zaabab A, Zhang QJ, and Nakhla M. A neural network modeling approach to circuit optimization and statistical design. IEEE Transactions on Microwave Theory and Techniques. 1995;43(6):1349–1358.

[84] Mandal S, Sural S, and Patra A. ANN- and PSO-based synthesis of on-chip spiral inductors for RF ICs. IEEE Transactions on Computer-Aided Design of Integrated Circuits and Systems. 2008;27(1):188–192.

[85] Cao Y, Ding R, and Zhang QJ. State-space dynamic neural network technique for high-speed IC applications: modeling and stability analysis. IEEE Microwave and Wireless Components Letters. 2006;54(6):2398–2409.

[86] Cao Y, Erdin I, and Zhang Q. Transient behavioral modeling of nonlinear I/O drivers combining neural networks and equivalent circuits. IEEE Microwave and Wireless Components Letters. 2010;20(12):645–647.

[87] Sridhar A, Vincenzi A, Ruggiero R, *et al.* Neural network-based thermal simulation of integrated circuits on GPUs. IEEE Transactions on Computer-Aided Design of Integrated Circuits and Systems. 2012;31(1):23–36.

[88] Stevanovic I, and McAndrew C. Quadratic backward propagation of variance for nonlinear statistical circuit modeling. IEEE Transactions on Computer-Aided Design of Integrated Circuits and Systems. 2009;28(9):1428–1432.

[89] Viraraghavan J, Pandharpure S, and Watts J. Statistical compact model extraction: a neural network approach. IEEE Transactions on Computer-Aided Design of Integrated Circuits and Systems. 2012;31(12):1920–1924.

[90] Magerl M, Ceperic V, and Baric A. Echo state networks for black-box modeling of integrated circuits. IEEE Transactions on Computer-Aided Design of Integrated Circuits and Systems. 2016;35(8):1309–1317.

[91] Maidon Y, Jervis B, Dutton N, and Lesage S. Diagnosis of multifaults in analogue circuits using multilayer perceptrons. IEE Proceedings – Circuits, Devices and Systems. 1997;144:149–154.

[92] Yang ZR, and Musgrave G. Applying artificial neural network to fault diagnosis of analogue circuit. In: Proc. 3rd International Conference on VLSI and CAD; 1993.

[93] Yu Q, and Sechen C. Efficient approximation of symbolic network functions using matroid intersection algorithms. IEEE Transactions on Computer-Aided Design of Integrated Circuits and Systems. 1997;16(10):1073–1081.

[94] Wambacq P, Gielen G, Kinget P, *et al.* High-frequency distortion analysis of analog integrated circuits. IEEE Transactions on Circuits & Systems – Part II. 1999;46(3):335–345.

[95] Shi G, Hu B, and Shi CR. On symbolic model order reduction. IEEE Transactions on Computer-Aided Design of Integrated Circuits and Systems. 2006;25(7):1257–1272.

[96] Katzenelson J, and Unikovski A. Symbolic-numeric circuit analysis or symbolic circuit analysis with online approximations. IEEE Transactions on Circuits and Systems I: Fundamental Theory and Applications. 1999;46(1): 197–207.

Chapter 7
On the usage of machine-learning techniques for the accurate modeling of integrated inductors for RF applications

Fábio Passos[1], Elisenda Roca[1], Rafael Castro-López[1] and Francisco V. Fernández[1]

This chapter describes an inductor modeling strategy based on machine-learning techniques. The model developed is based on Kriging functions and uses a novel modeling technique based on a two-step strategy, which is able to obtain an extremely accurate model with less than 1% error when compared to electromagnetic (EM) simulations. Due to its extreme accuracy and efficiency, the model can be used in inductor synthesis processes using single- or multi-objective optimization algorithms in order to obtain a single design or a Pareto-optimal front. Also, the model can describe the inductor behavior in frequency and therefore can also be used in circuit design using modern electrical simulators. This chapter discusses both applications (inductor synthesis and circuit design), performing several single- and multi-objective inductor optimizations, using different inductor topologies and operating frequencies. Furthermore, the model is also used in order to accurately model inductors during the design of a voltage-controlled oscillator (VCO) and a low-noise amplifier (LNA).

7.1 Introduction

Nowadays radio-frequency (RF) integrated circuits, passive components play a key role in circuit design for impedance matching, tuning, filtering, or biasing. Integrated inductors are one of the most important components in RF applications, where they are used in many circuits and applications. Inductors have historically been considered as one of the biggest bottlenecks of RF design due to its hard-to-predict performances and effects such as eddy currents and proximity effects [1].

Integrated inductor design is highly complex due to its peculiar features, which the RF designer has to take into account during the design stage. Through the variation of geometrical parameters that characterize the inductor (e.g., number of

[1]Instituto de Microelectrónica de Sevilla (CSIC and Universidad de Sevilla), Seville, Spain

turns, inner diameter and turn width), the designer usually tries to achieve a given inductance, while several design decisions have to be made in order to design the inductor that meets the inductance requirements with the highest quality factor and the smallest area. These decisions highly depend on the expertise of the RF designer and the technology process. While designing a circuit where inductors are needed, RF designers usually rely on three different methods in order to predict the inductor performances: inductor libraries (previously fabricated and measured) provided by the foundry, EM simulations and analytical/physical models.

In the past, some approaches, such as [2], used the foundry inductor libraries during the circuit design stage. However, this approach is very limited because the foundry usually provides very few measured inductors, which therefore decreases the possibilities of finding an optimal inductor for a given application. In order to overcome this limitation, EM simulators can be used, as [3,4], in order to predict the inductor performances based on the inductor geometrical parameters. EM simulations have the benefit of predicting inductors' performances with the highest accuracy possible; however, they have the disadvantage of being computational expensive, critically slowing down the RF design cycle. Furthermore, the accuracy of such simulations is completely dependent on an accurate process technology description, which is not easily provided by the foundries. Therefore, in order to overcome the problematic issues of EM simulations, a possible solution is to model integrated inductors using analytical models based on the physical properties of inductors and the substrate where they are integrated [5–8]. This type of model is usually very efficient to use and easy to implement. However, most times they are based on rather simple equations that fail to predict the exact inductor behavior, especially at high frequencies. Therefore, on the one hand, we have EM simulators which are very accurate but very inefficient, and on the other hand, we have analytical/physical models that are very efficient but also inaccurate. Thus, there is an accuracy–efficiency gap, as shown in Figure 7.1.

Nowadays, optimization algorithms are widely used for the synthesis of integrated inductors. Therefore, it is easy to understand that when these two inductor performance evaluation alternatives (EM simulations and analytical/physical models) are embedded into an iterative optimization loop for inductor synthesis, the accuracy–efficiency trade-off shown in Figure 7.1 is even more evident. When EM simulations are used, the highest accuracy is achieved at the price of the highest computation time, usually in the range of days. However, despite this limitation, they have shown their potentialities especially within multi-objective optimization-based synthesis, because the multi-objective optimization only has to be performed once for a given operating frequency (since inductors have a specification-independent use at circuit-level) [3]. The inductor synthesis using physical/analytical models for performance evaluation lies at the opposite end of the accuracy–efficiency trade-off [9]. The synthesis times are short, typically in the range of minutes, but with the lowest accuracy, because, generally, equivalent circuit models do not allow an accurate modeling of RF-passive components [8].

In the last few years, surrogate models are being used to efficiently and accurately model complex structures that are computationally expensive to

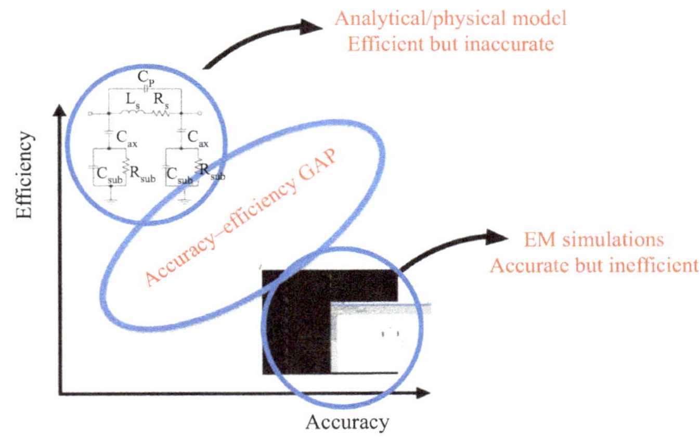

Figure 7.1 Denoting the existence of an accuracy–efficiency gap. On the one hand, EM simulators are very accurate but very inefficient, and on the other hand, analytical/physical models are very efficient but also inaccurate

evaluate, and therefore, they are good candidates to fill the accuracy–efficiency gap shown in Figure 7.1 [10–12]. However, most of these surrogate models still show relatively high errors when trying to model the inductors' entire design space and therefore its use in optimization-based approaches, where the entire design space is explored, is limited. With the development of an accurate surrogate model over the entire design space, the accuracy–efficiency trade-off can be significantly improved by the use of such models [13]. However, the difficulties in generating surrogate models that allow an accurate modeling of the entire design space have hampered the development of these approaches. In this chapter, a surrogate modeling strategy able to model the entire design space for inductors is described and, therefore, its usage in optimization-based methodologies is highly encouraged.

The rest of this chapter is organized as follows. Section 7.2 briefly describes the inductor geometrical parameters and gives some design insights for the design of these passive components. Section 7.3 explains how to build a surrogate model of inductors and the strategy developed by the authors in order to ensure such high accuracy over the entire design space. In Section 7.4, the developed model is used in several applications of RF design; first, the model is used in single- and multi-objective inductor optimization processes and second, the model is used in order to design two common RF circuits: a VCO and an LNA. Finally, in Section 7.5, conclusions are drawn.

7.2 Integrated inductor design insight

Integrated inductors are typically built using two metal layers, with an intermediate dielectric layer. In Figure 7.2, it is possible to observe the layouts of an octagonal

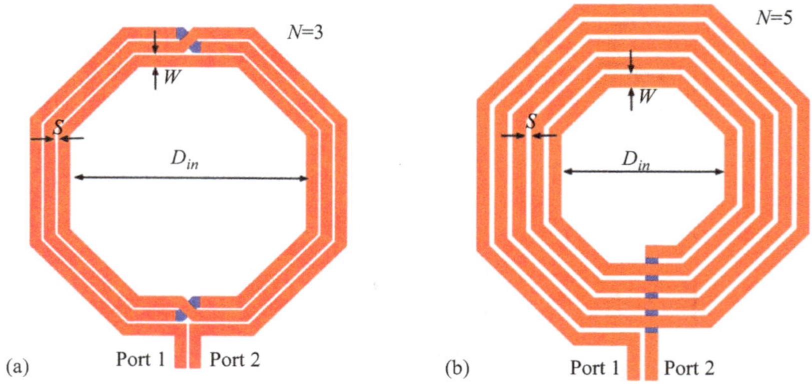

Figure 7.2 (a) Octagonal symmetric inductor and (b) octagonal asymmetric inductor

symmetric inductor and an asymmetric spiral inductor. The geometry of these inductors is usually defined by four geometric parameters: number of turns (N), inner diameter (D_{in}) (or alternatively the outer diameter (D_{out})), turn width (W) and spacing between turns (S).

Usually the designer is interested in the equivalent inductance, L_{eq}, and the quality factor, Q_{eq}. These performance parameters can be easily obtained from the S-parameters of the two-port structure representation of the inductor. For the symmetric topology, the inductor is measured in differential mode (DM) and, therefore, the following formulas can be used in order to calculate L_{eq} and Q_{eq} [14]:

$$S_{DM} = \frac{S_{11} - S_{12} - S_{21} + S_{22}}{2} \tag{7.1}$$

$$Z_{eq} = 2Z_0 \frac{1 + S_{DM}}{1 - S_{DM}} \tag{7.2}$$

where Z_{eq} is the equivalent input impedance and Z_0 is the characteristic impedance of the device (usually 50 Ω). For the asymmetric topology, the inductor is measured in single-ended mode (SEM) and the following formulas can be used [14]:

$$S_{SEM} = S_{11} - \frac{S_{12} \cdot S_{21}}{1 + S_{22}} \tag{7.3}$$

$$Z_{eq} = Z_0 \cdot \frac{1 + S_{SEM}}{1 - S_{SEM}} \tag{7.4}$$

After obtaining the equivalent input impedance, Z_{eq}, (either for the asymmetric or symmetric inductor), the inductance L_{eq}, and the quality factor, Q_{eq}, are calculated as follows:

$$L_{eq}(f) = \frac{\text{Im}\left[Z_{eq}(f)\right]}{2\pi f} \tag{7.5}$$

$$Q_{eq}(f) = \frac{\text{Im}\left[Z_{eq}(f)\right]}{\text{Re}\left[Z_{eq}(f)\right]} \tag{7.6}$$

where f stands for the frequency variable. In Figure 7.3, it is possible to observe a typical plot of the inductance and quality factor as a function of the frequency. Another important parameter of integrated inductors is the self-resonance frequency (SRF) which is defined as the frequency at which the imaginary part of Z_{eq} is zero, or, equivalently, the frequency at which the behavior of the inductor changes from inductive to capacitive (see (7.5)).

Furthermore, while designing an inductor, it is important to take into account the frequency behavior of such passive components. Ideally, an integrated inductor must be designed for an operating frequency such that the SRF is sufficiently above that frequency. Also, apart from the SRF, the inductor must also be designed for an operating frequency that is in the flat-bandwidth (BW) zone.

In order for the inductor to be in this flat-BW zone, the inductance value must be sufficiently flat from DC (L_{DC}) to slightly above the operating frequency ($L_{eq}(f_0 + \Delta_f)$, where f_0 is the operating frequency) and the SRF must lie sufficiently above this frequency. Furthermore, some inductors exhibit a "valley" behavior in inductance closely to the abrupt rise (between $L_{eq}(f_0 - \Delta_f)$ and

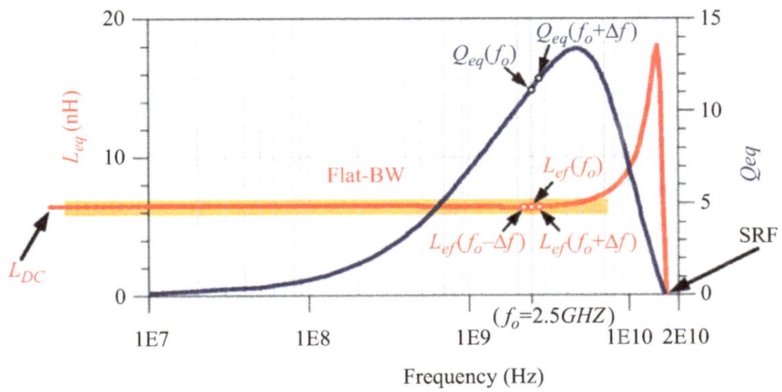

Figure 7.3 Illustrating inductance and quality factor as a function of frequency and the SRF and the flat-bandwidth zone of an inductor, which is the ideal zone for good inductor design

$L_{eq}(f_0 + \Delta_f)$ in Figure 7.3). This behavior is unwanted and, therefore, the inductance value must be flat in this region. Also, the quality factor at the operating frequency $Q_{eq}(f_0)$ must be near its maximum and always with a positive slope around it ($Q_{eq}(f_0 + \Delta_f) - Q_{eq}(f_0) > 0$). By ensuring these design requirements, the designer can ensure a "good behavior" of its inductor at the operating frequency [3].

7.3 Surrogate modeling

Surrogate models are commonly used when an outcome of interest of a complex system cannot be easily, or cheaply, measured by experiments or simulations [10]. Therefore, an approximate model of the output is developed and used instead. In this section, the basic steps for the generation of a surrogate model are described. Generating a surrogate model usually involves four steps that are briefly described next.

1. Design of experiments: The objective of surrogate models is to emulate the output response of a given system. Therefore, being them a type of machine learning technique, the model has to learn how the system responds to a given input. So, the first step in generating surrogate models is to select the input samples from which the model is going to learn. These samples should evenly cover the design space, so that it can be accurately modeled. In order to perform this sampling, different techniques are available, from classical Monte Carlo (MC) to quasi-MC (QMC) or Latin hypercube sampling [10]. In this chapter, QMC is used.
2. Accurate expensive evaluation: Surrogate models are usually built by learning from the most accurate evaluation possible for a given structure. However, most times these evaluations are time-consuming. In our work, these accurate evaluations are EM simulations, which are performed with ADS *Momentum* [15]. Depending on the size of the training set (samples from which the model is going to learn), these simulations could even last for weeks. However, these simulations are only performed once for a given technology node, therefore being useful for several years, as technology nodes do not become obsolete in months. Any modeling technique can later be used in order to build a new model using the same training set.
3. Model construction: This concerns the core functions used to build a surrogate model. The literature reports approaches based on artificial neural networks, support vector machines, parametric macromodels, Gaussian-process or Kriging models, etc. [16]. In this chapter, ordinary Kriging models are used. Different MATLAB® toolboxes like SUMO [17] or DACE [18] support this type of models. In the examples reported in this chapter, DACE was used. The motivations to use DACE were practical reasons, such as the fact that it is freely available, simple to use and it runs over a widely used software platform, which is MATLAB.
4. Model validation: Many different techniques may be used in order to validate the model and assess its accuracy, e.g., cross-validation, bootstrapping and

subsampling [19]. In this chapter, in order to validate the model, a set of points was generated independently of the training samples. These samples will be referred to as test samples and were also generated using QMC.

7.3.1 Modeling strategy

The modeling of integrated inductors has always been considered as one of the biggest bottlenecks in RF design. As previously mentioned, through times, several models were developed in order to accurately predict the behavior of such passive components [5–9,11,12]. However, most of them have not been successful in modeling the complete design space of the inductors. Therefore, in order to attain such goal, new modeling strategies have to be developed.

It is important to recall that inductance and quality factor are functions of the frequency, as seen in Figure 7.3. Usually, when developing models for integrated inductors, the usual efforts are in the direction of building frequency-dependent models [20]. However, surrogate models become much more complex with the increase in the number of parameters. This exponential complexity growth is also valid if the number of training samples increases. In order to alleviate this problem, problem-specific knowledge can be used in order to alleviate the model creation complexity.

In this chapter, the modeling of inductors is performed in a frequency-independent fashion. Hence, an independent model is created for each frequency point. This allows to increase the accuracy and to highly reduce the complexity of the models and also the time to generate them [13]. The initial strategy to build the model was to create a surrogate model valid in the complete design space. The model was created using 700 inductors generated with the QMC technique. Two different models were developed: one for predicting the inductance and the other for the quality factor (denoted as L and Q models, for the sake of simplicity). Remember that, since the models are frequency-independent, L and Q models have to be created for each frequency point. The technology selected for the model creation was a 0.35 μm CMOS technology and two different inductor topologies were modeled: asymmetric and symmetric topologies (as shown in Figure 7.2). The selection of the technology process in these experiments was only motivated by the available foundry data for EM simulation needed to build the inductor surrogate model. The created L and Q models were valid for inductors with any given number of turns, inner diameter and turn width within the ranges shown in Table 7.1.

Since a validation set of inductors is needed in order to test the model, 210 test inductors were generated, also using a QMC sampling technique. Models for three

Table 7.1 Inductor design variables

Parameter	Minimum	Grid	Maximum
N	1	1	7
W	5 μm	0.05 μm	25 μm
D_{in}	10 μm	0.05 μm	300 μm
S	2.5 μm	0.05 μm	2.5 μm

different frequencies were generated (20 MHz, 900 MHz and 2.45 GHz). It is possible to observe in Tables 7.2 and 7.3 that the mean relative errors of the model are larger at higher frequencies (e.g., 2.45 GHz), for both the asymmetric and the symmetric inductors topologies, respectively.

Therefore, a more advanced modeling strategy is needed in order to minimize the modeling error as much as possible. It is well known that the number of inductor turns can only take a few discrete values (in general, just integer values, depending on the implementation of the parameterized layout template). Therefore, it becomes clear that by creating several surrogate models, one for each number of turns (e.g., one model for inductors with two turns, another for inductors with three turns, etc.), the model accuracy can be increased, because the complexity of the modeled design space dramatically decreases (for each model). The generation of separate surrogate models for each number of turns instead of considering the number of turns as an input parameter of the surrogate model brings several benefits: not only is the accuracy significantly enhanced but also the computational cost is significantly decreased as the computational complexity of the training process grows exponentially with the number of samples. The number of models to create is manageable as the number of turns is typically between 1 and 7. This strategy increases the overall accuracy and efficiency of the model, as shown for the average errors of 210 test inductors in Table 7.4 for the asymmetric inductors and in Table 7.5 for the symmetric inductors. However, inductors with many turns still present large L and Q errors, specifically at high frequencies (e.g., 2.45 GHz).

The high error obtained at higher number of turns can be explained by the fact that some inductors from the training set have their SRF below or around the operating frequency. Kriging surrogate models assume continuity: if an input variable changes by a small amount, the output varies smoothly. However, in the

Table 7.2 Average error (in %) of inductance and quality factor for 210 test asymmetric octagonal inductors: single model for all N

20 MHz		**900 MHz**		**2.45 GHz**	
ΔL	ΔQ	ΔL	ΔQ	ΔL	ΔQ
0.04	0.17	0.15	0.37	63.60	16.20

Table 7.3 Average error (in %) of inductance and quality factor for 210 test symmetric octagonal inductors: single model for all N

20 MHz		**900 MHz**		**2.45 GHz**	
ΔL	ΔQ	ΔL	ΔQ	ΔL	ΔQ
0.04	0.23	0.08	0.23	155.19	1.70

Table 7.4 Inductance and quality factor average error for 210 test asymmetric octagonal inductors (in %): one model for each N

	20 MHz		900 MHz		2.45 GHz	
N	ΔL	ΔQ	ΔL	ΔQ	ΔL	ΔQ
1	0.07	0.34	0.07	0.43	0.11	0.55
2	0.04	0.28	0.07	0.33	0.09	0.49
3	0.02	0.24	0.06	0.36	0.10	0.21
4	0.02	0.20	0.06	0.36	0.10	0.31
5	0.02	0.18	0.06	0.32	0.60	0.79
6	0.01	0.16	0.15	0.32	32.44	29.25
7	0.01	0.18	0.06	0.37	13.85	1.61

Table 7.5 Inductance and quality factor average error for 210 test symmetric octagonal inductors (in %): one model for each N

	20 MHz		900 MHz		2.45 GHz	
N	ΔL	ΔQ	ΔL	ΔQ	ΔL	ΔQ
1	0.15	0.65	0.16	0.71	0.18	0.82
2	0.01	0.18	0.02	0.14	0.02	0.14
3	0.01	0.11	0.02	0.10	0.03	0.13
4	0.01	0.15	0.02	0.13	0.08	0.13
5	0.01	0.12	0.02	0.11	0.39	0.29
6	0.01	0.16	0.02	0.11	122.75	24.17
7	0.02	0.41	0.03	0.09	41.14	2.38

adopted technology node, for operating frequencies such as 2.45 GHz, some inductors with more than five turns, have their SRF close to (or even below) the operating frequency, where the inductance does not change smoothly. Therefore, the use of these inductors during the model construction severely degrades the accuracy of the model, because they present a sharp behavior and blur the model creation. Therefore, the accuracy of L and Q of the useful inductors is dramatically increased if only inductors with SRF sufficiently above the desired operating frequency are used for model training. However, this option is only feasible if we can detect which inductors have their SRF sufficiently above the frequency of operation and are, hence, useful.

Therefore, in the proposed strategy, the construction of the model is based on a two-step method:

1. Generate surrogate models for the SRF (for each number of turns) using all training inductors.
2. In order to generate highly accurate surrogate models for L and Q, only those inductors from the training set whose SRF is sufficiently above the operating

frequency are used. For example, if the operating frequency is 2.45 GHz, only inductors with, e.g., SRF > 2.95 GHz (using 500 MHz of margin) are used to generate L and Q models.

Consequently, with this methodology, whenever a test inductor is going to be evaluated, its SRF value is predicted first. If the predicted SRF is below 2.95 GHz, the inductor is discarded since it is not useful for the selected operating frequency. Otherwise, its inductance and quality factor are calculated using the L/Q models. The flowchart of the proposed modeling strategy is shown in Figure 7.4.

In order to evaluate the validity of the new proposed strategy, the 210 test inductors were evaluated for three different frequencies. The model errors for L and Q are shown in Tables 7.6 and 7.7 and for the SRF model in Tables 7.8 and 7.9. It can be concluded that by following this modeling strategy, the average error for inductance and quality factor is usually below 1% for L and Q (even for the model at 2.45 GHz), which is a very accurate model for integrated inductors.

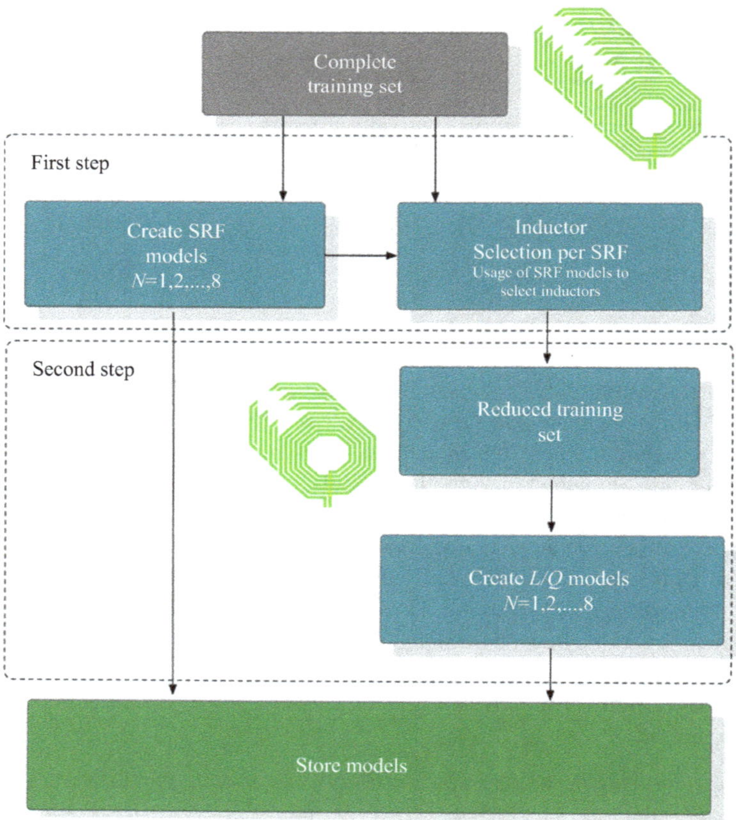

Figure 7.4 Flowchart of the proposed modeling strategy

Table 7.6 Inductance and quality factor average error for 210 test asymmetrical octagonal inductors (in %): one model for each N filtered by SRF

Operating frequency	**20 MHz**		**900 MHz**		**2.45 GHz**	
Filtering frequency	–		**1.1 GHz**		**2.95 GHz**	
N	**Δ*L***	**Δ*Q***	**Δ*L***	**Δ*Q***	**Δ*L***	**Δ*Q***
1	0.07	0.34	0.07	0.43	0.11	0.55
2	0.04	0.28	0.07	0.33	0.09	0.49
3	0.02	0.24	0.06	0.36	0.10	0.21
4	0.02	0.20	0.06	0.36	0.10	0.31
5	0.02	0.18	0.06	0.32	0.24	0.47
6	0.01	0.16	0.15	0.32	0.33	0.64
7	0.01	0.18	0.06	0.37	0.19	0.60

Table 7.7 Inductance and quality factor average error for 210 test symmetrical octagonal inductors (in %): one model for each N filtered by SRF

Operating frequency	**20 MHz**		**900 MHz**		**2.45 GHz**	
Filtering frequency	–		**1.1 GHz**		**2.95 GHz**	
N	**Δ*L***	**Δ*Q***	**Δ*L***	**Δ*Q***	**Δ*L***	**Δ*Q***
1	0.15	0.65	0.16	0.71	0.18	0.82
2	0.01	0.18	0.02	0.14	0.02	0.14
3	0.01	0.11	0.02	0.10	0.03	0.13
4	0.01	0.15	0.02	0.13	0.08	0.13
5	0.01	0.12	0.02	0.11	0.28	0.21
6	0.01	0.16	0.02	0.11	0.23	0.24
7	0.02	0.41	0.03	0.09	1.14	0.30

Table 7.8 SRF average error for 210 test asymmetrical octagonal inductors (in %)

	***N* = 1**	***N* = 2**	***N* = 3**	***N* = 4**	***N* = 5**	***N* = 6**	***N* = 7**
SRF	1.28	1.61	0.69	0.58	0.35	0.34	0.38

Table 7.9 SRF average error for 210 test symmetrical octagonal inductors (in %)

	***N* = 1**	***N* = 2**	***N* = 3**	***N* = 4**	***N* = 5**	***N* = 6**	***N* = 7**
SRF	0.76	0.34	0.33	0.29	0.27	0.17	0.25

7.4 Modeling application to RF design

The accurate developed models in the previous section can be used by the RF designer as an inductor performance evaluator in optimization processes in order to synthesize inductors. Furthermore, such models can also be used in order to model inductors in a circuit design stage. In this section, both model applications will be exemplified.

7.4.1 Inductor optimization

In this subsection, the model is used as a performance evaluator in several inductor synthesis processes. A typical synthesis process is shown in Figure 7.5. The single or multi-objective optimization algorithm is linked with an inductor performance evaluator and automatically varies the inductor variables in order to achieve the optimal inductor performances (optimization objectives), while meeting several design constraints.

During the inductor optimization/design stage, the designer is usually interested in obtaining a given inductance at the frequency of operation, with the largest quality factor, and with the smallest area occupation since fabrication cost grows linearly with area, and inductors represent a large percentage of the RF circuit area. These three parameters: inductance, quality factor and area are mutually conflicting. In addition, the inductor must be designed in the flat-BW zone and the SRF should be sufficiently above the frequency of operation [3]. Consequently, the inductor optimization problem can be posed as a constrained optimization problem:

$$\begin{aligned} &\text{maximize } \boldsymbol{F}(\boldsymbol{x}); && \boldsymbol{F}(\boldsymbol{x}) = \{f_1(\boldsymbol{x}), \ldots, f_n(\boldsymbol{x})\} \in \mathbb{R}^n \\ &\text{such that: } \boldsymbol{G}(\boldsymbol{x}) \geqslant 0; && \boldsymbol{G}(\boldsymbol{x}) = \{g_1(\boldsymbol{x}), \ldots, g_m(\boldsymbol{x})\} \in \mathbb{R}^m \\ & && \text{where } x_{Li} \leqslant x_i \leqslant x_{Ui}, i \in [1, p] \end{aligned} \tag{7.7}$$

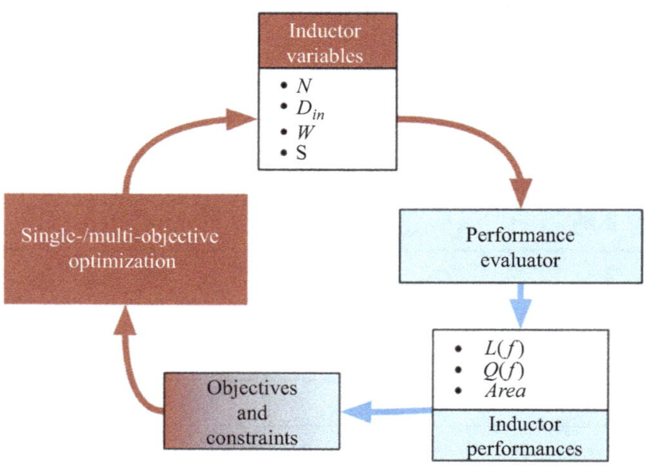

Figure 7.5 Optimization-based inductor synthesis loop

where $\boldsymbol{x}$ is a vector with p geometric parameters, each design parameter being restricted between a lower limit (x_{Li}) and an upper limit (x_{Ui}). The functions $f_j(\boldsymbol{x})$, with $1 \leq j \leq n$, are the objectives that will be optimized, where n is the total number of objectives. The functions $g_k(\boldsymbol{x})$, with $1 \leq k \leq m$, are design constraints that can be defined independently for each optimization problem.

If $n = 1$ then the optimization problem is *single-objective*, and, if $n > 1$, it is *multi-objective*. Whereas the solution to the former is a single design point, the solution to the latter is a set of solutions in the search space, the Pareto set, exhibiting the best trade-offs between the objectives, i.e., the Pareto-optimal front (POF).

In this chapter, the selection-based differential evolution (SBDE) algorithm [21] is used for single-objective optimization. As in evolutionary algorithms, the system is initialized with a population of random solutions and searches for optimal solutions along a set of iterations. The population-based evolutionary optimization algorithm NSGA-II [22] was selected as multi-objective optimization algorithm for the experiments in this chapter. NSGA-II is based on the concept of Pareto dominance and non-dominated ranking of solutions, and the output of the algorithm is a non-dominated set of points in the feasible objective space, i.e., the Pareto-optimal front. The inductor synthesis results reported in this chapter do not exploit any specific characteristics of SBDE and NSGA-II; hence, they can be replaced by any other single- and multi-objective optimization algorithm, respectively.

Also, as previously said, it has to be taken into account that the inductors synthesized must be in the flat-BW zone, as described in Section 7.2. Therefore, in all optimizations, constraints are applied in order to guarantee that the selected inductors can operate at the chosen working frequency (WF). These constraints are specified in the following set of equations:

$$\begin{cases} \text{area} < 400\ \mu\text{m} \times 400\ \mu\text{m} \\ \left|\dfrac{L_{@WF} - L_{@WF+0.05\ \text{GHz}}}{L_{@WF}}\right| < 0.01 \\ \left|\dfrac{L_{@WF} - L_{@WF-0.05\ \text{GHz}}}{L_{@WF}}\right| < 0.01 \\ \left|\dfrac{L_{@WF} - L_{@0.1\ \text{GHz}}}{L_{@WF}}\right| < 0.05 \\ Q_{@WF+0.05\ \text{GHz}} - Q_{@WF} > 0 \end{cases} \tag{7.8}$$

where $L_{@WF}$ and $Q_{@WF}$ are the inductors' inductance and quality factor at the WF, respectively, and $L_{@WF\pm0.05\ \text{GHz}}$ and $Q_{@WF\pm0.05\ \text{GHz}}$ are the inductance and quality factor at $WF\pm0.05$ GHz, respectively. The first constraint in (7.8) is a reasonable upper size of integrated inductors to be used in practice. The second to fourth constraint ensures that the inductance is sufficiently flat from DC to slightly above the WF. Finally, the last constraint guarantees that the SRF is sufficiently above the WF.

As a first synthesis example, a single-objective optimization, using the SBDE algorithm, is performed where the objective is to achieve a symmetrical octagonal inductor with $L = 2.25$ nH (± 0.025 nH) at 0.9 GHz, while maximizing the quality factor. The optimization was performed with 300 individuals and 100 generations and took just 22.14 s of CPU time. The geometrical parameters of the obtained inductor were $N = 2$, $D_{in} = 286$ µm, $W = 25$ µm and its performance parameters are $L = 2.268$ nH and $Q = 8.921$. The SRF of this inductor is around 13.31 GHz. The inductor layout and frequency behavior over frequency can be observed in Figure 7.6. However, in order to assess the model accuracy, the obtained inductor was EM-simulated and compared with the performances achieved by the surrogate model. The accuracy evaluation can be observed in Table 7.10, where it is possible to conclude that the model is extremely accurate.

Afterward, the same type of single-objective optimization is performed in order to obtain an asymmetric inductor with $L = 1.5$ nH (± 0.05 nH) at 2.45 GHz. Again, the optimization was performed with 300 individuals and 100 generations and took only 23.44 s of CPU time, which, as in the previously example, is an impressive time. The geometrical parameters of the obtained inductor were $N = 2$, $D_{in} = 187$ µm, $W = 17.05$ µm and its performance parameters are $L = 1.548$ nH and $Q = 11.697$. The SRF of this inductor is around 16.94 GHz. The inductor

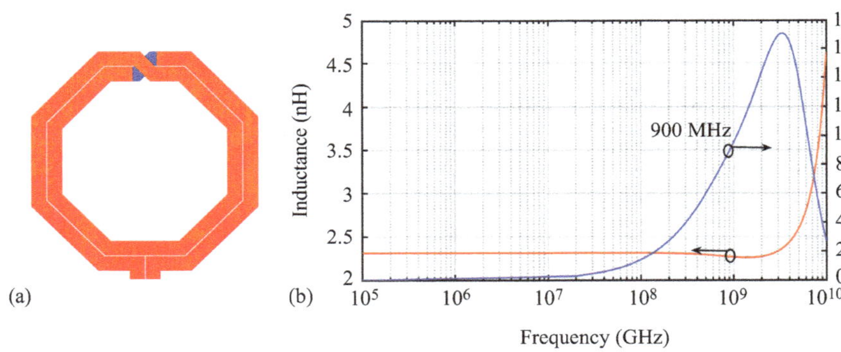

Figure 7.6 (a) Layout of the symmetric inductor obtained with the single-objective optimization. (b) Illustrating the inductance (red curve and left y-axis) and quality factor (blue curve and right y-axis) as a function of frequency of the same inductor

Table 7.10 Comparison of inductance and quality factor values obtained with the surrogate model and with EM simulation for the inductor shown in Figure 7.6

Geometric parameters			***WF* = 900 MHz**					
N	D_{in} (µm)	W (µm)	L_{EM} (nH)	L_{SUR} (nH)	Error (%)	Q_{EM}	Q_{SUR}	Error (%)
2	286	25	2.267	2.268	0.044	8.918	8.921	0.033

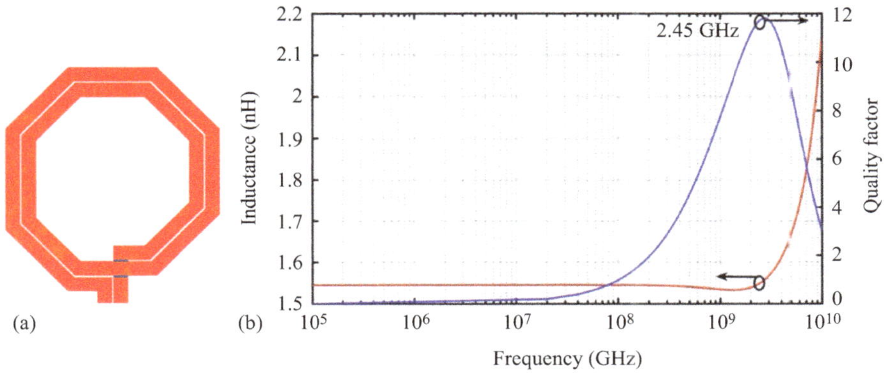

Figure 7.7 (a) Layout of the asymmetric inductor obtained with the single-objective optimization. (b) Illustrating the inductance (red curve and left y-axis) and quality factor (blue curve and right y-axis) as a function of frequency of the same inductor

Table 7.11 Comparison of inductance and quality factor values obtained with the surrogate model and with EM simulation for the inductor shown in Figure 7.7

Geometric parameters			***WF* = 2.45 GHz**					
N	D_{in} (µm)	*W* (µm)	L_{EM} (nH)	L_{SUR} (nH)	**Error (%)**	Q_{EM}	Q_{SUR}	**Error (%)**
2	187	17.05	1.548	1.548	0.001	11.683	11.697	0.119

layout and frequency behavior over frequency are shown in Figure 7.7. Again, as performed for the previous obtained inductor, the accuracy evaluation of the model can be observed in Table 7.11, where the inductor obtained in the optimization was EM-simulated and its error assessed.

After the single-objective optimizations were performed, it is interesting to perform some multi-objective optimizations in order to generate Pareto-optimal fronts (POFs) of inductors. As previously said, for this purpose, NSGA-II is used. The first optimization is performed with the objectives of maximizing inductance and quality factor, while minimizing the area of the inductors. The inductor topology considered for this optimization is the symmetric topology and the WF is 900 MHz. The entire optimization with 400 individuals and 100 generations takes 178.24 s of CPU time, and the obtained POF can be seen in Figure 7.8. Due to the efficiency of the model, in just 3 min it is possible to generate a POF comprising the best trade-offs between objectives for this inductor topology, technology and frequency of operation.

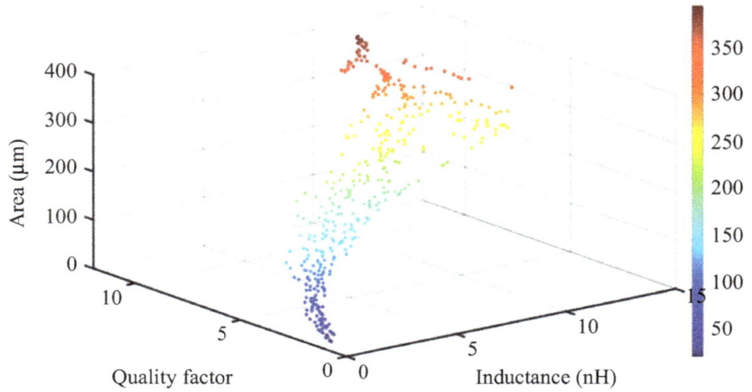

Figure 7.8 POF obtained of the optimization maximizing inductance and quality factor, while minimizing the area for a symmetric inductor topology and working frequency of 900 MHz. The color bar represents the third objective of the optimization, which is the area in μm

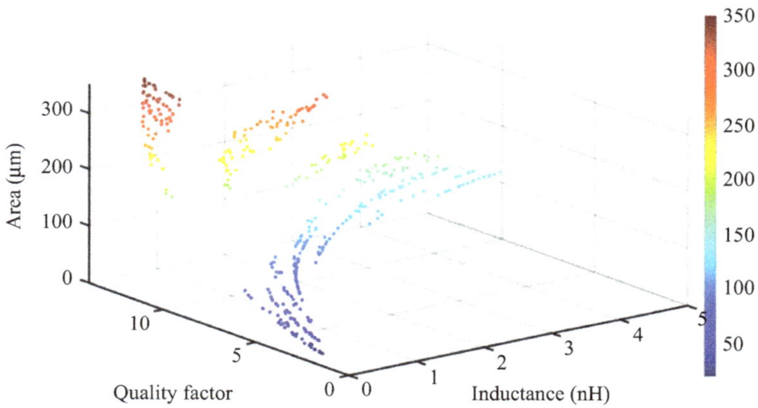

Figure 7.9 POF obtained of the optimization maximizing inductance and quality factor, while minimizing the area for an asymmetric inductor topology and working frequency of 2.45 Hz. The color bar represents the third objective of the optimization, which is the area in μm

Afterward, for illustration purpose, another optimization was performed with the same objectives of maximizing inductance and quality factor, while minimizing the area of the inductors, but considering an asymmetric topology and WF of 2.45 GHz. The entire optimization with 400 individuals and 100 generations takes 202.40 s of CPU time, and the obtained POF can be seen in Figure 7.9.

As previously said, the POFs in Figures 7.8 and 7.9 show the best inductor designs for the selected objectives, operating frequencies and topologies.

By obtaining these POFs, the RF designer can then use them as inductor-optimized databases during the circuit design stages. The idea is that instead of searching for an inductor in the entire design stage, the designer only has to search the POF for the inductor that best suits his/her needs.

7.4.2 Circuit design

In this section, the model will be used during the design stage of two well-known RF circuits. The model is extremely accurate and therefore its usage in circuit design is highly endorsed. A VCO and an LNA will be manually designed and the surrogate model is used in order to design the inductors used.

7.4.2.1 Voltage-controlled oscillator

The first circuit to be designed is a LC double-differential cross-coupled VCO, as shown in Figure 7.10, intended to oscillate at 900 MHz.

The oscillation frequency, f_{osc}, of an LC VCO is given by:

$$f_{osc} = \frac{1}{2\pi\sqrt{L\,C}} \tag{7.9}$$

where L is the inductance and C is the tank capacitance, which can be varied by using the varactors (C_{var}). Since we are designing the VCO to oscillate at 900 MHz, from (7.9) it is possible to fix the capacitance and find the needed inductance or vice versa. In this example, the VCO is designed using a capacitance bank of six parallel capacitances and two varactors (in series, in order to have symmetry in the circuit), as shown in Figure 7.10. Each capacitance has a value of 100 fF and each varactor has a maximum capacitance, $C_{var\text{MAX}}$, of 2.8 pF. All of the component

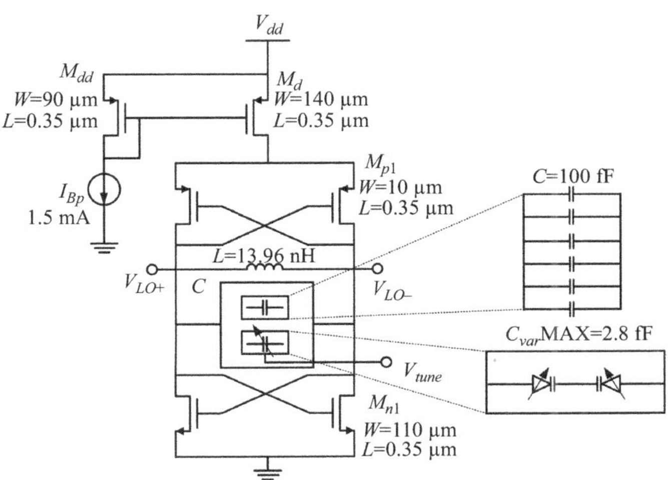

Figure 7.10 Topology of the LC double-differential cross-coupled differential VCO used and sizes of all components used in the design

values are also shown in Figure 7.10. The parameter V_{tune} represents an independent voltage source used to vary the capacitance of the varactors. In this example, we have performed a single-objective optimization in order to obtain an inductor with approximately 14 nH (as performed as in Section 7.4.1). The objective was to achieve a symmetrical octagonal inductor with $L = 14$ nH (± 0.1 nH) at 0.9 GHz, while maximizing the quality factor. The geometrical parameters of the inductor were $N = 5$, $D_{in} = 252$ μm, $W = 5.2$ μm and its performance parameters are $L = 13.960$ nH and $Q = 6.921$. The SRF of this inductor is around 3.923 GHz. The inductor layout and frequency behavior over frequency can be observed in Figure 7.11. The obtained inductor was EM-simulated later on and the error of the model was assessed. The results are shown in Table 7.12, where it is possible to observe that the model has negligible errors. The inductor optimization was performed with 300 individuals and 100 generations and took 20.81 s of CPU time.

The transient response of the designed VCO can be observed in Figure 7.12 (blue curve). Since an inductor model is being used, it is expected that the performances shift due to the usage of this model. Therefore, in order to inspect the shifts that occur in the VCO performances, the used inductor was EM-simulated and the VCO re-simulated. The differences in the transient response can also be observed in

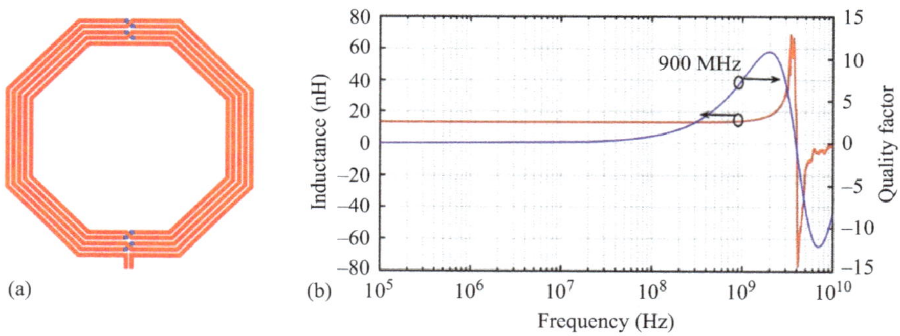

Figure 7.11 (a) Layout of the 13.960 nH symmetric inductor obtained with the single-objective optimization. (b) Illustrating the inductance (red curve and left y-axis) and quality factor (blue curve and right y-axis) as a function of frequency of the same inductor

Table 7.12 Comparison of inductance and quality factor values obtained with the surrogate model and with EM simulation for the inductor shown in Figure 7.11

Geometric parameters			***WF* = 900 MHz**					
N	D_{in} **(μm)**	*W* **(μm)**	L_{EM} **(nH)**	L_{SUR} **(nH)**	**Error (%)**	Q_{EM}	Q_{SUR}	**Error (%)**
5	252	5.2	13.958	13.960	0.014	6.918	6.921	0.043

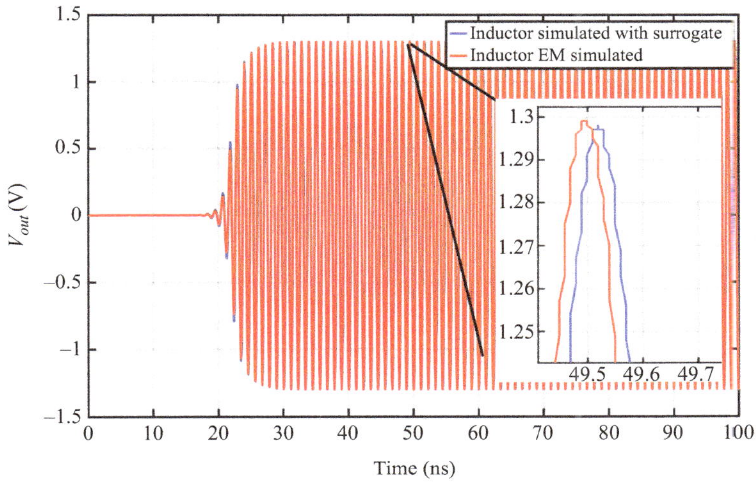

Figure 7.12 Transient response of the designed VCO. A zoom over the signals is performed in order to illustrate the small difference between both transient responses

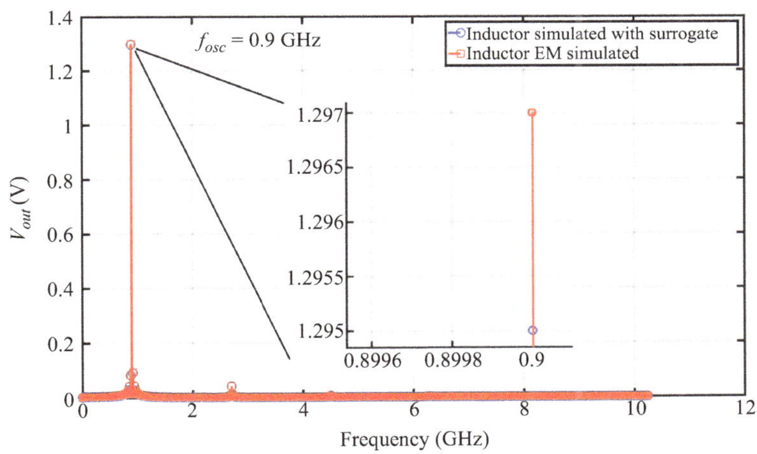

Figure 7.13 DFT of the transient signal shown in Figure 7.12. A zoom over the signals is performed in order to illustrate the small difference between both DFT signals

Figure 7.12 (red curve), where it is possible to conclude that negligible performance shifts occur (f_{osc} does not shift), confirming once again the model accuracy. Furthermore, the discrete Fourier transform (DFT) of the transient signal can be observed in Figure 7.13. The phase noise of the VCO can be observed in

Figure 7.14, and the tuning range in Figure 7.15, where for these performances, the differences are negligible even at centesimal level. It can be concluded that since the model is so accurate it is extremely useful in RF circuit design.

7.4.2.2 Low-noise amplifier

In this subsection the design of another well-known RF circuit, an LNA, is considered. The circuit topology considered in this chapter is the source-degenerated topology shown in Figure 7.16. The LNA is manually designed and is intended to operate at the ISM band of 2.4–2.5 GHz.

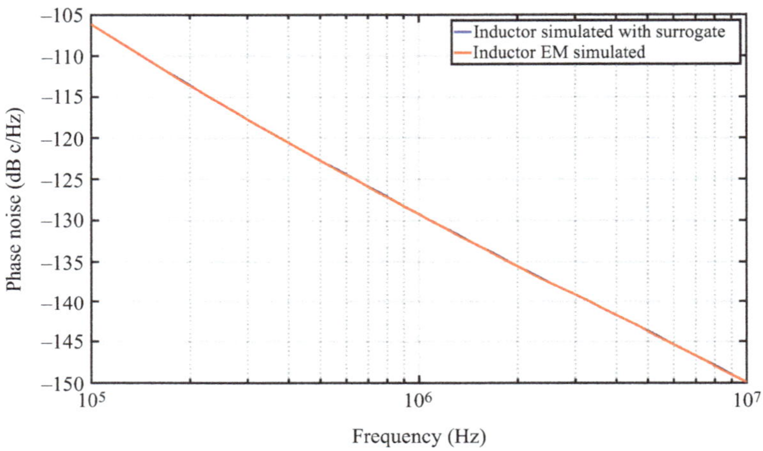

Figure 7.14 Phase noise of the designed VCO

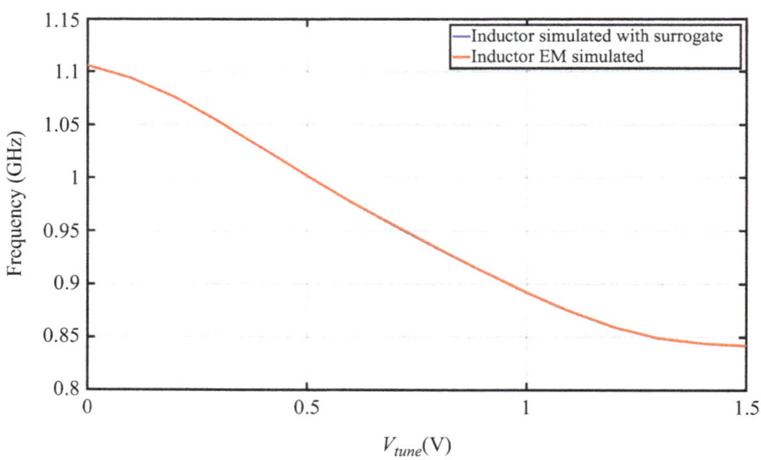

Figure 7.15 Tuning range of the designed VCO

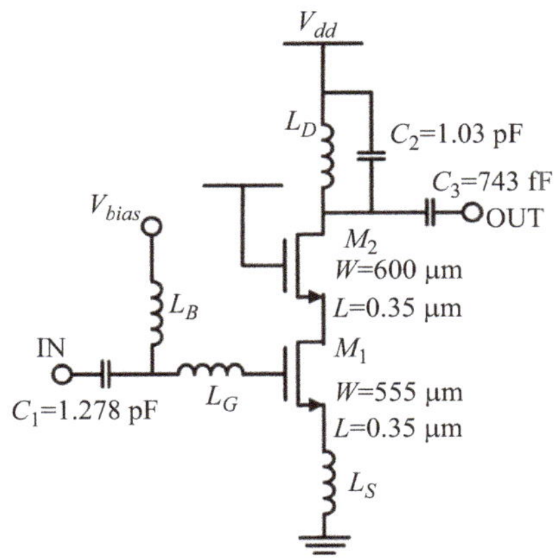

Figure 7.16 Topology of the source-degenerated LNA used and sizes of all components (apart from the inductors) used in the design

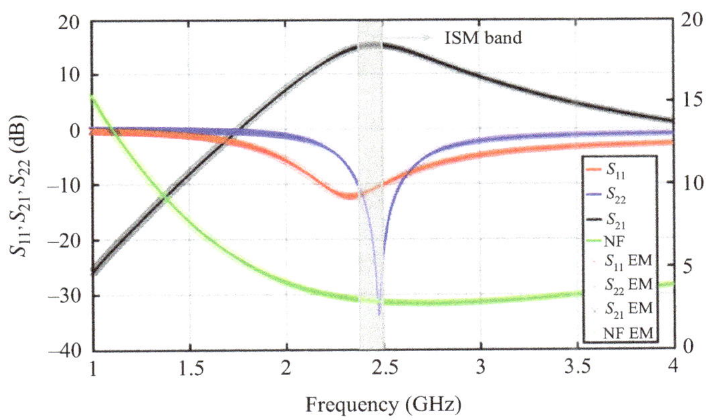

Figure 7.17 Performances of the designed LNA

The strategy used in order to design the LNA is different from the used in the previously designed VCO. In this example, instead of using a single-objective optimization algorithm in order to size each inductor (L_S, L_G, L_B and L_D), the POF shown in Figure 7.9 is used as an optimized inductor library from which inductors can be selected to design the LNA. The needed inductors are therefore selected from the library and the LNA is manually designed. The values of each component are shown in Figure 7.16 and the most important LNA performances are shown in Figure 7.17. The gray shaded area corresponds to the ISM band, where the designed LNA is supposed to work. It is possible to observe that at the center frequency of the ISM band (2.45 GHz) the LNA has a gain (S_{21}) of 15.23 dB, an input matching (S_{11}) of −11.01 dB, an output matching (S_{22}) of −21.12 dB and a noise figure of 2.917 dB.

Table 7.13 Comparison of inductance and quality factor values obtained with the surrogate model and with EM simulation for the inductors used in the LNA design of Figure 7.16

	Geometric parameters			WF = 2.45 GHz					
Inductor	N	D_{in} (μm)	W (μm)	L_{EM} (nH)	L_{SUR} (nH)	Error (%)	Q_{EM}	Q_{SUR}	Error (%)
L_S	1	22	11.40	0.041	0.041	0.851	3.720	3.682	1.025
L_G	2	15	6.90	0.138	0.137	0.154	3.154	3.169	0.466
L_B	6	23	7.00	2.484	2.484	0.004	8.204	8.233	0.356
L_D	2	216	13.45	1.905	1.903	0.054	11.51	11.55	0.317

Since the surrogate model yields an associated error, the inductors were EM-simulated and the LNA re-simulated. The performance deviations are also shown in Figure 7.17, where it is possible to observe that the performance deviations are negligible compared to the ones obtained when the surrogate model was used. The error associated with the inductor performances can be seen in Table 7.13.

7.5 Conclusions

This chapter presents a surrogate model for integrated inductors based on machine-learning techniques and also an enhanced modeling strategy that is able to increase the model accuracy. The model can be used for the simulation of a single inductor and also for different applications inherent to RF design, such as the use of optimization algorithms and usage in circuit simulation during circuit design. In this chapter, several models were created for different operating frequencies and inductor topologies. Afterward, the model was used in both single- and multi-objective optimizations and has proved to be very accurate and efficient in several optimizations, when compared to EM simulations. Furthermore, the model has been used in circuit design experiments (design of VCOs and LNAs) and has exhibited to be an accurate and efficient model that can assist the RF designer in many different design examples.

Acknowledgment

This work was supported by TEC2016-75151-C3-3-R Project (AEI/FEDER, UE).

References

[1] R. C. Li, RF Circuit Design, 2nd Ed. Hoboken, NJ: Wiley, 2012.

[2] G. Zhang, A. Dengi, and L. R. Carley, "Automatic synthesis of a 2.1 GHz SiGe low noise amplifier," in *IEEE Radio Frequency Integrated Circuits (RFIC) Symposium*, Seattle, WA, USA, 2002, pp. 125–128.

[3] R. González-Echevarría, R. Castro-López, E. Roca, *et al.*, "Automated generation of the optimal performance trade-offs of integrated inductors," IEEE Transactions on Computer-Aided Design of Integrated Circuits and Systems, vol. 33, no. 8, pp. 1269–1273, 2014.

[4] C. De Ranter, De Ranter, Geert Van der Plas, *et al.*, "CYCLONE: automated design and layout of RF LC-oscillators," in *Proceedings 37th Design Automation Conference*, Los Angeles, CA, USA, 2000, pp. 11–14.

[5] C. P. Yue and S. S. Wong, "Physical modeling of spiral inductors on silicon," IEEE Transactions on Electron Devices, vol. 47, pp. 560–568, 2000.

[6] C. Wang, H. Liao, C. Li, *et al.*, "A wideband predictive double-π equivalent-circuit model for on-chip spiral inductors," IEEE Transactions on Electron Devices, vol. 56, no. 4, pp. 609–619, 2009.

[7] Y. Cao, R.A. Groves, Xuejue Huang, *et al.*, "Frequency-independent equivalent circuit model for on-chip spiral inductors," IEEE Journal of Solid-State Circuits, vol. 38, no. 3, pp. 419–426, 2003.

[8] F. Passos, M. Kotti, R. González-Echevarría, *et al.*, "Physical vs. surrogate models of passive RF devices," in *IEEE International Symposium on Circuits and Systems (ISCAS)*, Lisbon, Portugal, 2015, pp. 117–120.

[9] F. Passos, M. H. Fino, and E. Roca, "Single-objective optimization methodology for the design of RF integrated inductors," in Camarinha-Matos L. M., Barrento N. S., and Mendonça R. (eds.), Technological Innovation for Collective Awareness Systems. DoCEIS 2014. IFIP Advances in Information and Communication Technology, vol 423, 2014. Springer, Berlin, Heidelberg.

[10] A. I. J. Forrester, A. Sobester, and A. J. Keane, Engineering Design via Surrogate Modelling – A Practical Guide. Wiley, 2008.

[11] S. K. Mandal, S. Sural, and A. Patra, "ANN- and PSO-based synthesis of on-chip spiral inductors for RF ICs," IEEE Transactions on Computer-Aided Design of Integrated Circuits and Systems, vol. 27, no. 1, pp. 188–192, 2008.

[12] I. Couckuyt, F. Declercq, T. Dhaene, H. Rogier, and L. Knockaert, "Surrogate-based infill optimization applied to electromagnetic problems," International Journal of RF and Microwave Computer-Aided Engineering, vol. 20, no. 5, pp. 492–501, 2010.

[13] F. Passos, E. Roca, R. Castro-López, and F. V. Fernández, "Radio-frequency inductor synthesis using evolutionary computation and Gaussian-process surrogate modeling," Applied Soft Computing, vol. 60, pp. 495–507, 2017.

[14] K. Okada and K. Masu, "Modeling of spiral inductors," in Advanced Microwave Circuits and Systems. InTech, 2010. doi: 10.5772/8435.

[15] Keysigth, ADS Momentum. http://www.keysight.com/en/pc-1887116/momentum-3d-planar-em-simulator?cc=ES&lc=eng [accessed, 2019].

[16] M. B. Yelten, T. Zhu, S. Koziel, P. D. Franzon, and M. B. Steer, "Demystifying surrogate modeling for circuits and systems," IEEE Circuits and Systems Magazine, vol. 12, no. 1, pp. 45–63, 2012.

[17] D. Gorissen, I. Couckuyt, P. Demeester, T. Dhaene, and K. Crombecq, "A surrogate modeling and adaptive sampling toolbox for computer based design," The Journal of Machine Learning Research, vol. 11, pp. 2051–2055, 2010.

[18] DACE – A Matlab Kriging Toolbox. http://www2.imm.dtu.dk/pubdb/views/publication_details.php?id=1460 [Accessed November, 2019].
[19] B. Bischl, O. Mersmann, H. Trautmann, and C. Weihs, "Resampling methods for meta-model validation with recommendations for evolutionary computation," Evolutionary Computation, vol. 20, no. 2, pp. 249–275, 2012.
[20] F. Ferranti, L. Knockaert, T. Dhaene, and G. Antonini, "Parametric macromodeling based on amplitude and frequency scaled systems with guaranteed passivity," International Journal of Numerical Modelling: Electronic Networks, Devices and Fields, vol. 25, no. 2, pp. 139–151, 2012.
[21] K. Zielinski and R. Laur, "Constrained single-objective optimization using differential evolution," in *IEEE Congress on Evolutionary Computation*, Vancouver, BC, Canada, 2006, pp. 223–230.
[22] K. Deb, A. Pratap, S. Agarwal, and T. Meyarivan, "A fast and elitist multiobjective genetic algorithm: NSGA-II," IEEE Transactions on Evolutionary Computation, vol. 6, pp. 182–197, 2002.

Chapter 8

Modeling of variability and reliability in analog circuits

Javier Martin-Martinez[1], Javier Diaz-Fortuny[1], Antonio Toro-Frias[2], Pablo Martin-Lloret[2], Pablo Saraza-Canflanca[2], Rafael Castro-Lopez[2], Rosana Rodriguez[1], Elisenda Roca[2], Francisco V. Fernandez[2], and Montserrat Nafria[1]

One of the main challenges in the design of integrated circuits (ICs) in recent ultrascaled complementary metal oxide semiconductor (CMOS) technologies is how to deal with the large variability of the electrical characteristics of devices. In the last years, large progress has been made in understanding the metal oxide semiconductor field effect transistors (MOSFETs) variability associated with fabrication processes, namely, the time-zero variability (TZV) introduced by phenomena, such as random dopant distribution or line edge roughness [1–7], and their impact on circuits [8,9]. However, the time-dependent variability (TDV), though can strongly limit the IC reliability, is still a controversial issue. TDV is typically associated with processes, such as random telegraph noise (RTN) [10–12], and wear-out mechanisms, such as bias temperature instabilities (BTI) [13–17] or hot-carrier injection (HCI) [18–21]. At device level, during circuit operation, these phenomena lead to stochastic and time-dependent variations of several CMOS device parameters (e.g., the threshold voltage, V_{th}), negatively impacting circuit parameters, as the delay in digital circuits [22], small-signal parameters in low noise amplifiers (LNAs) [23,24], retention times and stability in memories [25,26] and eventually causing the failure of the IC. Consequently, TDV must be taken into account during the design of digital and analog ICs.

Aging (also known as degradation or wear out) of MOSFETs has been extensively studied in the last years and physical models for the BTI and HCI mechanisms have been proposed [27–29]. Nowadays, it is broadly accepted that the aging of the device during its operation is related to the generation of defects (in the bulk of the gate oxide and/or close to its interfaces) that can trap/detrap charges, leading to the observed threshold voltage shifts [30–32]. Because of the complexity of the phenomena, taking into account the details of the related physics during circuit simulation would be

[1]Electronic Engineering Department, Universitat Autònoma de Barcelona (UAB), Barcelona, Spain
[2]Instituto de Microelectrónica de Sevilla, IMSE-CNM, CSIC and Universidad de Sevilla, Sevilla, Spain

computationally unaffordable, so that a more simplistic, though accurate enough, description of the aging effects must be formulated, implementable into commercial circuits simulators. In this regard, physics-based compact models are required, which accurately describe the TDV observed in the device electrical properties and allow the evaluation of the aging of devices during their operation in the circuit within reasonable computation times. These compact models are actually keys to bridge the gap between device and circuit levels, but, unfortunately, a limited number is still available [33,34]. In parallel to the models, suitable model parameter extraction procedures must be developed to take the particularities of the aging of the underlying technology into account. However, this second step also poses several challenges when dealing with ultrascaled technologies. First, suitable test structures and a complete and advanced instrumentation system are required to characterize TDV in a large number of transistors [35–40], to get statistically significant results in reasonable testing times. Second, smart post-processing tools [41–44] are needed to handle the huge amount of data coming out of the characterization, to get the parameters of the compact models. Once the devices have been fully modeled, a simulation methodology must be developed, which, considering the physics-based device TDV compact models with the suitable model parameter set, provides an accurate estimation of the impact of device TDV on the circuit performance and reliability [45–47].

In this chapter, some of the most recent approaches adopted in all these fields will be presented, with especial emphasis on analog circuits. The chapter is divided into four sections, covering each of the aforementioned issues. In Section 8.1, the probabilistic defect occupancy (PDO) model, a physics-based compact model, will be introduced, which can be easily implemented into circuit simulators. Section 8.2 describes a purposely designed IC which contains suitable test structures, together with a full instrumentation system for the massive characterization of TZV and TDV in CMOS transistors, from which aging of the technology under study can be statistically evaluated. Section 8.3 is devoted to a smart methodology, which allows extracting the statistical distributions of the main physical parameters related to TDV from the measurements performed with our instrumentation system. Finally, Section 8.4 describes CASE, a new reliability simulation tool that accounts for TZV and TDV in analog circuits, covering important aspects, such as the device degradation evaluation, by means of stochastic modeling and the link between the device biasing and its degradation. As an example, the shifts of the performance of a Miller operational amplifier related to the device TDV is evaluated using CASE. Finally, in Section 8.5 the main conclusions are summarized.

8.1 Modeling of the time-dependent variability in CMOS technologies: the PDO model

During the operation of devices in a circuit, mechanisms, such as BTI and/or HCI, will lead to the aging of the device, i.e., to a progressive change of its electrical properties, which could end in its final failure, limiting the IC reliability. Fortunately, under nominal operation conditions, the device/IC failure can take years, so that accelerated tests are used in the laboratory to reduce the evaluation of the lifetime to reasonable testing times. In these accelerated tests, temperature and/or drain voltage and/or gate voltage (namely, the stress conditions) are raised above their nominal values during a certain time, i.e., the stress time. After this stress phase, the electrical properties of the device are measured again, during the so-called measurement phase, at lower voltages/temperatures, to evaluate the shift of some device electrical parameter (with respect to that measured on the unstressed or pristine device), as a consequence of the aging induced by the previous stress. These consecutive stress-measurement (SM) phases are repeated several times, so that the temporal evolution of the device aging can be evaluated, in what is known as SM testing scheme [14]. Physical models allow extrapolating the results obtained under accelerated test conditions to normal operation conditions [33,48].

In the case of BTI, a voltage is applied to the gate (source and drain grounded) and/or eventually at high temperature. To trigger HCI, in addition to the gate voltage, a drain voltage has to be applied. Independently of the aging mechanism (BTI or HCI), the aging of the device leads to an increment of the threshold voltage (or, equivalently, to a decrease of the current). Figure 8.1 shows an example of I_D–V_G curves recorded during the measurement phases of a 3-cycle SM test on a pMOS device. The stress voltage was 2.1 V and the stress times, t_s, were 100, 1,000 and 10,000 s. The I_D–V_G curve measured on the fresh device is also shown. Note that, after the stress, the curve shifts toward the right (smaller currents), which is interpreted as an increase of the threshold voltage. Figure 8.2(a), left-hand side, shows a typical increase of the threshold voltage, ΔV_{th}, as a function of the stress time. However, it is well known that during the measurement of BTI effects, the threshold voltage of the stressed device starts recovering toward their original values immediately after the removal/decrease of the stress [14], in what is known as recovery/relaxation phase, with duration t_r. An example is shown on the right-hand side of Figure 8.2(a). Note, however, that the recovery is only partial (i.e., the

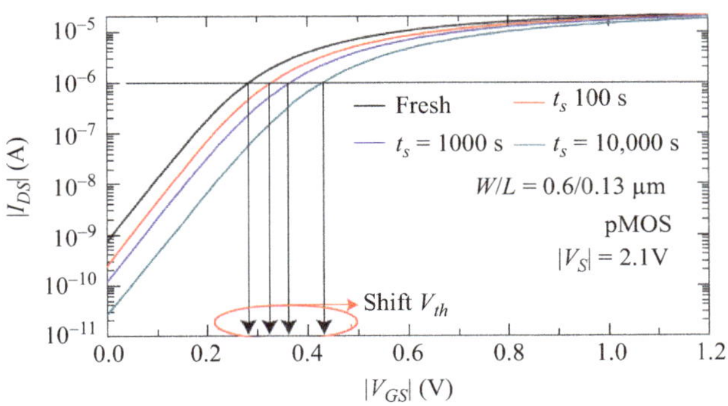

Figure 8.1 Characteristic I_D–V_G curve shift after several BTI stresses on a pMOS after different stress times (100, 1,000 and 10,000 s) at a constant stress voltage V_s = 2.1 V. The black curve corresponds to the one measured on the fresh (as-grown) device

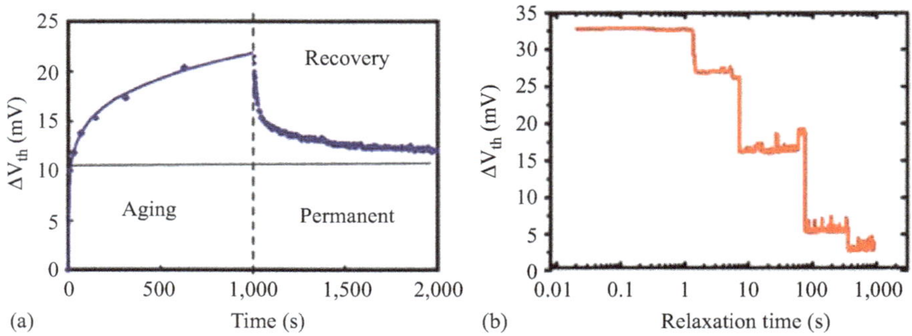

Figure 8.2 (a) Example of the threshold voltage shift caused by BTI during 1,000 s of stress. After that when the stress is removed, the threshold voltage is partially recovered. In this large-area device, the recovery is continuous with time. (b) In small-area devices, the V_{th} recovery is observed as discrete drops associated to the discharge of defects that were previously charged during the stress

value of the threshold voltage previous to the stress is not reached); so it is said that BTI has actually two components: a permanent and a recoverable component [49]. In the recovery trace, shown in Figure 8.2(a), the threshold voltage continuously recovers (i.e., decreases), but in ultrascaled devices, this recovery is discrete, being observed as sudden V_{th} drops, as shown in Figure 8.2(b). This kind of evolution introduces an intrinsic statistical nature of the BTI aging [31,32].

From data like those shown in Figures 8.1 and 8.2, physical models have been developed to explain the device aging and its dependences on time (stress and recovery), voltage and temperature during BTI [14,15,17] and HCI tests [18–21]. In spite of their differences, all of them associate aging to the generation of defects in the oxide bulk and/or its interfaces, which can be charged/discharged. When these defects are charged, they create a barrier that can hinder the electron transport along the channel, reducing the drain current, or equivalently increasing the threshold voltage [50]. Then, V_{th} is the parameter that is extensively considered in the models to describe the BTI degradation [27–29]. It must be emphasized that a similar physical framework explains the phenomenon of RTN. Actually, the observation of RTN and/or BTI is strongly dependent on the operation conditions (voltage and temperature) of the device [31]. RTN and BTI effects can be observed simultaneously, the first one as fast increase/decrease events in the device current, the second one as slower or permanent changes in the current.

This general picture must be translated into compact models, which provide equations that (i) accurately describe the change of V_{th} as a function of the device operation and (ii) can be implemented into circuit simulators with a reasonable increase of the computation time. Several attempts have been already presented [51] but, as an example, this work will focus on the PDO model [29], which was initially developed to describe the BTI aging. However, it has been recently demonstrated that it is able to describe in a unified way the effects related to aging (BTI and HCI) and RTN transients [16,31].

The PDO model describes the shift of the V_{th} (ΔV_{th}) in a MOSFET when a stress voltage is applied to the gate (V_s) during a certain stress time (t_s), followed by a relaxation/recovery phase at a lower voltage, V_r, of duration t_r (usually $V_r \approx V_{th}$). This model assumes that each device has a number of defects (N), which in a set of devices is Poisson distributed [52]. When each of these defects is charged/discharged, a shift of the V_{th}, η, is registered, different for each defect [32]. Figure 8.3 shows a schematic representation of a pristine device ($t = 0$) during a BTI stress. When a stress voltage (V_s) is applied to the gate contact, the occupancy probability of traps in the oxide rises leading to a device V_{th} increase. After this perturbation, gate voltage is reduced to V_r (to measure the effect of the stress) and the charged traps in the oxide are gradually discharged (occupancy probability of traps in the oxide decreases) and the system returns to the previous state, recovering partially or totally the previous V_{th} level. Figure 8.4(a) represents the energy band diagram of a metal oxide semiconductor (MOS) structure, where the tunneling

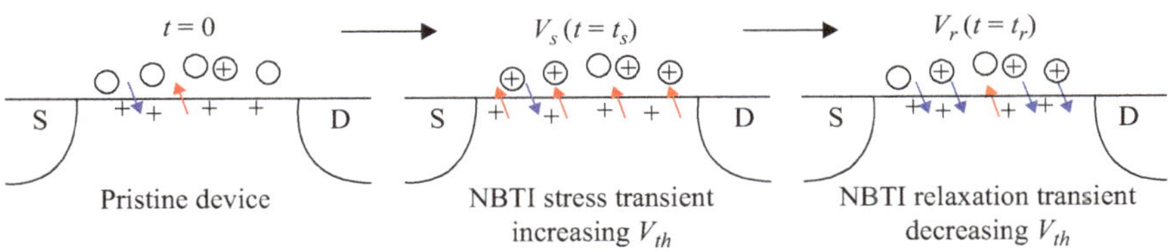

Figure 8.3 Schematic representation of NBTI phenomenon. Defects are charged during the stress and discharged during the recovery (measurement) phase

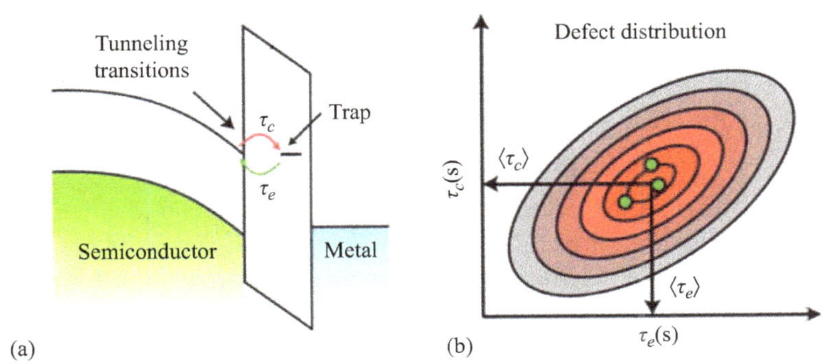

Figure 8.4 (a) Energy band diagram of an MOS structure, where the tunneling transitions of the electrons from the semiconductor to a trap are shown. (b) Example of defect distribution in a bidimensional graph (τ_c–τ_e)

transitions of the electrons from the semiconductor to a trap that leads to the charge/discharge events that cause the device V_{th} shifts are sketched.

Each of the N defects in the device has associated an emission time (τ_e) and a capture time (τ_c), which are the average times that a defect needs to be discharged or charged, respectively. Given these assumptions, the PDO model equation that describes ΔV_{th}, for a given set of t_r, t_s and operation conditions is (8.1):

$$\Delta V_{th} = N\langle\eta\rangle \int_0^{\infty}\int_0^{\infty} D(\tau_e,\tau_c)\cdot P_{occ}(\tau_e,\tau_c;t_s,t_r)d\tau_e d\tau_c + P_p(t_s) \tag{8.1}$$

where N is the number of defects, $\langle\eta\rangle$ is the average shift of the V_{th} when the defects change their state. $D(\tau_e, \tau_c)$ describes the distribution of the defects in the τ_e and τ_c space and depends on voltage and temperature. P_p (also dependent on V and T) is the BTI permanent part, which is associated with charged defects whose emission time is much larger than the typical experimental window and therefore they appear permanently charged. $D(\tau_e, \tau_c)$ and P_p are characteristics of the technology and implicitly include the voltage and temperature dependences of the device aging. Note, however, that in ultrascaled devices, N will be small, so that actually few individual traps (statistically distributed following $D(\tau_e, \tau_c)$) will have to be considered, leading to device-to-device variability of the aging. Figure 8.4(b) shows a typical $D(\tau_e, \tau_c)$ defect distribution, which must be introduced into (8.1) and integrated to evaluate its impact on the device V_{th}. Green dots represent the particular case for an ultrascaled device, where only three defects are active ($N = 3$).

Finally, P_{occ} in (8.1) is the occupancy probability for each defect with given τ_e and τ_c, for a particular stress/relaxation conditions, t_s, t_r, V_r and V_s, and can be calculated using the following equation:

$$P_{occ}(t_{r/s}) = P_{occ}(t_i) + \left(\frac{\tau_e(V_{r/s})}{\tau_e(V_{r/s}) + \tau_c(V_{r/s})} - P_{occ}(t_i)\right)\left(1 - \exp\left(\frac{t_i - t_{r/s}}{\tau_{eff}(V_{r/s})}\right)\right) \tag{8.2}$$

where (t_i) is the initial time and $\tau_{eff}(V_{r/s})$ is given by the following equation:

$$\left(\tau_{eff}(V_{r/s})\right)^{-1} = \left(\tau_e(V_{r/s})\right)^{-1} + \left(\tau_c(V_{r/s})\right)^{-1} \tag{8.3}$$

the term $V_{r/s}$ indicates that the values τ_{eff}, τ_c and τ_e correspond to either the stress voltage (V_s) or the relaxation voltage (V_r). Then, P_{occ} depends on the operation conditions of the device (voltage, temperatures and times) and must be evaluated after a transient simulation of the circuit.

N, $\langle\eta\rangle$, $D(\tau_e, \tau_c)$ and P_p are the model parameters, which are technology dependent. Then, to evaluate the TDV effects using the PDO model, the values of the model parameters, together with their voltage and temperature dependences, must be evaluated for the technology under study. It has been shown that they can be obtained from the relaxation traces measured after the stress of the device [53]. Note that, in the case of ultrascaled devices, where device-to-device variability can be very important, it is necessary to measure many devices in order to get enough statistical data under different stress conditions (to get the corresponding dependences), which is a very challenging issue, as will be discussed in the next section.

8.2 Characterization of time zero variability and time-dependent variability in CMOS technologies

Traditionally, TDV effects are analyzed in isolated MOSFETs on a wafer (non-encapsulated). However, as described in Section 8.1, the stochastic nature of the aging phenomena and the need to evaluate the voltage/temperature dependences of the model parameters require stressing thousands of samples under several stress conditions and long stress times. Then, extensive statistical on-wafer characterization of MOSFET aging effects is hardly affordable, because applying aging tests to a large number of devices results in long measurement times that can take months and even years. In order to perform a statistical characterization of the device TDV, an area-efficient solution is the use of on-chip array structures that contain thousands of devices under test (DUT) [35–40]. Together with the device array, with a well-suited architecture, the complete characterization system should fulfill the following requirements:

- Individual DUT terminal access and accurate control of applied voltages (since the model parameters strongly depend on voltage).
- Possibility to perform parallel stress of devices with accurate timing and accurate control of the stress and recovery times for all transistors during aging tests, minimizing at the same time the time gaps between both phases.
- Possibility to perform electrical characterization of process variability, RTN, BTI and HCI aging.

The design of such system is not easy, which explains that most of the developed systems only partially fulfill these requirements. As an alternative, we have designed and fabricated a device array chip, called ENDURANCE, whose design satisfactorily addresses all of them [54]. It is equipped with an internal full custom digital control circuitry (for DUT selection) and Force & Sense architecture (to get a precise control of the actual voltages applied to the DUT terminals). The chip has been fabricated in a 1.2 V, 65-nm CMOS technology and occupies an area of $1,800 \times 1,800\ \mu m^2$. To allow statistical characterization, the chip includes 3,136 MOS test transistors, distributed in two electrically isolated nMOS and pMOS DUT blocks, with eight different geometries.

For the electrical characterization of the chip, a test setup has been implemented (Figure 8.5), which includes a printed circuit board (PCB), where the chip is inserted; a semiconductor parameter analyzer (SPA), used for voltage and current measurements and DUTs biasing; and a power supply for chip biasing. The Force & Sense voltage system supported by the SPA and incorporated into the PCB and the chip minimizes the negative effects of series resistance and parallel capacitance of cables, connectors and PCB routing, as well as voltage drops due to metal chip lines,

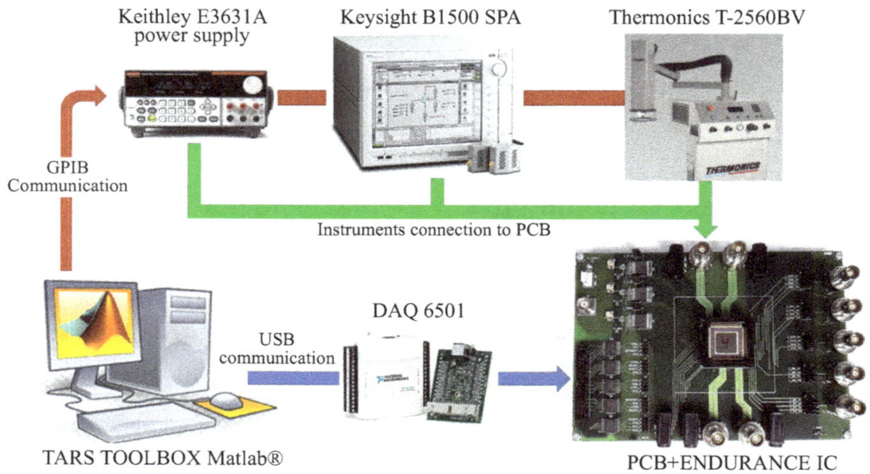

Figure 8.5 Complete system for the characterization of TZV and TDV in CMOS transistors. The devices are included in a purposely designed array chip

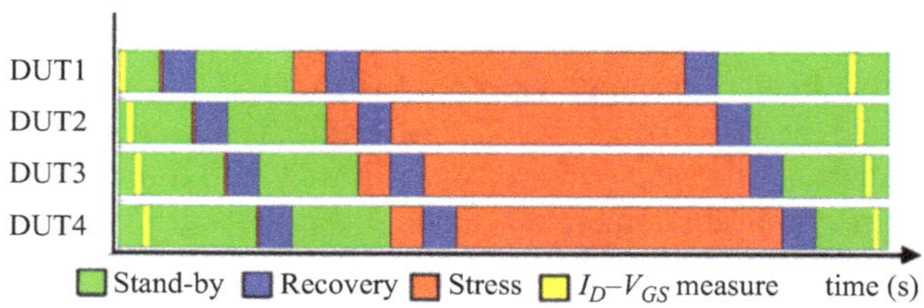

Figure 8.6 Illustration of a parallel 4-DUT 4-cycle SM scheme where the stress phases overlap. The introduction of stand-by periods avoids the overlap of the measurement phases

i.e. IR-drops, and other series resistances. A temperature system allows the control of the test temperature. The GPIB bus is used for the communication between the controller (a personal computer) and the instrumentation. The digital control of the chip is carried out with a USB digital acquisition system. A dedicated software tool automatically controls the digital inputs of the chip and all the instrumentation for automatic execution of the TDV tests [55].

The architecture of our chip together with a new stress parallelization algorithm allows a strong reduction of the testing time, so that aging tests of many devices can be carried out in reasonable testing times. The concept behind this algorithm is the overlap of the stress phases of multiple devices, while only one DUT can be measured at a time. Figure 8.6 shows an example of the parallelization of four DUTs under a typical aging test. Our algorithm avoids overlapping of the measurement phases (i.e., I_D–V_{GS} measurement and/or recovery phases), i.e., when the device is operated at low voltage and the device current is being registered by the system, but overlapping between stress phases in many transistors is allowed, leading to a strong reduction of the test time, because stress phase is typically the most time-consuming part of any aging test. This is done through the addition of suitable stand-by phases in the test (where the same constant voltage is applied to all the terminals of the DUT), which also grants equal recovery and stress times in all of the devices. To give some figures as example of the reduction of the testing time with this new parallelization capability, a test on a set of 784 DUTs that would last approximately 104 days if performed serially would be completed in 4 days with the new algorithm.

To show the potentiality of the system, some examples of the TZV and TDV characterization results obtained with our chip will be presented.

Before applying any TDV test, an initial characterization of all the DUTs in the ENDURANCE chip is performed, to evaluate the process variability (TZV) and obtain the threshold voltage of the fresh devices, which will be the reference value from which the aging-induced shifts will be evaluated. The I_D–V_{GS} curves of 784 80/60 nm nMOS are shown in Figure 8.7. The I_D–V_{GS} curves were measured by sequentially selecting each of the DUTs in the array. To measure the curves, V_{GS} was ramped from 0 to 1.2 V, $V_{DS} = 100$ mV and bulk and source terminals were grounded. As it can be seen in the figure, a high variability can be observed. Using these measured curves and the constant current parameter extraction procedure as a first approach [14], the threshold voltage (V_{th}) of each DUT can be obtained.

As a second example, full RTN characterization can be automatically done on the 3,136 DUTs of the ENDURANCE chip. Two different tests are applied to each DUT for RTN characterization: a ramped voltage test, to measure an initial I_D–V_{GS} curve (with $V_{DS} = 0.1$ V), followed by a constant voltage test to measure the I_D–t curve, during 100 s, with fixed $V_{GS} = 0.4$ V and $V_{DS} = 0.1$ V (absolute value). As an illustrative example, Figure 8.8 shows a set of seven current traces displaying

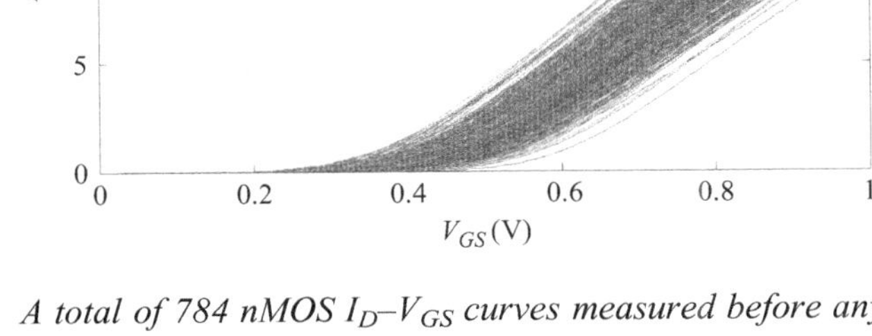

Figure 8.7 A total of 784 nMOS I_D–V_{GS} curves measured before any aging test, demonstrating process variability

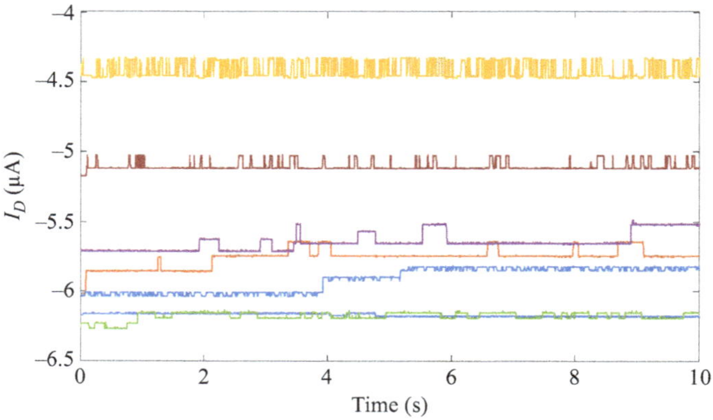

Figure 8.8 Experimental results of RTN tests on seven pMOS transistors of the ENDURANCE chip

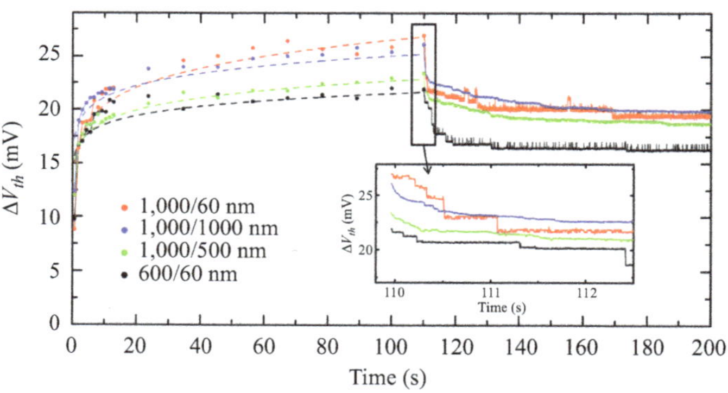

Figure 8.9 Threshold voltage shifts extracted from the currents measured during a BTI aging test on pMOS devices with different geometries. $V_{GS} = 2.5$ V and $V_{DS} = 0$ V were applied during the stress phase, at room temperature. The inset shows zooms of the recovery phase of the device

RTN effects. This figure demonstrates that the system is capable of capturing current fluctuations associated to charge trapping/detrapping in/from defects in the analyzed devices, with different emission and capture times. Multilevel signals, corresponding to more than one defect in the transistor, can also be observed [55].

Finally, Figure 8.9 shows the results of four BTI tests on four pMOS with different geometries. During the test, a stress with $V_{GS} = 2.5$ V and $V_{DS} = 0$ V has been applied for 100 s, which was periodically interrupted to measure the stress effects. A final recovery phase of 100 s, during which the current was measured at $V_{GS} = 0.5$ V and $V_{DS} = 0.1$ V was considered for each DUT. The corresponding V_{th}

shifts were extracted from the measured currents. As expected, during the stress phase V_{th} increases, whereas during the recovery phase V_{th} decreases. Note, however, that for the smallest devices, the evolution of the threshold voltage shift is no longer continuous, but discrete, as shown in the inset of Figure 8.9, because a small number of defects are present in the device.

8.3 Parameter extraction of CMOS aging compact models

During circuit operation, the variability effects related to the trapping/detrapping in/from oxide defects (as a consequence of TDV) could result in circuit malfunction, due to the shift of some transistor parameters, such as the threshold voltage (V_{th}) [10–21]. Thus, to implement reliability-aware circuits, it is critical for IC designers to take TDV effects into account [22–26]. To this end, appropriate TDV compact models, like the PDO model [29], described in Section 8.1, are essential, together with suitable model parameters extraction procedures. In the context of the PDO model, the stochastic impact of aging can be modeled through the analysis of the emission time (τ_e) and corresponding impact on the variation of the threshold voltage (η) of each defect in the transistor. With these parameters, and their dependences with the operation conditions, the corresponding voltage/temperature dependent $D(\tau_e, \tau_c)$ can be built. These parameters can be "visually" evaluated from the recovery traces of small-area devices (Figure 8.10(a)). But, since thousands of recovery traces have to be analyzed in massive aging tests, an automatic

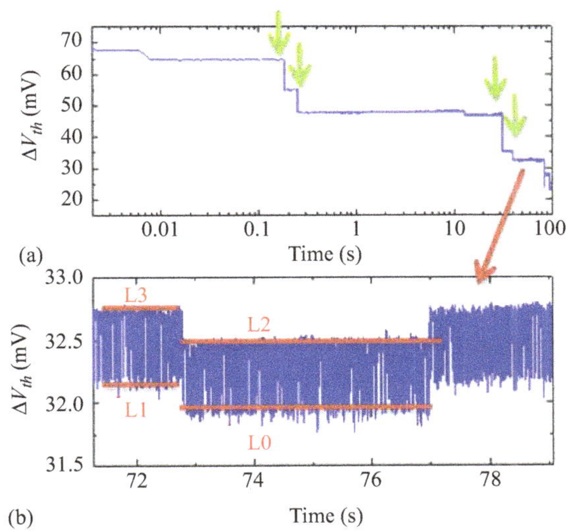

Figure 8.10 (a) Typical relaxation trace after a BTI stress. (b) Zoom in this trace shows fast defect transitions (RTN)

model parameter extraction procedure must carefully analyze recovery traces to obtain these parameters. However, during BTI testing, RTN transients could simultaneously appear together with other noise sources (Figure 8.10(b)), so that "artifacts" could mask or significantly increase the current increment determined during the extraction procedure. Then, for accurate parameter extraction, one must clearly distinguish the "slow" defects, responsible for aging-induced degradation, from the "fast" defects causing the RTN transient variations.

Here we introduce a smart method of defect parameters extraction, which identifies the BTI-related τ_e and η values from a large number of experimental transistor recovery traces, where fast defect capture/emissions (i.e., RTN) and background noise are present [42].

8.3.1 Description of the method

The defect parameter extraction method analyses the corresponding recovery traces individually. A five-step procedure is followed in order to remove the background noise and RTN (if any) from the trace, so that the τ_e and η parameters can be accurately evaluated. The procedure is described in detail as follows:

1. I_{DS} conversion to ΔV_{th}. For each device, the measured I_{DS}–t recovery traces are converted into the equivalent ΔV_{th} vs. time trace, using information obtained from the I_{DS}–V_{GS} curve measured on the device before the stress was applied, as those in Figure 8.7. For instance, Figure 8.10(a) shows the results of the conversion of a measured I_{DS}–t curve to ΔV_{th}–t, where several charge emissions (denoted with green arrows) can be clearly observed in the 100 s measurement window. In addition to these emissions, which contain the information of interest for parameter extraction, fast transitions associated to RTN are superposed. Figure 8.10(b) reveals fast capture/emission transitions, switching between four different ΔV_{th} levels (i.e., L0–L3) and mixed with background noise. A visual inspection reveals the joint contribution of two individual RTN signals, one switching between levels L0 and L1 and the other one switching with lower capture/emission times between L1 and L3 and between L0 and L2.
2. Identification of ΔV_{th} levels. The next step consists in the application of the weighted time lag plot (WTLP) method [41] to each recovery trace, in order to identify the number and magnitude of the ΔV_{th} levels. For instance, Figure 8.11 (b) shows the WTLP resulting from the analysis of the trace in Figure 8.11(a). By analyzing the diagonal of the WTLP, four groups of populated data regions, separated in the figure by red dashed arrows, can be distinguished. Each of the data groups corresponds to a different ΔV_{th} level present in the recovery trace. Transitions from one data group to the next are considered BTI slow emissions at a specific τ_e and the difference between two ΔV_{th} levels corresponds to the η of the discharged defect. Furthermore, when present, fast capture/emissions on top of each ΔV_{th} level can also be clearly distinguished as red-colored regions in the diagonal (i.e., ΔV_{th} levels L0–L9) with other less populated regions located outside of the diagonal (i.e., ΔV_{th} level transitions).

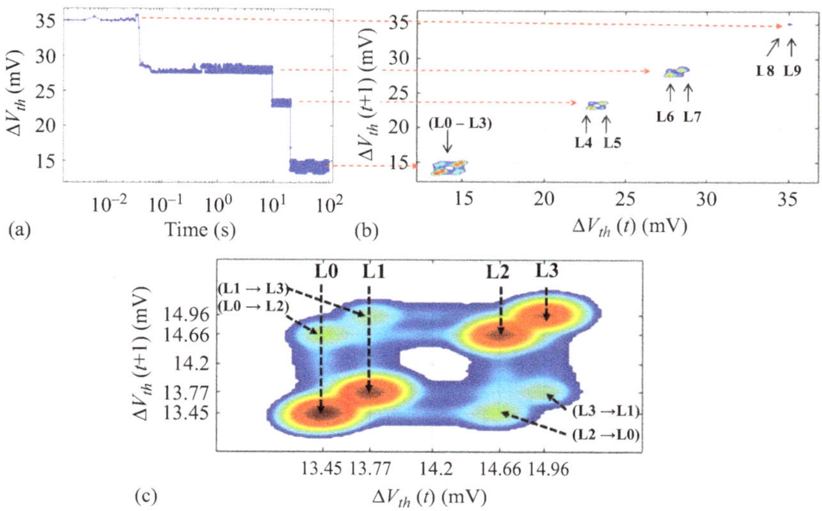

Figure 8.11 (a) Typical recovery trace measured during a BTI test and (b) the corresponding WTLP. (c) Zoom-in in (b) showing the combination of two fast defects

As an example, Figure 8.11(c) shows a zoom-in of the WTLP plotted in Figure 8.11(b), where the four red regions across the diagonal, which correspond to the L0, L1, L2 and L3 ΔV_{th} levels, can be clearly observed. The blue regions out of the diagonal in the WTLP indicate the transitions between levels (e.g., L1→L3). These levels result from the joint combination of the two individual RTN signals from the fast charges/discharges of two oxide defects. Figure 8.11 clearly demonstrates that the application of the WTLP to the ΔV_{th} recovery trace allows an accurate location of all ΔV_{th} levels and is able to distinguish between RTN and BTI contributions.

3. Background noise removal. Once the ΔV_{th} levels are identified in the recovery trace, the procedure assigns the closest ΔV_{th} level to each sample in the ΔV_{th} trace. This step quantizes the trace, removing the background noise and keeping only ΔV_{th} levels associated to captures/emissions of RTN and BTI contributions. For instance, Figure 8.12(a) displays three ΔV_{th} traces (in blue), showing high ΔV_{th} degradation due to the previous stress, and also the ΔV_{th} trace reconstruction (in red) with the background noise removed. During the recovery period, several ΔV_{th} levels can be observed because of defect emissions mixed with RTN phenomena, as shown in detail in the zoom-in plot in Figure 8.12(b).
4. Removal of RTN-related transients. In order to distinguish between RTN and BTI contributions, we define a square matrix, named transition matrix (TM), to store the transitions between different ΔV_{th} levels in the recovery trace. The dimension of the TM is the total number of ΔV_{th} levels in the trace, ten in this

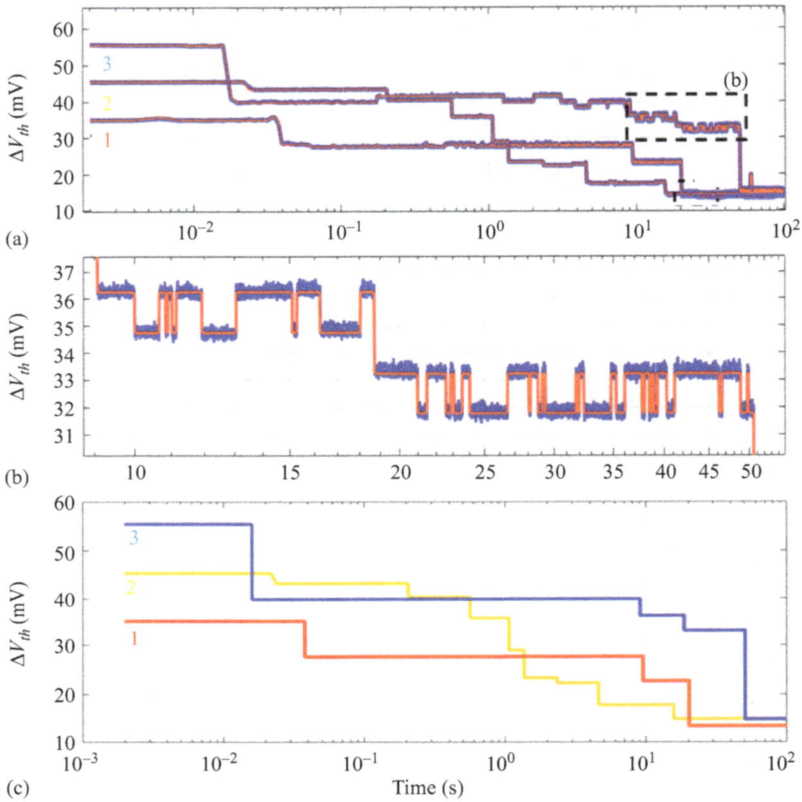

Figure 8.12 (a) Several examples of experimental relaxation traces (blue) and clean traces after the background noise removal (red). (b) Detail of the dashed square in (a). (c) Traces after the RTN removal; transitions here are only due to relaxation of BTI stress

particular case. Each ΔV_{th} level is denoted as LN, where N ranges from 0 to 9. The rows of the TM are defined as the initial ΔV_{th} levels (i.e., ith ΔV_{th} level), while the columns are defined as the final ones (i.e., $(i + 1)$th ΔV_{th} level). During the ΔV_{th} trace analysis, three different ΔV_{th} level transitions can be distinguished:

(i) Case i: There is not any change of ΔV_{th} level, i.e., two consecutive ΔV_{th} samples stay at level LN.
(ii) Case ii: Defect charging, i.e., ΔV_{th} shifts to a larger level. For instance, ΔV_{th} switches from level L1 to L3.
(iii) Case iii: Defect discharging. ΔV_{th} shifts to a smaller level. For instance, ΔV_{th} switches from level L3 to L1.

The TM is constructed by analyzing all the sample values in the ΔV_{th} recovery trace and counting the number of cases i, ii and iii. Case i will lay in the main diagonal of the TM, while case ii and case iii transitions will be located above and below the TM main diagonal, respectively. For instance, in Figure 8.11(a), ten ΔV_{th} levels have been detected by the WTLP method (Figure 8.11(b)). The resulting 10×10 TM is filled with all ΔV_{th} transitions identified during the ΔV_{th} recovery trace sweep as shown in Figure 8.13. In order to remove fast RTN transitions from the traces, the method analyzes the data in the TM, distinguishing the following two cases:

(a) Slow emission recognition: This type of defect emissions is characterized by a unique defect discharge to a lower ΔV_{th} level. In the TM, this appears as a "1" in the corresponding element below the TM main diagonal, and a "0" in the TM symmetric position. Analyzing the TM in Figure 8.13, three different emissions, without any further capture, can be found marked with circles: from initial to final ΔV_{th} levels L4–L3, L6–L4 and L8–L7.

(b) Transients recognition: The method identifies multiple and consecutive transitions between two distinct ΔV_{th} levels. For instance, transitions from/to levels L0 to L3 and from L6 to L7 that are attributed to RTN signals are marked with dash and solid square box in Figure 8.13. Consequently, the construction of TM can be used to distinguish between transitions provoked by RTN or by BTI. As examples, Figure 8.12(c) displays three resulting traces after the application of the methodology showing a total of three, eight and four slow emissions, without artifacts coming from fast defect transitions and background noise, respectively.

	L0	L1	L2	L3	L4	L5	L6	L7	L8	L9
L0	12658	127	297	0	0	0	0	0	0	0
L1	125	9650	144	213	0	0	0	0	0	0
L2	299	144	8825	107	0	0	0	0	0	0
L3	0	212	109	6391	0	0	0	0	0	0
L4	0	0	0	1	3255	88	0	0	0	0
L5	0	0	0	0	88	1958	0	0	0	0
L6	0	0	0	0	1	0	2702	109	0	0
L7	0	0	0	0	0	0	110	1809	0	0
L8	0	0	0	0	0	0	0	1	13	2
L9	0	0	0	0	0	0	0	0	2	1

Figure 8.13 TM corresponding to the trace of Figure 8.11(a). Dashed and solid boxes indicate transitions attributed to RTN, circles indicate the location of slow defect emissions

5. Defect parameters extraction. The last step consists in obtaining the τ_e and η parameters of the slow emissions. This is done by locating the elements below the TM main diagonal showing single discharge that have a symmetrical zero above the TM main diagonal. The η value of each detected defect is obtained by subtracting the final ΔV_{th} from the initial ΔV_{th}. To allow the evaluation of the τ_e associated to the defect, when the TM is constructed, also the times at which $(i, i + 1)$ transitions occurred are saved. When a slow discharge is encountered, the τ_e is assigned to the computed η value so that, the defect is characterized by the tuple $\{\tau_e, \eta\}$.

8.3.2 Application examples

As example of application of the method, the model parameters obtained from a particular BTI test applied on devices within our chip are described. BTI experiments were performed on a set of 248 pMOS, which were tested using a 6-cycle SM scheme. The duration of the stress phases is increased exponentially (i.e., 1, 10, 100, 1,000, 10,000 and 100,000 s), while the measurement (i.e., recovery) phase time always lasts 100 s. During the stress period, a gate-source voltage V_{GS} has been applied maintaining the drain-source voltage V_{DS} at 0 V. Four different V_{GS} voltages have been considered (1.2, 1.5, 2.0 and 2.5 V), while for measurements $V_{GS} \approx V_{th}$ and $V_{DS} = 100$ mV. The total test time required for a 6-cycle SM test on a single device takes 111,111 s + (6 · 100 s) = 111,711 s ≈ 31 h. Thus, for the four BTI tests involving 992 devices, the total test time, with conventional serial testing procedures, would be ≈3.5 years, a considerably large and prohibitive test time. However, thanks to the array-based IC chip ad hoc design, to significantly reduce the total testing time, the stress phases of each BTI test have been parallelized, without overlapping the measurement phases of any device. In this case, the testing time is reduced to only 64 h per BTI test, resulting in a 4-BTI test that only lasts 10 days, a significant test time reduction when compared to conventional serial testing.

From the automated analysis of the data, slow emissions have been identified. Figure 8.14 shows the histogram plots of the extracted τ_e from the defects found

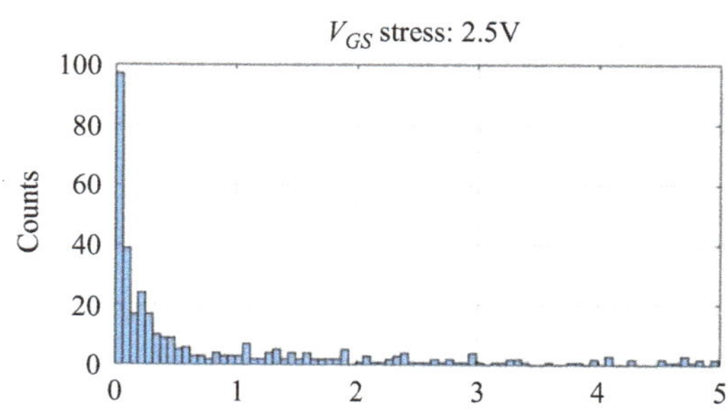

Figure 8.14 τ_e Histogram plots found during recovery analysis for the 80/60 nm transistor geometry after 10,000 s of pure BTI stress

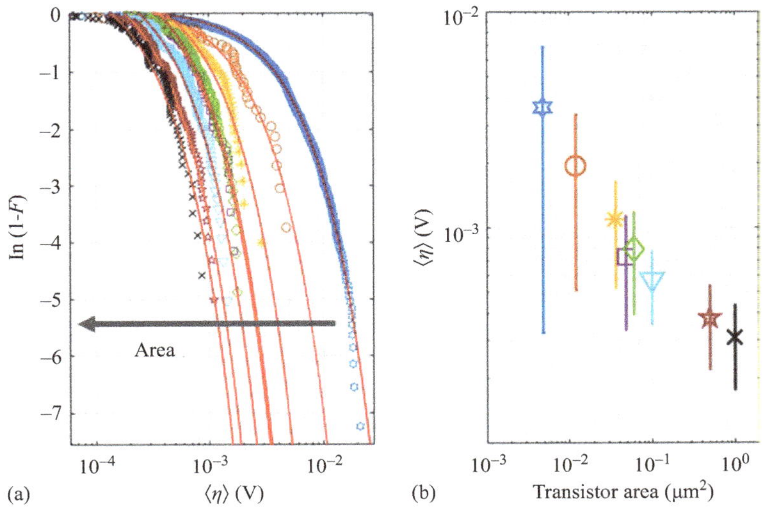

Figure 8.15 (a) Exponential distribution of the η values (symbols), with exponential fitting (lines) extracted from the four 6-cycle BTI tests for each geometry (device area increases from right to left). (b) Distribution of the ⟨η⟩ values of the slow emissions as a function of the transistor area

during the recovery traces, showing that most charges are emitted during the very first moments of the recovery phase.

Figure 8.15 shows the statistical results obtained from the analysis of all $\{\tau_e, \eta\}$ tuples extracted by using the method detailed in this work. Figure 8.15(a) shows that η is exponentially distributed. Figure 8.15(b) shows the dependence of the $\langle\eta\rangle$ value on the transistor area. The results demonstrate that $\langle\eta\rangle$ decreases with the transistor area. The results are in agreement with those in the literature [32], which supports the validity of the methodology. Note that if the RTN-related transients were not removed from the recovery traces before the slow emissions identification, a large number of "false" events (with equal ΔV_{th} value as the RTN amplitude) would have been taken into account during the $\langle\eta\rangle$ calculation. Therefore, the resulting $\langle\eta\rangle$ value, for all tested geometries, would have been close to the ΔV_{th} of the fastest RTN, masking the actual $\langle\eta\rangle$ of the BTI-related defects.

8.4 CASE: a reliability simulation tool for analog ICs

Many challenges have to be faced during the co-optimization of performance and reliability of ICs during the initial design phase. As shown in the previous sections, the characterization and modeling of TDV impact on ultrascaled devices are two of the aspects to be considered. The implementation of the stochastic TDV models into circuit simulation tools is a third one. In the case of digital circuits, the aging of many devices, which are operated under different conditions, will have to be evaluated, which could take unaffordable computation time, so that approximations of the models [33] or dedicated computing architectures will have to be considered [51]. For analog circuits, the problem of the number of devices could not be so relevant as in digital circuits, but accuracy of the model predictions will become a must. In any case, the number of available circuit reliability simulation tools is still very limited [45–47]. Focusing on analog circuits, available commercial tools that deal with TDV only offer limited simulation solutions. First, they are based on deterministic aging models, which do not account for the stochastic nature of the wear-out phenomena and, therefore, accuracy errors occur. Second, they cannot carry out a complete reliability analysis taking into account both TZV and TDV (each source of variability is considered uncorrelated, losing accuracy along the way). Third, the bidirectional link between biasing and stress [56,57] is taken into account inefficiently (in the trade-off CPU vs. accuracy), with linear or, at best, logarithmic scales defining the steps where the biasing conditions of each device in the circuit are updated.

This section presents CASE, a reliability simulator that tackles the above-mentioned deficiencies with a streamlined simulation flow, an underlying stochastic physics-based model (i.e., the PDO model), and a user-friendly interface. As it will be shown later, CASE can carry out reliability simulations in a complete manner, the results of the analysis can be plotted easily, and several pieces of information can be extracted: the statistics of the circuit performances, the statistics of the variability of each device, or even the impact that the variability of each device has on a particular circuit performance.

8.4.1 Simulator features

This section provides a brief description of the main features of the tool, as well as its advantages over other existing solutions. The main features of the implemented tool, which allow overcoming the limitations of commercial ones, are as follows:

- Handling of the stochastic nature of aging.
- Inclusion of the combined effect of TZV and TDV.
- Use of an adaptive technique to efficiently consider the variability of each device.

In contrast with the solutions derived from deterministic models, the use of a stochastic model, based on the underlying physics of the aging phenomena, provides more accurate information. In this regard, the tool uses the foundry-provided Monte Carlo models (for TZV) and the PDO model (for TDV) characterized as described in previous sections. In any case, the simulator allows integrating any model to evaluate the aging effects easily.

Reliability simulators commonly use transient analysis to determine the stress conditions of the different devices. With this information, the device degradation at the *target time* (the time at which the degradation is to be calculated and the aged performances simulated, e.g., 10 years) is estimated to obtain the aged circuit, as depicted in Figure 8.16. A final simulation is carried out to obtain the aged performances of the aged circuit. To reduce errors in the extrapolation of the conditions at the *target time*, the calculation process is carried out (feedback loop in Figure 8.16) with several intermediate time steps, instead of a single time jump to the *target time*.

One of the improvements provided by CASE is that TZV and TDV are considered in each of the actions taken within the dashed line in Figure 8.16, whereas other reliability simulators only take TDV or, at the very best, TDV with TZV at the end of the flow.

The feedback loop between the calculation of the degradation and the stress conditions is also improved. The stress conditions change during the circuit operation, precisely because of the device degradation over its lifetime. As it is well known, there is a strong feedback between device biasing, stress conditions and device aging, as illustrated in Figure 8.17. For this reason, a number of intermediate steps are used to update the stress conditions. This option is already included in several commercial simulators. However, these tools use fixed scales where the steps are uniformly distributed along a linear or logarithmic scale. It is important to emphasize that there are two issues to take into account: the number of steps and the distribution of the steps. A higher number of steps implies a higher computational cost, but, if few steps are carried out, the required accuracy in the calculation of the degradation might not be achieved.

Fixed scales (such as a linear or logarithmic distribution of steps) are used because the degradation of a single device, under unalterable stress conditions, follows a power law distribution. This assumption is only correct for a single

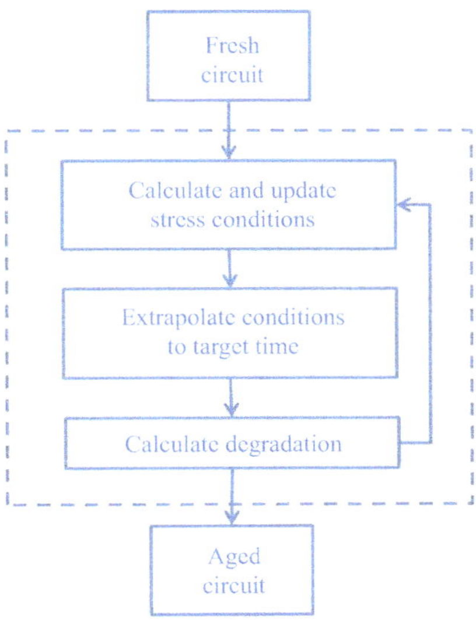

Figure 8.16 Simulation flow of the CASE simulator

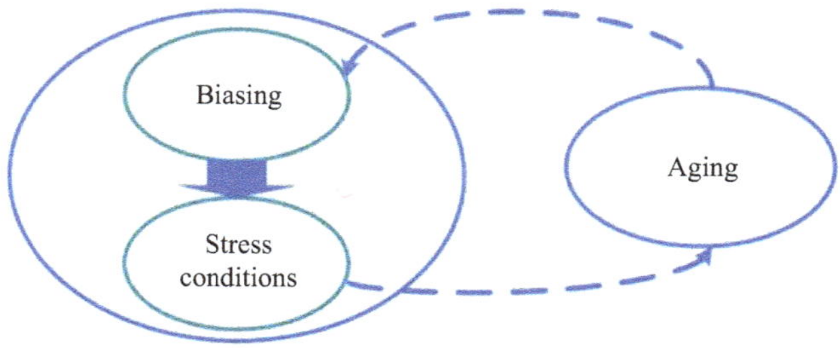

Figure 8.17 Bidirectional link between stress (i.e., biasing) and aging

device, operating alone, but in a circuit, where different devices coexist, the general behavior cannot be adjusted with these scales, merely for the fact that the degradation of each device can alter the biasing of any other device.

Therefore, in contrast to a predefined number of steps over a fixed scale, CASE provides an innovative algorithm [58] that adapts both the number and the distribution of steps with the progressive aging of the circuit. This algorithm can be used in two ways:

- Setting a maximum value for the degradation of the devices (e.g., V_{th}) at each intermediate step, up to which a new calculation (i.e., a new step where a transient analysis is performed to update the stress conditions) is not required. This approach allows reducing CPU time while maintaining the desired accuracy for the calculation of the circuit degradation.
- Setting a CPU budget (i.e., setting the number of times that a new transient analysis is carried out). In this case, the algorithm optimizes the distribution of the steps where stress conditions should be updated. In contrast to other adaptive solutions presented in the literature, the one used in CASE allows fixing the number of steps and, therefore, the CPU time can be bounded and controlled.

CASE links to commercial off-the-shelf electric simulators, like Spice or Spectre, to carry out all required analyses (transients, nominal or Monte Carlo simulations) to evaluate the circuit performance and get the stress conditions of each device.

8.4.2 TZV and TDV studied in a Miller operational amplifier

In this section, the operation of the reliability simulation tool will be illustrated through a case study: the TZV and TDV impact on a Miller operational amplifier (Figure 8.18).

The different configuration settings used for this case study are the following: (i) TZV (process and mismatch) and TDV are selected as variability sources. (ii) The *target time* is 1 year and the update of the stress conditions uses the adaptive scale with a fixed number of 20 steps. (iii) The temperature of the analysis is set to 25°C (iv) The number of samples for the Monte Carlo analysis is 1,000. The target

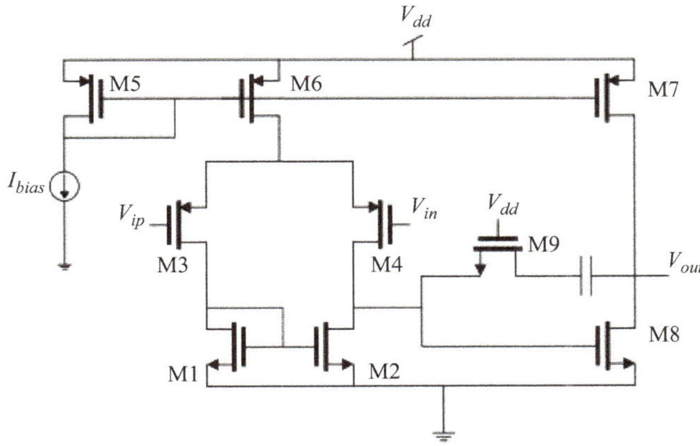

Figure 8.18 The Miller operational amplifier considered to illustrate the capabilities of the CASE simulator

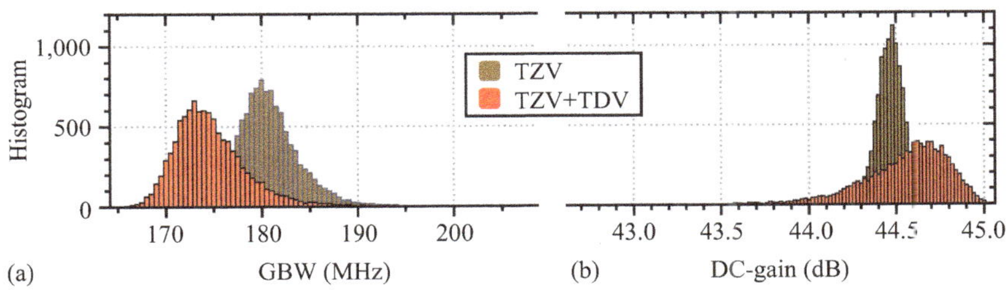

Figure 8.19 Histograms of the performances of the circuit: (a) GBW and (b) DC-gain

performances of the circuit are the DC-gain and the gain–bandwidth product (GBW) of the operational amplifier. The constraints established are a phase margin larger than 60° and that all transistors are in their correct region of operation (M1–M8 set in saturation and M9 operating in the linear region).

Once the simulation has been carried out, it is possible to plot the statistics of each performance, using the options provided by the GUI. In Figure 8.19, two examples of these plots for the case study circuit are shown. To analyze the wear-out of the devices, CASE uses the variation in the threshold voltage (as most of today's aging models). Nevertheless, CASE can handle any other varying parameter (e.g., mobility) as long as the underlying aging model takes that parameter into account. The variation in the threshold voltage of each transistor is shown in Figure 8.20. Each type of transistor can be represented individually, since the impact of aging can be very different for each nMOS and pMOS transistors. The variation of the threshold voltage is presented with the mean value (green for pMOS and red for nMOS), the standard deviation (blue error bars) and the maximum and minimum values (grey error bars).

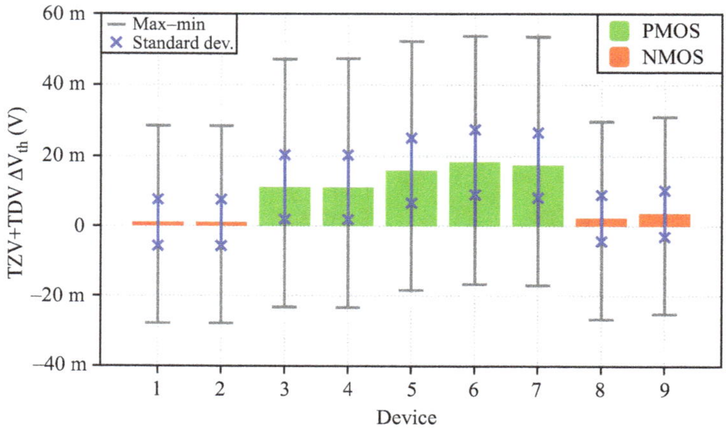

Figure 8.20 ΔV_{th} *of the transistors, including spatial and temporal variability*

Figure 8.21(a) shows the statistical distribution of the samples obtained in the Monte Carlo analysis of the performance of the operational amplifier. This plot represents the possible correlation between different performances. The *Yield* obtained in the reliability analysis is 0.993. The tool also provides the option of plotting a selected performance against the degradation of a selected device. An example of this is shown in Figure 8.21(b) and (c). As it can be seen, while variations in the threshold voltage of device M8 seem to be uncorrelated with variations in the DC-gain, lower values of its threshold voltage seem to follow a decrease of the GBW of the amplifier. In this way, this type of figure can provide information about the correlation between each performance and each device and let the user take appropriate measures (e.g., further analyzing the correlation by selecting device M8 as the only transistor to be aged in the *input file*).

8.5 Conclusions

TDV, due to stochastic processes as RTN, BTI and HCI, and TZV are identified as relevant sources of degradation of circuit functionality in scaled CMOS technological nodes. Reliability-aware circuit design requires accurate physics-based device compact models, together with an accurate computing time-efficient reliability simulation methodology. In this chapter, the challenging demands in these fields have been covered, focusing on analog circuits. As far as the compact modeling of the device TDV is concerned, the main features of the PDO model, a physics-based compact model, which is able to describe in a unified manner RTN transients and BTI/HCI aging has been described. In ultrascaled technologies, the extraction of the model parameters requires the measurement of a large number of devices, subjected to several stress conditions during long testing times. To allow such kind of massive statistical characterization, suitable tests structures and smart variability

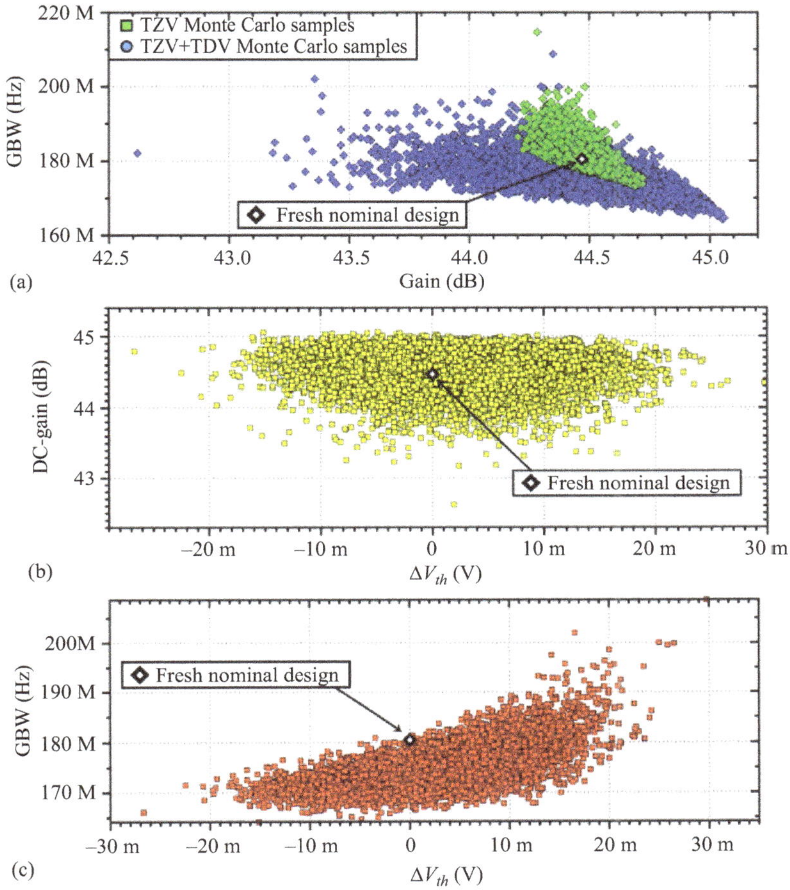

Figure 8.21 (a) Frequency vs. gain, (b) DC-gain vs. ΔV_{th} of device M8, (c) GBW vs. ΔV_{th} of device M8. A target time of 1 year has been considered during the simulation

characterization and analysis techniques are required. Thus, in Section 8.2, a full TZV and TDV characterization system for a large number of devices has been presented, whose key element is a purposely designed IC that contains thousands of devices in an array arrangement. The chip architecture together with a new stress parallelization scheme (which allows the simultaneous stress of a large number of devices while preserving identical degradation conditions) allow the reduction of the time needed for the statistical characterization of device TDV from years to days. In Section 8.3, a developed smart methodology to massively analyze the huge amount of measured TDV data has been presented. BTI-related events are automatically identified in the measured data by the use of the Weighted Time Lag method. Further analysis allows to distinguish between events associated to BTI degradation and RTN, separating the effects of two sources of TDV that inevitably are coupled. This methodology has been used to obtain the statistical distributions of parameters of the physics-based PDO compact model, which are the inputs to a circuit reliability simulation tool. Section 8.4 presents CASE, an example of reliability simulation tool, which has been optimized to investigate the effects of TZV and TDV on analog circuits functionality. CASE evaluates the degradation in each transistor of the circuit using the PDO model (and the suitable set of model parameters), accounting for the change of the stress conditions during the circuit operation. As an example, the effects of the device TDV in the performance of a Miller operational amplifier has been evaluated using CASE, identifying the most variability-sensitive transistors and showing that TDV has a strong impact on the circuit performance.

In summary, to accurately predict the reliability of analog circuits, strong efforts have to be carried out at different levels: at device level, on device physics (to develop accurate compact models for the device TDV) and characterization (to extract the model parameters that describe the device aging in a particular technology) and circuit level (to develop suitable simulation methodologies that evaluate the shifts of their performance and reliability linked to the device TDV). Many challenges rise on the way but several feasible solutions have been proposed, which can help designers to implement reliability-aware circuits, by accounting for the impact of device variability during their design.

Acknowledgments

This work has been supported in part by the TEC2013-45638-C3-R and TEC2016-75151-C3-R projects (funded by the Spanish AEI and ERDF). The work of J. Diaz-Fortuny and P. Saraza-Canflanca was supported by AEI under grants BES-2014-067855 and BES-2017-080160, respectively.

References

[1] A. Asenov, A. R. Brown, G. Roy, *et al.*, "Simulation of statistical variability in nano-CMOS transistors using drift-diffusion, Monte Carlo and non-equilibrium Green's function techniques," Journal of Computational Electronics, vol. 8, no. 3–4, pp. 349–373, 2009.

[2] A. Asenov, F. Adamu-Lema, X. Wang, and S. M. Amoroso, "Problems with the continuous doping TCAD simulations of decananometer CMOS transistors," IEEE Transactions on Electron Devices, vol. 61, no. 8, pp. 2745–2751, 2014.

[3] S. K. Saha, "Modeling process variability in scaled CMOS technology," IEEE Design & Test of Computers, vol. 27, no. 2, pp. 8–16, 2010.

[4] K. J. Kuhn, M. D. Giles, D. Becher, *et al.*, "Process technology variation," IEEE Transactions on Electron Devices, vol. 58, no. 8, pp. 2197–2208, 2011.

[5] K. Takeuchi, A. Nishida, and T. Hiramoto, "Random fluctuations in scaled MOS devices," IEEE International Conference on Simulation of Semiconductor Processes and Devices, 2009.

[6] C. M. Mezzomo, A. Bajolet, A. Cathignol, R. D. Frenza, and G. Ghibaudo, "Characterization and modeling of transistor variability in advanced CMOS technologies," IEEE Transactions on Electron Devices, vol. 58, no. 8, pp. 2235–2248, 2011.

[7] K. Bernstein, D. J. Frank, A. E. Gattiker, *et al.*, "High-performance CMOS variability in the 65-nm regime and beyond," IBM Journal of Research and Development, vol. 50, no. 4.5, pp. 433–449, 2006.

[8] J. K. Rangan, N. P. Aryan, J. Bargfrede, C. Funke, and H. Graeb, "Timing variability analysis of digital CMOS circuits," Reliability by Design 9. ITG/GMM/GI-Symposium, 2017.

[9] A. Ghosh, R. M. Rao, J. J. Kim, C. T. Chuang, and R. B. Brown, "Slew-rate monitoring circuit for on-chip process variation detection," IEEE Transactions on VLSI Systems, vol. 21, no. 9, pp. 1683–1692, 2013.

[10] C. Claeys, M. G. C. de Andrade, Z. Chai, *et al.*, "Random telegraph signal noise in advanced high performance and memory devices," Symposium on Microelectronics Technology and Devices, 2016.

[11] S. Dongaonkar, M. D. Giles, A. Kornfeld, B. Grossnickle, and J. Yoon, "Random telegraph noise (RTN) in 14 nm logic technology: High volume data extraction and analysis," IEEE Symposium on VLSI Technology, 2016.

[12] F. M. Puglisi, A. Padovani, L. Larcher, and P. Pavan, "Random telegraph noise: Measurement data analysis and interpretation," IEEE 24th International Symposium on the Physical and Failure Analysis of Integrated Circuits (IPFA) 2017, 2017.

[13] T. Grasser, B. Kaczer, W. Goes, *et al.*, "Recent advances in understanding the bias temperature instability," IEEE International Electron Devices Meeting, 2010.

[14] B. Kaczer, T. Grasser, J. Roussel, *et al.*, "Ubiquitous relaxation in BTI stressing-new evaluation and insights," IEEE International Reliability Physics Symposium, 2008.

[15] S. Mahapatra, and N. Parihar, "A review of NBTI mechanisms and models," Microelectronics Reliability, vol. 81, pp. 127–135, 2018.

[16] N. Ayala, J. Martin-Martinez, R. Rodriguez, M. Nafria, and X. Aymerich, "Unified characterization of RTN and BTI for circuit performance and variability simulation," IEEE European Solid-State Device Research Conference, 2012.

[17] T. Grasser, Bias Temperature Instability for Devices and Circuits, New York, NY: Springer, 2014.

[18] T. Aichinger, M. Nelhiebel, and T. Grasser, Hot Carrier Degradation in Semiconductor Devices, Switzerland AG: Springer Nature, 2014.

[19] C. Schlünder, J. Berthold, F. Proebster, A. Martin, W. Gustin, and H. Reisinger, "On the influence of BTI and HCI on parameter variability," IEEE International Reliability Physics Symposium, 2017.

[20] P. Magnone, F. Crupi, N. Wils, *et al.*, "Impact of hot carriers on nMOSFET variability in 45-and 65-nm CMOS technologies," IEEE Transactions on Electron Devices, vol. 58, no. 8, pp. 2347–2353, 2011.

[21] E. Amat, T. Kauerauf, R. Degraeve, *et al.*, "Channel hot-carrier degradation under static stress in short channel transistors with high-k/metal gate stacks," IEEE International Conference on Ultimate Integration of Silicon, pp. 103–106, 2008.

[22] M. Islam, T. Nakai, and H. Onodera, "Measurement of temperature effect on random telegraph noise induced delay fluctuation," IEEE International Conference on Microelectronic Test Structures, 2018.

[23] D. P. Ioannou, Y. Tan, R. Logan, *et al.*, "Hot carrier effects on the RF performance degradation of nanoscale LNA SOI nFETs," IEEE International Reliability Physics Symposium, 2018.

[24] A. Crespo-Yepes, E. Barajas, J. Martin-Martinez, *et al.*, "MOSFET degradation dependence on input signal power in a RF power amplifier," Microelectronic Engineering, vol. 178, pp. 289–292, 2017.

[25] A. Goda, C. Miccoli, and C. Monzio Compagnoni, "Time dependent threshold-voltage fluctuations in NAND flash memories: From basic physics to impact on array operation," IEEE International Electron Devices Meeting, 2015.

[26] F. Ahmed, and L. Milor, "Online measurement of degradation due to bias temperature instability in SRAMs," IEEE Transactions on VLSI Systems, vol. 24, no. 06, pp. 2184–2194, 2016.

[27] S. Mahapatra, A. Islam, S. Deora, *et al.*, "A critical re-evaluation of the usefulness of R-D framework in predicting NBTI stress and recovery," IEEE International Reliability Physics Symposium, 2011.

[28] T. Grasser, B. Kaczer, W. Goes, T. Aichinger, P. Hehenberger, and M. Nelhiebel, "A two-stage model for negative bias temperature instability," IEEE International Reliability Physics Symposium, 2009.

[29] J. Martin-Martinez, B. Kaczer, M. Toledano-Luque, *et al.*, "Probabilistic defect occupancy model for NBTI," International Reliability Physics Symposium, 2011.

[30] J. P. Campbell, P. M. Lenahan, A. T. Krishnan, and S. Krishnan, "NBTI: An atomic-scale defect perspective," IEEE International Reliability Physics Symposium, 2006.

[31] T. Grasser, "Stochastic charge trapping in oxides: From random telegraph noise to bias temperature instabilities," Microelectronics Reliability, vol. 52, no. 1, pp. 39–70, 2012.

[32] B. Kaczer, T. Grasser, Ph. J. Roussel, *et al.*, "Origin of NBTI variability in deeply scaled pFETs," IEEE International Reliability Physics Symposium, 2010.

[33] H. Amrouch, J. Martin-Martinez, V. M. van Santen, *et al.*, "Connecting the physical and application level towards grasping aging effects," IEEE International Reliability Physics Symposium, 2015.

[34] B. Kaczer, S. Mahato, V. Valduga de Almeida Camargo, *et al.*, "Atomistic approach to variability of bias-temperature instability in circuit simulations," IEEE International Reliability Physics Symposium, 2011.

[35] C. S. Chen, L. Li, Q. Lim, *et al.*, "A compact test structure for characterizing transistor variability beyond 3σ," IEEE Transactions on Semiconductor Manufacturing, vol. 28, no. 3, pp. 329–336, 2015.

[36] C. Schlünder, J. M. Berthold, M. Hoffmann, J. M. Weigmann, W. Gustin, and H. Reisinger, "A new smart device array structure for statistical investigations of BTI degradation and recovery," IEEE International Reliability Physics Symposium, 2011.

[37] T. Fischer, E. Amirante, K. Hofmann, M. Ostermayr, P. Huber, and D. Schmitt-Landsiedel, "A 65 nm test structure for the analysis of NBTI induced statistical variation in SRAM transistors," IEEE European Solid-State Device Research Conference, 2008.

[38] H. Awano, M. Hiromoto, and T. Sato, "Variability in device degradations: Statistical observation of NBTI for 3996 transistors," IEEE European Solid-State Device Research Conference, 2014.

[39] C. Schlunder, F. Proebster, J. Berthold, W. Gustin, and H. Reisinger, "Influence of MOSFET geometry on the statistical distribution of NBTI induced parameter degradation," IEEE International Integration Reliability Workshop. Final Report, vol. 2016, March 2016.

[40] M. Simicic, A. Subirats, P. Weckx, *et al.*, "Comparative experimental analysis of time-dependent variability using a transistor test array," IEEE International Reliability Physics Symposium, 2016.

[41] J. Martin-Martinez, J. Diaz, R. Rodriguez, M. Nafria, and X. Aymerich, "New weighted time lag method for the analysis of random telegraph signals," IEEE Electron Device Letters, vol. 35, no. 4, pp. 479–481, 2014.

[42] J. Diaz-Fortuny, J. Martin-Martinez, R. Rodriguez, *et al.*, "A noise and RTN-removal smart method for parameters extraction of CMOS aging compact models," Joint International EUROSOI Workshop and International Conference on Ultimate Integration on Silicon (EUROSOI-ULIS), 2018.

[43] T. Grasser, H. Reisinger, P.-J. Wagner, F. Schanovsky, W. Goes, and B. Kaczer, "The time dependent defect spectroscopy (TDDS) for the characterization of the bias temperature instability," IEEE International Reliability Physics Symposium, pp. 16–25, May 2010.

[44] T. Nagumo, K. Takeuchi, S. Yokogawa, K. Imai, and Y. Hayashi, "New analysis methods for comprehensive understanding of random telegraph noise," IEEE International Electron Devices Meeting, 2009.

[45] M. Nafria, R. Rodriguez, M. Porti, J. Martin-Martinez, M. Lanza, and X. Aymerich, "Time-dependent variability of high-k based MOS devices: Nanoscale characterization and inclusion in circuit simulators," IEEE International Electron Devices Meeting, 2011.

[46] C. Hu. "The Berkeley reliability simulator BERT: An IC reliability simulator," Microelectronics Journal, vol. 23, no. 2, pp. 97–102, 1992.

[47] P. Martin-Lloret, A. Toro-Frias, R. Castro-Lopez, *et al.*, "CASE: A reliability simulation tool for analog ICs," 14th International Conference on Synthesis, Modeling, Analysis and Simulation Methods and Applications to Circuit Design, 2017.

[48] S. F. Wan Muhamad Hattam, H. Hussin, F. Y. Soon, *et al.*, "Negative bias temperature instability characterization and lifetime evaluations of submicron pMOSFET," IEEE Symposium on Computer Applications & Industrial Electronics, 2017.

[49] T. Grasser, B. Kaczer, P. Hehenberger, *et al.*, "Simultaneous extraction of recoverable and permanent components contributing to bias-temperature instability," IEEE International Electron Devices Meeting, 2007.

[50] V. Velayudhan, J. Martin-Martinez, M. Porti, *et al.*, "Threshold voltage and on-current variability related to interface traps spatial distribution," IEEE European Solid-State Device Research Conference, 2015.

[51] V. M. van Santen, J. Diaz-Fortuny, H. Amrouch, *et al.*, "Weighted time lag plot defect parameter extraction and GPU-based BTI modeling for BTI variability," IEEE International Reliability Physics Symposium, 2018.

[52] S. E. Rauch, "Review and reexamination of reliability effects related to NBTI-induced statistical variations," IEEE Transactions on Device and Materials Reliability, vol. 7, no. 4, pp. 524–530, 2007.

[53] M. Moras, J. Martin-Martinez, R. Rodriguez, M. Nafria, X. Aymerich, and E. Simoen, "Negative bias temperature instabilities induced in devices with millisecond anneal for ultra-shallow junctions," Solid-State Electronics, vol. 101, pp. 131–136, 2014.

[54] J. Diaz-Fortuny, J. Martin-Martinez, R. Rodriguez, *et al.*, "A versatile CMOS transistor array IC for the statistical characterization of time-zero variability, RTN, BTI and HCI," IEEE Journal of Solid-State Circuits, vol. 54, no. 2, pp. 476–488, 2019.

[55] J. Diaz-Fortuny, J. Martin-Martinez, R. Rodriguez, *et al.*, "TARS: A toolbox for statistical reliability modeling of CMOS devices," International Conference on Synthesis, Modeling, Analysis and Simulation Methods and Applications to Circuit Design, 2017.

[56] E. Maricau, and G. Gielen, Analog IC Reliability in Nanometer CMOS, New York: Springer, 2013.

[57] E. Afacan, G. Berkol, G. Dundar, A. E. Pusane, and F. Baskaya, "A deterministic aging simulator and an analog circuit sizing tool robust to aging phenomena," International Conference on Synthesis Modeling Analysis and Simulation Methods and Applications to Circuit Design, 2015.

[58] A. Toro-Frias, P. Martin-Lloret, J. Martin-Martinez, *et al.*, "Reliability simulation for analog ICs: Goals, solutions and challenges," Integration the VLSI Journal, vol. 55, pp. 341–348, 2016.

Chapter 9

Modeling of pipeline ADC functionality and nonidealities

Enver Derun Karabeyoğlu[1] *and Tufan Coşkun Karalar*[1]

During the design of mixed signal circuits and systems, engineers often begin with high-level behavioral descriptions of the target system. The main appeal of this approach is to reveal theoretical limits and impact of nonidealities before starting transistor level design. This method becomes especially valuable during the design of high-performance and high-resolution circuits. Therefore, behavioral models of mixed signal systems, such as analog-to-digital converters (ADCs), have become a popular research topic. Using such models, the design parameters can be explored based on fast, high-level simulations. As will be described in this chapter, behavioral models of the circuit nonidealities can reveal many issues early in the design. Hence, nonidealities of the circuits should be described and modeled carefully. Sections 9.1 and 9.2 briefly explain the structure of the pipeline ADC and flash ADC, respectively. Section 9.3 describes an ideal model for a pipeline ADC. Next, in Section 9.4, circuit nonidealities are analyzed and modeled in MATLAB® and Simulink™ environments. Finally, the model of whole ADC with nonideality models are simulated and results are presented in Section 9.5.

9.1 Pipeline ADC

Many contemporary applications, such as 5G mobile systems, broadband communications, digital imaging, etc., require high-resolution data converters that can reach 10–12-bit data resolution while sampling data at in excess of 1 Gsps [1]. ADCs in pipeline topology are commonly employed to attain such performance levels. By simply adding extra stages, resolution can be increased at the expense of larger chip area.

Each pipeline stage consists of a sub-ADC and a multiplying digital-to-analog converter (MDAC). Sub-ADC compares the sampled signal against reference levels. MDAC is composed of a sub-DAC and a switched capacitor circuit that

[1]Electronics and Communications Engineering Department, Istanbul Technical University, Istanbul, Turkey

performs subtraction and amplification. The low-resolution digital output is converted back to analog value via the sub-DAC. This reconstructed signal is subtracted from the held signal and amplified by 2^n where n is the number of output bits in the stage. Output of the MDAC is also known as the residue. The residue is passed to the next pipeline stage input. The residue from the next stage is fed to the subsequent stage and this arrangement repeats itself until the last stage as seen in Figure 9.1. To compute the overall ADC output, digital output codes from all stages are combined after proper timing alignment and weightings. The throughput of the pipeline ADC is at the full sampling rate while the latency is m clock cycle long where m is the number of pipeline stages.

To illustrate the basic principle of pipeline ADCs, an input at 60% of full-scale voltage (V_{FS}) is applied to an ideal 6-bit pipelined ADC with 1-bit stages. The comparator reference level is half of the full scale voltage. If the sampled signal is

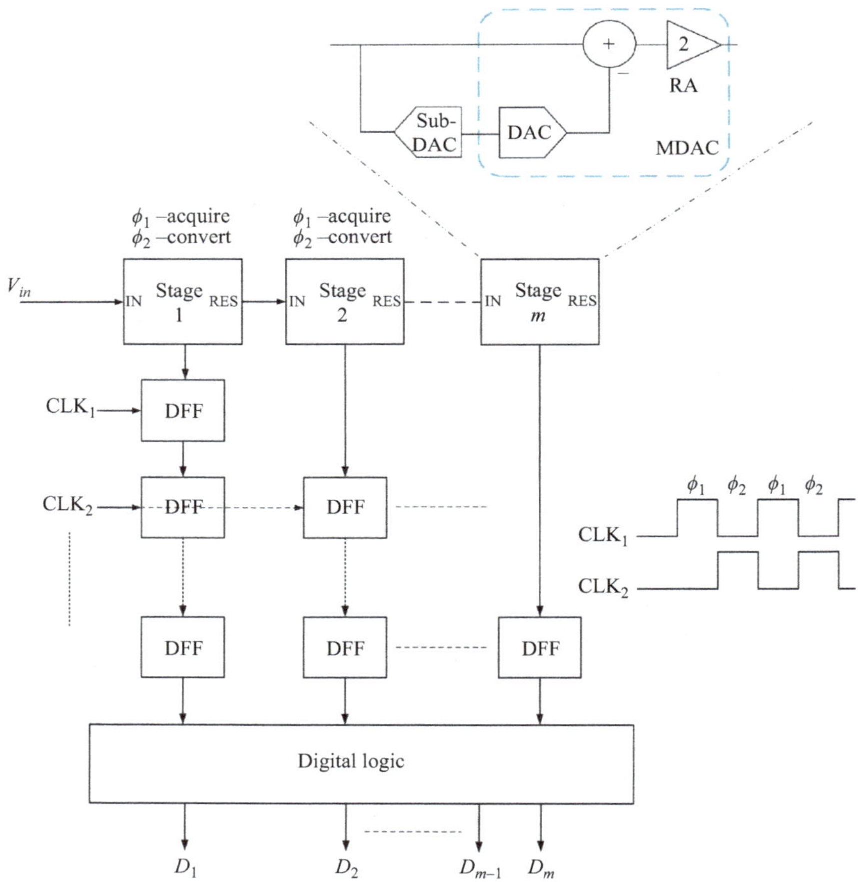

Figure 9.1 Generic pipeline ADC

greater than the comparator reference level, the digital output becomes high, otherwise low. Residue voltage of a 1-bit pipeline stage can be expressed as:

$$V_{res} = 2V_{in} - DV_{FS} \tag{9.1}$$

where V_{in} and D are the input voltage and the digital output of the stage, respectively. For a 1-bit stage, D can be 0 or 1. According to (9.1), residue voltage and digital output of each stage are demonstrated in Figure 9.2.

One problem with 1-bit pipeline stages is its vulnerability to comparator offset and residue amplifier gain errors. Here, missing codes or missing decision may appear as a result of small offset voltage or small gain error in the stage. Consider the transfer function when the comparator offset or residue amplifier offset are explicitly added in Figure 9.3, the residue voltage exceeds full-scale voltage near the center of the input range. Assuming that the subsequent stages are ideal, the output codes of the following stages remain all high resulting in missed decision levels.

This problem can be eliminated by using 1.5-bit pipeline stages. Instead of one comparator, two comparators are used where reference levels are set to $3V_{FS}/8$ and $5V_{FS}/8$. As a result, comparator offsets up to $V_{FS}/8$ could be tolerated without

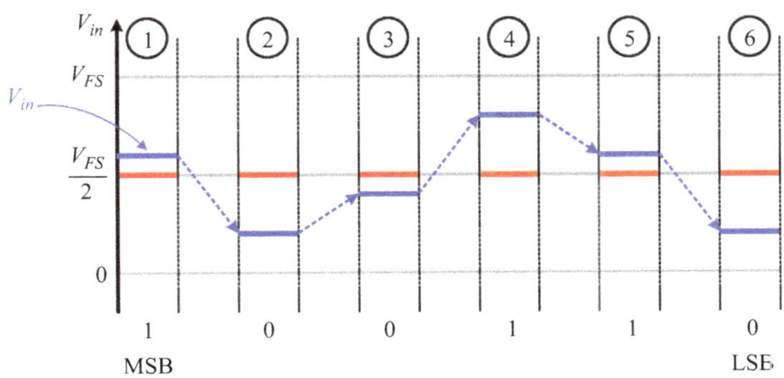

Figure 9.2 Residue voltages and final digital code of an ideal 6-bit pipelined ADC [2]

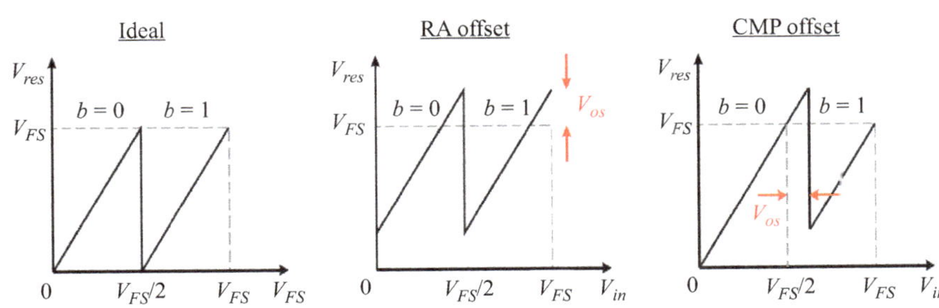

Figure 9.3 One-bit pipeline stage transfer function with offset of comparator and residue amplifier [2]

residue saturation. The digital outputs determine DAC output that can assume V_{FS}, $V_{FS}/2$ and 0. Residue voltage of a 1.5-bit stage can be ideally written as:

$$V_{res} = \begin{cases} 2V_{in} - V_{FS}, & \text{if } D_0 = 1 \text{ and } D_1 = 1 \\ 2V_{in} - \dfrac{V_{FS}}{2}, & \text{if } D_0 = 1 \text{ and } D_1 = 0\,. \\ 2V_{in}, & \text{if } D_0 = 0 \text{ and } D_1 = 0 \end{cases} \tag{9.2}$$

It should be noted that in the case of a fully differential implementation, the input signal can vary between $-V_{FS}$ and $+V_{FS}$. Then, the comparator thresholds should be set to $\pm V_{FS}/4$ and the DAC values can assume $-V_{FS}/2$, 0 and $+V_{FS}/2$.

9.2 Flash ADC

The flash architecture is the fastest ADC architecture. It is implemented by comparing input signals with a set of reference voltages. These reference levels are generated by using a resistor ladder. The comparator output is thermometer encoded and a decoding operation is employed to obtain the binary outputs. An n-bit flash ADC requires an array of $2^n - 1$ comparators, $2^n + 1$ resistors and decoding logic. However, the structure becomes very complex and power consuming for increasing resolutions. Also, nonidealities such as comparator offset and resistor mismatch prevent achieving high resolutions. Therefore, low-resolution flash ADCs are generally used as sub-ADC in addition to digitizing the last pipeline stage residue.

9.3 Behavioral model of pipeline ADCs

For the discussion regarding the behavioral model of a pipeline stage, we need to make some assumptions to define the system clearly. Here we will be considering an implementation that targets an 11-bit pipeline ADC that consists of eight cascaded pipeline stages and a last stage with a 3-bit flash ADC. Each pipeline stage has 1.5-bit configuration.

9.3.1 A 1.5-bit sub-ADC model

Since sub-ADC is implemented using two comparators whose reference levels are $-V_{FS}/4$ and $+V_{FS}/4$, its behavioral model can be created by using compare blocks followed by sample and hold blocks as illustrated in Figure 9.4. Each sub-ADC outputs 2-bit signals that are thermometer encoded. However, after the sub-ADC, an encoder can convert these values into any other encoding that is necessary.

9.3.2 Multiplier DAC

MDAC is the combination of a sub-DAC, a subtractor and an amplifier. The sub-DAC converts digital bits into proper analog voltage. Current steering DACs are

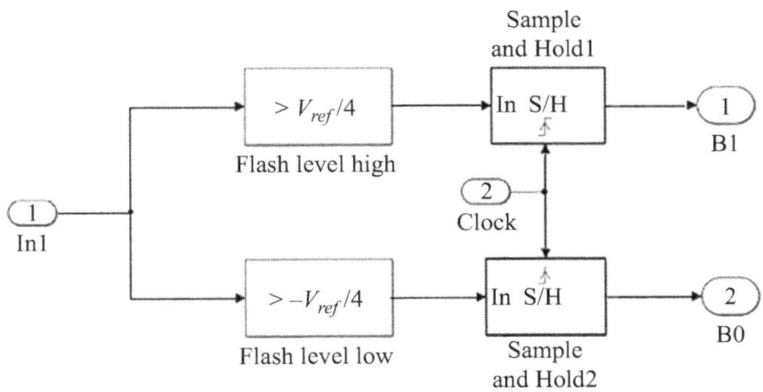

Figure 9.4 Model of ideal sub-ADC

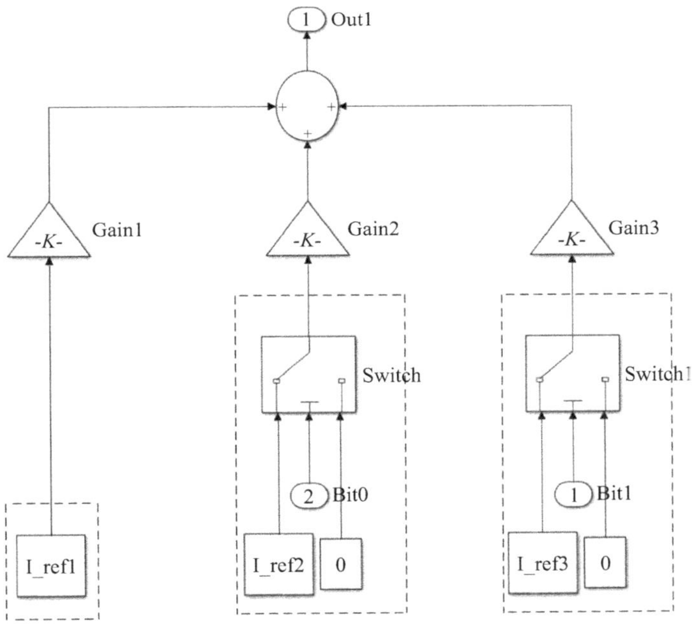

Figure 9.5 Model of ideal sub-DAC

generally preferred for high-speed and high-resolution applications. Each data bit controls a current source. Figure 9.5 displays the model of the 1.5-bit DAC. Models of the current sources are encircled with dashed lines. Depending on the digital bits at the input of the switches, the current value is multiplexed with 0 before being multiplied by gain block to obtain desired DAC output voltage.

Figure 9.6 shows a flip-around switched capacitor circuit that is capable of subtracting the DAC output voltage from the input signal and amplifying the

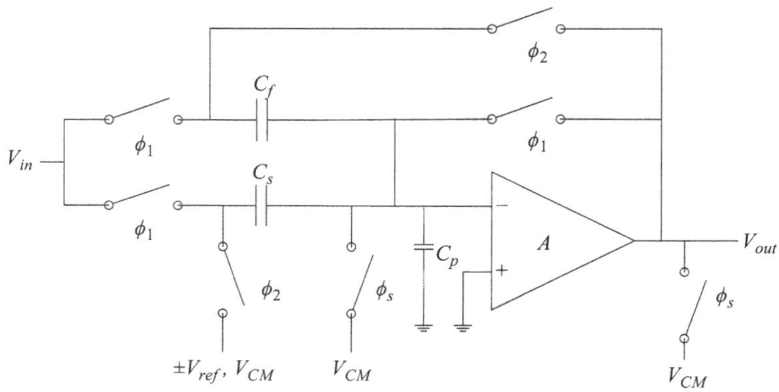

Figure 9.6 Flip-over switched capacitor circuit

difference to obtain the residue voltage of the pipeline stage. During its operation, in sampling phase, switches controlled by ϕ_1 are closed. Both capacitors are charged by the input signal. In the amplifying phase, when switches controlled by ϕ_2 are closed, residue voltage and DAC output voltage charge the feedback and the sampling capacitors, respectively. Hence, the output voltage of the residue amplifier can be expressed as:

$$V_{out} = V_{in}\frac{C_f + C_s}{C_f} - V_{DAC}\frac{C_s}{C_f} \tag{9.3}$$

where V_{DAC}, C_s and C_f are DAC output voltage, sampling and feedback capacitors, respectively.

In pipeline ADCs, MDAC of a stage transitions into amplifying phase as soon as its residue is sampled by the latter stage. So, in the worst case, during transition from hold to track mode, the charge due to the previous sample should be discharged before the current input can be sampled. This increased output swing is problematic especially at high-speed operation, as it can induce inter-sample interference (ISI) causing performance loss. To reduce the voltage swing during the track phase, a short reset phase ϕ_S can be added to discharge the previous cycle's charge on the capacitor [3,4]. And the track phase can start from the reset value. As a consequence, effects of previous samples are reduced and ISI performance loss can be mitigated.

9.3.3 A 3-bit flash ADC

After the cascaded pipeline stages, a 3-bit flash ADC, which consists of seven comparators and an encoder, is placed as shown in Figure 9.7. Comparator threshold levels increase with $+V_{FS}/4$ from bottom to top. Finally, the outputs of the compare blocks are converted to 3-bit binary code by an encoder.

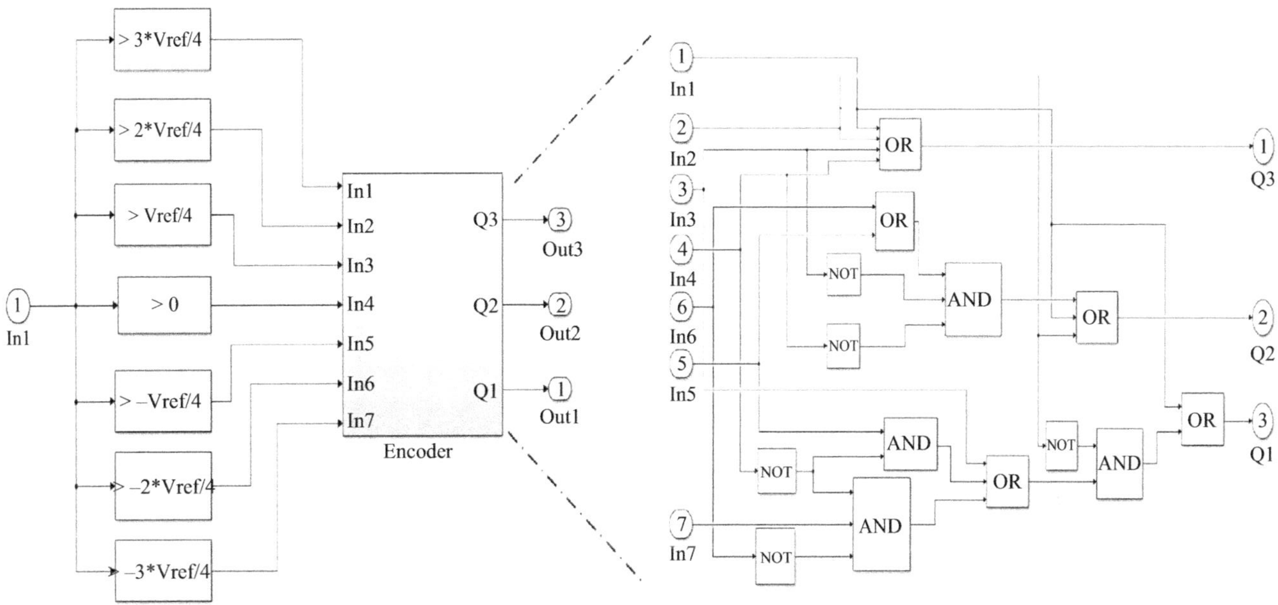

Figure 9.7 Model of ideal 3-bit flash ADC

9.4 Sources of nonidealities in pipeline ADCs

9.4.1 Op-amp nonidealities

Op-amps are often used and are important blocks for switched capacitor implementation in pipeline ADCs. Op-amps realized in deep-submicron complementary metal oxide semiconductor (CMOS) technology suffer from significant nonidealities. The impact of these should be analyzed precisely to get an accurate model.

9.4.1.1 Finite DC gain

Ideally, the DC gain of an op-amp is infinite; however, the actual open loop gain A is limited by circuit parameters. Including the finite DC gain of the op-amps, total charge stored on both capacitors during sampling phase becomes:

$$Q_s = V_{in}(C_f + C_s) \tag{9.4}$$

By taking C_p input capacitance of the op-amp into account, the total charge in amplifying phase is:

$$Q_a = (V_x - V_{DAC})C_s + (V_x - V_{out})C_f + V_x\, C_p \tag{9.5}$$

Based on conservation of total charge law:

$$V_{out} = V_{in}\left(\frac{C_s + C_f}{C_f}\right) + V_x\left(\frac{C_s + C_f + C_p}{C_f}\right) - V_{DAC}\frac{C_s}{C_f} \tag{9.6}$$

So, feedback factor β, which defines how much of the output voltage is fed back to the input, can be written as:

$$\beta = \frac{C_f}{C_s + C_f + C_p} \tag{9.7}$$

Then, considering the finite DC gain of the op-amp, the output can be written as:

$$V_{out} = G\, V_x \frac{1}{1 - (1/A \cdot \beta)} \tag{9.8}$$

where G is the ideal gain of the switched-capacitor circuit.

Furthermore, this output characteristic of the residue amplifier can also be approximated as a Taylor series expansion of the input voltage assuming a fifth-order polynomial [5,6]:

$$V_{out} = G\left(V_x + \sum_{n=1}^{5} V_x^n a_n\right) \tag{9.9}$$

where a_n denotes linear and harmonic distortion terms of the amplifier.

9.4.1.2 Bandwidth and slew rate

Ideally, an amplifier has infinite bandwidth. However, in reality, its bandwidth is limited by the gain-bandwidth product. The gain starts to reduce and beyond the point where the op-amp loop gain falls to 0 dB, the amplifier is not useful.

Slew-rate can be considered as the rate of change in the output voltage in the presence of large input signals. In large signal conditions, op-amp provides current to charge the capacitors of the circuit and the charging rate is called slew rate. If the current is insufficient or the op-amp does not have enough time to charge the capacitors, the output will not be able to reach the required level. At high speeds, this condition can be exasperated in the presence of ISI-defined earlier.

These two effects cause nonlinearity at the output and must be analyzed for two different conditions [7,8].

1. The slope of the curve is lower than the slew rate. In such a case, there is no slew-rate limitation. The output response is linear.

$$V_{out} = GV_{in}\left(1 - e^{-(t/\tau)}\right) \tag{9.10}$$

   where $\tau = 1/\beta 2\pi f_u$ is the time constant of the circuit where f_u is the unity gain frequency.
2. The slope of the curve is higher than the slew rate. In slewing mode, output of the op-amp behaves as a constant current source charging a capacitor, hence yielding ramp voltage during slewing time. Later on, as the charging current reduces below the slew current, it resumes linear behavior where the feedback takes control of the output voltage.

$$V_{out} = SR\, t \quad t < t_{slew} \tag{9.11}$$

$$V_{out} = V_{out}(t_{slew}) + (GV_{in} - SR\,\tau)\left(1 - e^{-[(t-t_{slew})/\tau]}\right) \quad t > t_{slew} \tag{9.12}$$

   where *SR* refers to slew rate and t_{slew} is the time in slewing mode. Imposing the condition for the continuity of the derivatives of (9.11) and (9.12) in t_{slew}, we obtain:

$$t_{slew} = \frac{V_{in}\,G}{SR} - \tau \tag{9.13}$$

9.4.1.3 Input offset of an operational amplifier

Ideally, the output voltage of an op-amp is zero when two input terminals are shorted. However, mismatch of the input transistors during fabrication of the silicon die and stresses placed on the die during the packaging process cause unequal current flowing through input transistors, resulting in nonzero output [9]. So, the input offset voltage is defined as required input voltages that zeroes the output voltage. The transfer function of a stage with offset voltage can be expressed as:

$$V_{out} = 2(V_{in} + V_{off}) - D\,V_{ref} \tag{9.14}$$

where V_{off} is the op-amp input offset which is modeled as a constant value addition.

9.4.1.4 Noise sources of an operational amplifier

Thermal noise is one of the dominant noise sources in op-amps. It occurs in the passive and active elements due to random vibration of the charge carriers. AC model of the switched-capacitor circuit in MDAC is shown in Figure 9.8. Here, the amplifier noise is modeled as an equivalent noise current source at the amplifier output. With this model, the transfer function can be expressed as:

$$H(s) = \frac{V_{out}}{i_n} = \frac{r_O}{(1 + g_m \cdot r_O \cdot \beta)(1 + (s \cdot C_o \cdot r_O/(1 + g_m \cdot r_O \cdot \beta)))} \tag{9.15}$$

where C_{load} is the load capacitor and $C_o = C_{load} + \beta(C_s + C_p)$. Input-referred noise power can be written as:

$$\overline{V_{in}^2} = \frac{\overline{V_{out}^2}}{G^2} = \frac{\int_0^\infty |H(s)|^2\,\overline{i_n^2}}{G^2} = \lambda\,kT\frac{1}{\beta}\frac{1}{C_o}\frac{C_f}{C_s + C_f} \tag{9.16}$$

where noise current source $i_n = \lambda\,4kT\,g_m\,\Delta f$, λ is a coefficient equal to 2/3 for long channel transistors, Δf is a small bandwidth at frequency f, G is the gain of the MDAC, k is the Boltzmann constant and T is the temperature in terms of Kelvin.

Another important type is flicker noise also called as $1/f$ noise. It is present in all active devices. It is highly related to random trapping and releasing of the electrons or holes at the interface and doping profile. It causes the noise power spectral density to increase 3 dB/octave at low frequencies. The flicker noise power can be calculated as:

$$\overline{V_n^2} = \frac{K_f}{C_{ox}\,W\,L}\frac{1}{f} \tag{9.17}$$

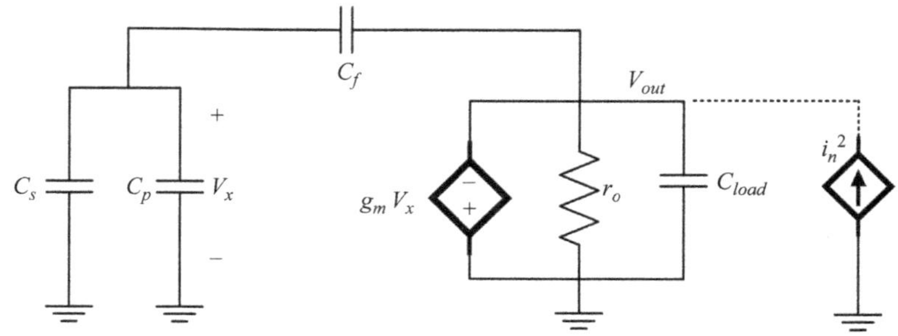

Figure 9.8 AC model for noise calculation

where K_f is the flicker noise coefficient, C_{ox} is the oxide capacitance and W and L are the width and length of the metal oxide semiconductor field effect transistor (MOSFET) transistor. Figure 9.9 illustrates the model to simulate the thermal noise and flicker noise of the op-amp.

9.4.2 Switch nonidealities

Switches are one of the major components in the switched-capacitor circuits. Ideally, they have zero resistance when they are on, on the contrary, they have infinite resistance when they are off. However, NMOS and PMOS transistors are used as switches in CMOS technology. Their parasitics can adversely affect the performance of the pipeline ADC. They can be categorized as switch thermal noise, charge injection, clock feedthrough and nonlinear switch resistance [8].

9.4.2.1 Switch thermal noise

Thermal noise associated with the sampling switches and the op-amps appears as white noise. When the input is sampled, thermal noise is also sampled on the input capacitor. Figure 9.10 provides the equivalent circuit for noise estimation. Its spectral power and the transfer function of the circuit can be calculated as:

$$\overline{V_{n,Rs}^2} = 4kT\,Rs\,\Delta f \tag{9.18}$$

$$A(jw) = \frac{1}{1 + jwRsC} \tag{9.19}$$

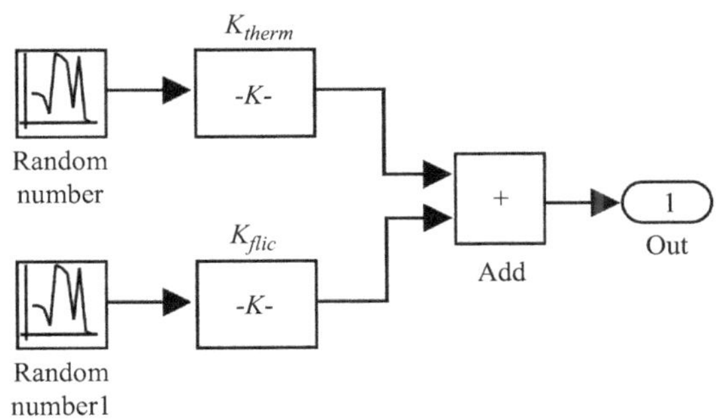

Figure 9.9 Model of thermal and flicker noise

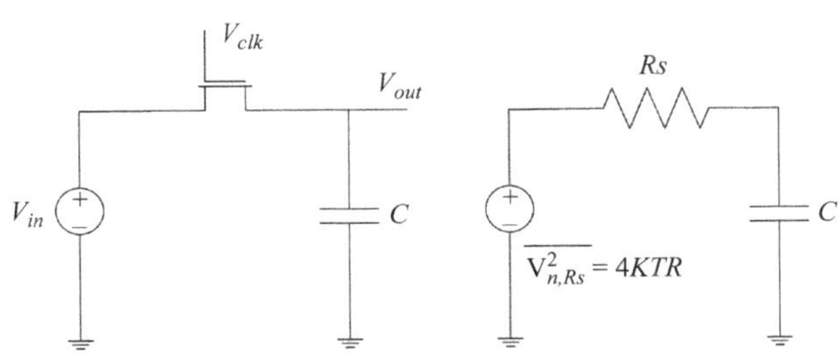

Figure 9.10 MOS sampling circuit and its equivalent

When the switch is off, total noise power stored on the sampling capacitor is:

$$\overline{V_n^2} = \frac{1}{2\pi}\int_0^b \overline{V_{n,Rs}^2}\, A(jw)^2 df = \frac{kT}{C} \tag{9.20}$$

where R_s is the on-resistance of the MOS switch and C is the sampling capacitor value.

The total noise power of MDAC is given by:

$$\overline{\sigma_{in,MDAC}^2} = \frac{kT(C_{s,mdac} + C_{f,mdac} + C_{p,mdac})}{(C_{s,mdac} + C_{f,mdac})^2} \tag{9.21}$$

Since thermal noise is random, a random number generator is used and it is multiplied by (9.21) to realize the switch thermal noise as in the Simulink model given in Figure 9.11.

9.4.2.2 Charge injection

Charge injection in MOS switches is caused by forcing mobile channel charge to escape when the switch turns off. This charge flows through each terminal depending on the ratio of the terminal's impedance, the switch parameters and the slope of the clock. This nonideality manifests itself as a voltage step at the output. The error for an NMOS switch can be expressed as:

$$\Delta V_{inj} = -\frac{Q_{ch}}{C_{load}} = -\frac{\alpha WLC_{ox}(V_{GS} - V_{th})}{C_{load}} \tag{9.22}$$

where W, L, C_{ox}, V_{GS} and V_{th} are the channel width, channel length, gate-to-oxide capacitance, gate-to-source voltage and threshold voltage of the NMOS, respectively. The factor α is the fraction of how much charge leaks into the sampling capacitor rather than the input source.

Model of charge injection effect is developed in Simulink as illustrated in Figure 9.12. It performs (9.22) by using a random variable block to define uncertainty of charge quantity.

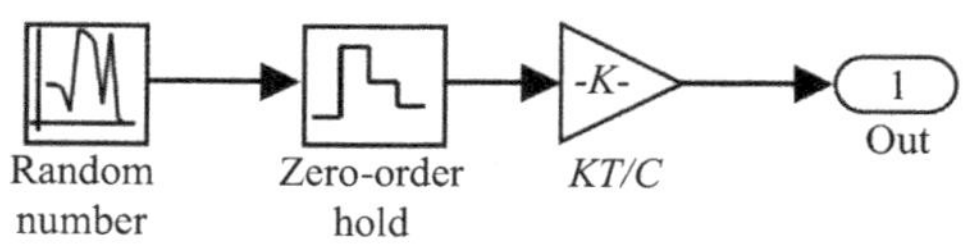

Figure 9.11 Model of switch thermal noise

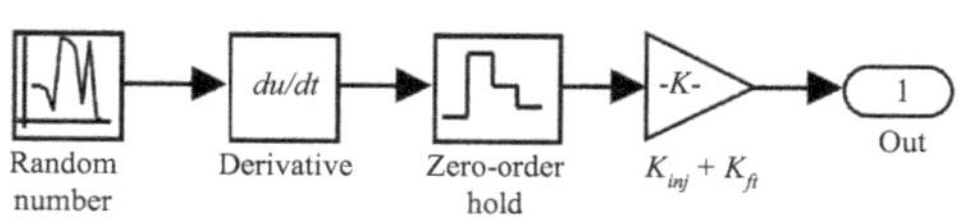

Figure 9.12 Model of the charge injection and clock feedthrough effects

9.4.2.3 Clock feedthrough

Besides drain current of the switch, there is an extra current flowing through the two-gate capacitances, C_{gd} and C_{gs}, when the switch turns off. In other words, clock transition is coupled to sampling capacitor through parasitic capacitances of MOS switches. This phenomenon is called clock feedthrough. Assuming that the overlap capacitances are fixed, the effect on the output voltage can be calculated as [8,10]:

$$\Delta V = -\frac{Q_{out}}{C_H + C_{ov}} = -\frac{V_{clk} \cdot C_{ov}}{C_H + C_{ov}} \tag{9.23}$$

where Q_{out} is the error charge due to the clock feedthrough, V_{clk}, C_{ov} and C_H represent the clock signal voltage, the overlap capacitor and the hold capacitor, respectively. As is seen in (9.23), clock feedthrough is independent of input signal. It can be modeled as shown in Figure 9.12.

9.4.2.4 Nonlinear switch resistance

Unlike ideal conditions, switches have nonzero resistance that is a nonlinear function of the input voltage. Nonlinear on-resistance causes harmonic distortions especially for high-frequency and high-swing signals. It can be calculated as:

$$R_{on} = \frac{1}{(1/2)\mu C_{ox}(W/L)(V_{in,s} - V_{th} - V_{in})} \tag{9.24}$$

where μ is the NMOS mobility, C_{ox} is the gate oxide capacitance per unit area, W, L are transistor dimensions, and $V_{in,s}$ is the sampling voltage that can be calculated as:

$$V_{in,s} = V_{in}\left(1 - e^{t_s/(R_{on}(j)C)}\right) \tag{9.25}$$

where j is the number of time intervals of the sampling phase. The switch on-resistance from (9.24) can be approximated by using a polynomial function of the input voltage [8]. For the Simulink model, a polynomial calculation is realized with addition and product blocks.

9.4.3 Clock jitter and skew mismatch

Clock jitter is the uncertainty in clock period because of thermal and power supply noises, crosstalk and reflections. These uncertainties can cause an increase in noise power that is proportional to the signal bandwidth and degrade ADC performance [11]. Consider a sinusoidal signal $V_{in}(t)$ with amplitude A, and frequency f_{in} as input. Sampling error due to jitter can be expressed as:

$$[V_{in}(t + \gamma) - V_{in}(t)] \approx \gamma \frac{dV_{in}(t)}{dt} \tag{9.26}$$

with the assumption that:

$$2\pi f_{in}\, \sigma \ll 1 \tag{9.27}$$

where γ is the sampling instance error characterized by a jitter process with $N(0, \sigma^2)$ error statistics and σ is the rms value of the jitter [12].

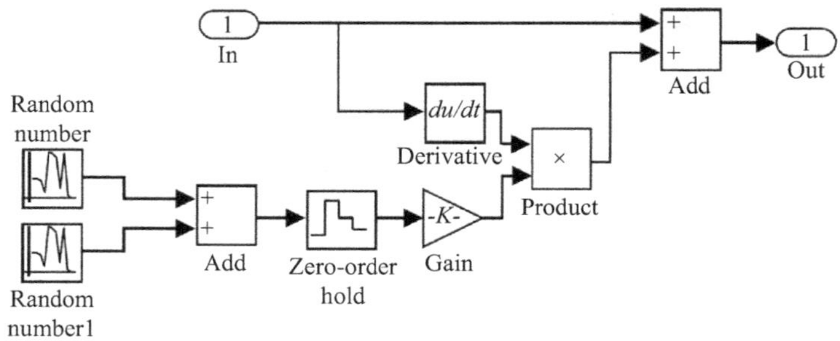

Figure 9.13 Model of the clock jitter and skew effects

Another performance-limiting factor for high-speed ADCs is the skew in the clock distribution network. This problem arises from unequal clock path lengths to different pipeline stages. Even though, H-tree structure is a common way to balance delay variations among stages, process variations in clock buffers and unequal coupling from neighboring interconnect can introduce uneven delays within the H-tree introducing noise due to clock skew [13]. These effects can be modeled in Simulink as shown in Figure 9.13. While one of the random sources generates a random number for each cycle of the ADC to model jitter, the other one generates a random number once for the simulation to model the clock skew effect.

9.4.4 Capacitor mismatches

The ratio between sampling capacitor and feedback capacitor determines the gain of the MDAC. Any deviation in capacitor values would introduce an error from the ideal value of the residue. Therefore, an adequate matching is needed between the capacitors. The capacitance value can vary due to systematic mismatch errors caused by layout engineers as well as random mismatch from lithographic non-idealities such as over etching and oxide thickness gradients [14–16]. For MDAC configuration, what matters is the ratio of capacitances. In our system, ideally, sampling and feedback capacitances are equal to each other. Therefore, including mismatch, the relationship between capacitances can be written as:

$$C_{s,k} = C_{f,k} + \Delta C_k \quad \text{and} \quad m_k = \frac{\Delta C_k}{C_{f,k}} \tag{9.28}$$

where $C_{s,k}$ and $C_{f,k}$ are sampling and feedback capacitors of the kth pipeline stage. Thus, the residue output becomes:

$$V_{out} = V_{in}(2 + m_k) - D\, V_{ref}(1 + m_k) \tag{9.29}$$

Accuracy of capacitors can be improved during design and layout. The most straight forward way is to use large capacitors, which has limitations due to silicon area. The usage of large unit sized capacitors and dummy capacitors to avoid asymmetric loading decreases the effect of systematic mismatch. Furthermore, gradient mismatches can be minimized by applying common centroid and inter-digitation layout techniques. However, even after these improvements, remaining mismatches can still limit the ADC performance that may necessitate the use of gain calibration methods.

9.4.5 Current sources matching error

As shown in Figure 9.1, the DAC value subtracted from input for residue calculation is generated by a sub-DAC. For this pipeline ADC, the sub-DAC is implemented as a three-level current steering DAC. Therefore, the mismatch between the current sources becomes a potential error source. Figure 9.14 shows a unit CMOS current mirror where the reference and output currents are equal. To the first-order output, current of the mirror source is proportional to the W/L ratios between the diode and current source devices. However, secondary effects such as channel length modulation, threshold mismatch and W/L mismatch can cause deviation from the ideal current source output.

The drain current in device $M0$ can be written as:

$$I_{ref} = I_{out,0} = \frac{1}{2}\mu C_{ox}\frac{W}{L}(V_{GS} - V_{th})^2 \tag{9.30}$$

where μ is the NMOS mobility, C_{ox} is the gate oxide capacitance per unit area and W, L are transistor dimensions. Including mismatch errors, the drain current of $M1$ can be written as following:

$$I_{out} = \frac{1}{2}\mu C_{ox}\left(\frac{W}{L} + \Delta\frac{W}{L}\right)(V_{GS} - (V_{th} + \Delta V_{th}))^2(1 + \lambda\, V_{DS}) \tag{9.31}$$

From (9.30) and (9.31), it is seen that the output current includes additional dependencies than W/L of $M1$. After evaluating the products and ignoring high-order terms the resulting current can be shown to include a current error ΔI_{out} [17,18]:

$$I_{out} = I_{out,0} + \Delta I_{out} \tag{9.32}$$

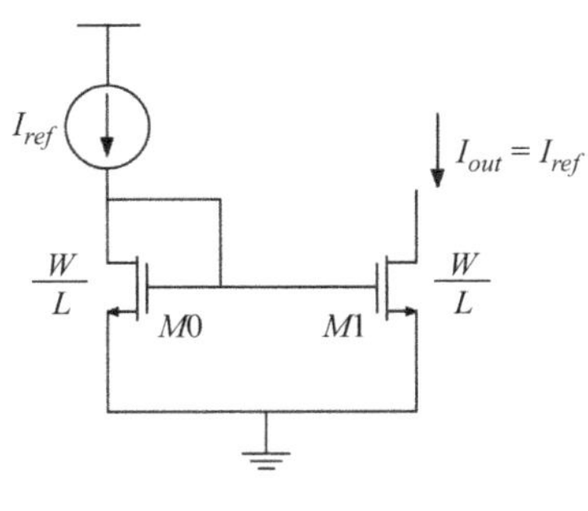

Figure 9.14 Basic CMOS current mirror

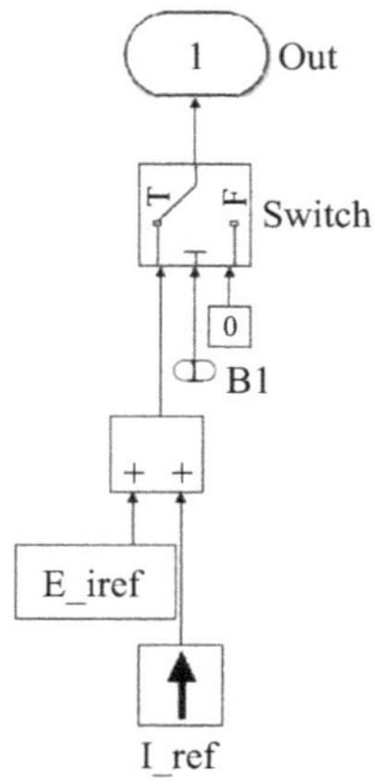

Figure 9.15 A current source model with matching error

Figure 9.15 illustrates the Simulink model of the current source with mismatch errors. Error value that can be extracted from Monte Carlo simulations is added to current value to model given nonideality.

9.4.6 Comparator offset

Comparator offset is the main source of error in the sub-ADC. When comparing two input voltages, input offset may affect the output decision. For the Simulink model, dynamic offset is implemented as a random number and is sampled at every cycle and the static offset is added to input signal as a constant number [16,19].

9.5 Final model of the pipeline ADC and its performance results

Simulink representation of the pipeline ADC models the basic ADC functionality in addition to the nonidealities that were defined in previous sections. For the sub-ADC, these include the clock jitter, skew, as well as static and dynamic comparator offsets. Next, for the sub-DAC, a current mismatch model is included. Figure 9.16 presents subblock models in pipeline ADCs with error models.

For the switch capacitor-based MDAC, first, charge injection, clock feed-through, switching thermal noise and nonlinear switch resistance effects are captured. Next, input signal and DAC output are amplified with gains that include the errors due to capacitive mismatches. Residue amplifier nonidealities such as input offset, finite gain error, thermal noise, flicker noise and bandwidth as well as slew rate limitations are included in our model that is illustrated in Figure 9.17. It should be noted that during the Taylor series expansion of the finite amplifier gain only odd-order terms were employed. This is due to the well-known fact that a fully differential implementation would cancel the even-order terms. Finally, bandwidth

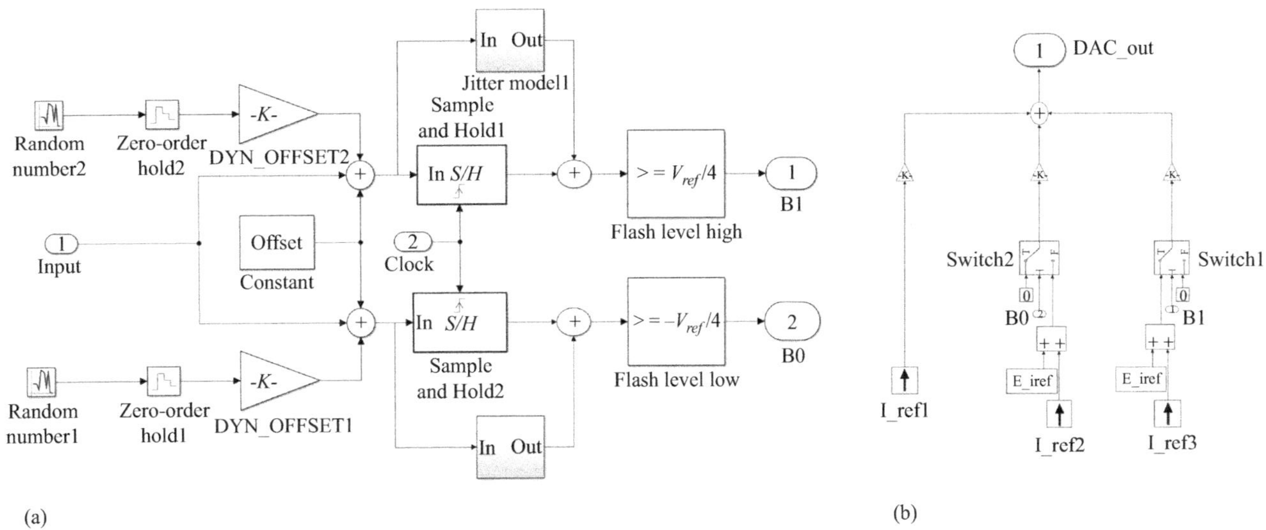

Figure 9.16 (a) Sub-flash ADC model and (b) sub-DAC model

and slew rate limitations are implemented via a MATLAB function code and near the output of the model. The resulting Simulink representation of the pipeline stage is illustrated in Figure 9.17.

To confirm the functionality of the model, an 11-bit ideal pipeline ADC model is simulated in MATLAB/Simulink. A sine wave with a frequency of 7.26 MHz is applied to the input. The sampling rate of the ADC is selected as 250 MHz. As is seen from Figure 9.18 and the left plot in Figure 9.19, the reconstructed digital output validates that the conversion is completed successfully.

Furthermore, simulation of the ADC with modeled nonidealities is performed under the same conditions. In order to extract signal-to-noise ratio (SNR) and spurious-free dynamic range (SFDR), the output spectrum is plotted. Degradation of dynamic performance can be observed by comparing ideal and nonideal spectra in Figure 9.19. SNR decreases to 44.55 from 68.09 dB, which means the effective number of bits reduces from 11 to 7.1 bits. It can also be seen that the SFDR approximately lost 40 dB because of harmonic distortions. Additionally, Figure 9.20 shows simulation results of the static performance tests, including differential nonlinearity (DNL) and integral nonlinearity (INL). The result indicates that DNL and INL are within the range of −1 to +1.26 and −13.71 to +15.5 LSB, respectively.

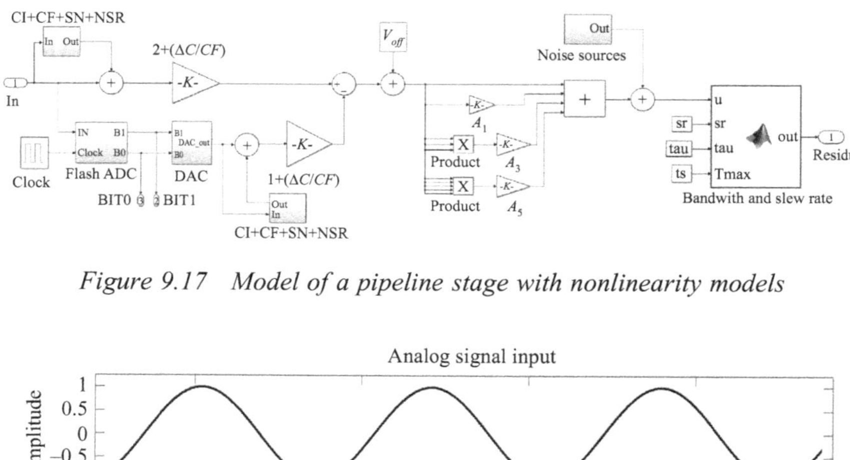

Figure 9.17 Model of a pipeline stage with nonlinearity models

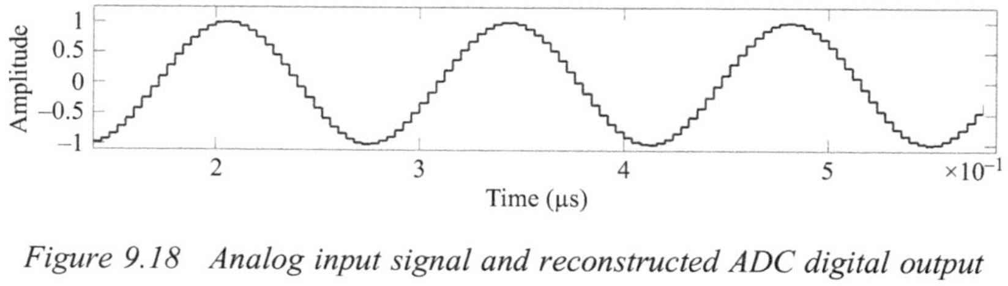

Figure 9.18 Analog input signal and reconstructed ADC digital output

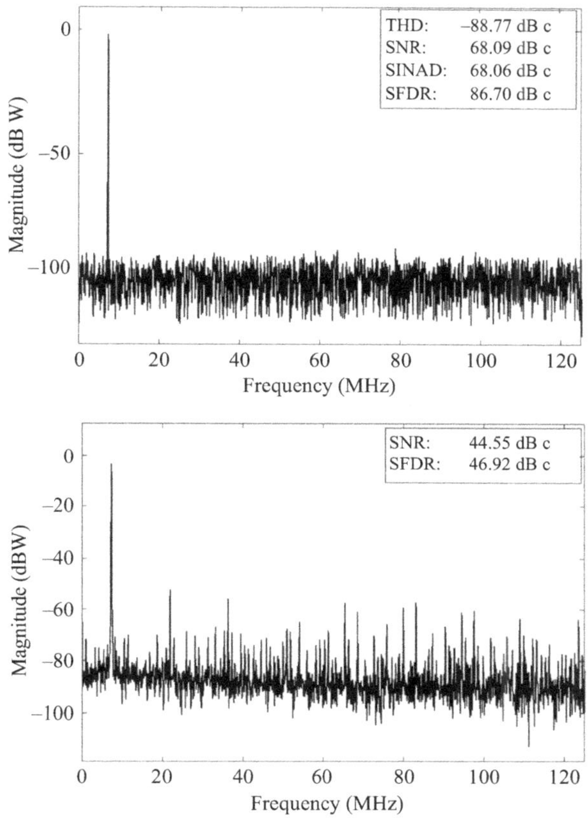

Figure 9.19 FFT spectrum of ideal and nonideal 11-bit pipeline ADC Simulink model

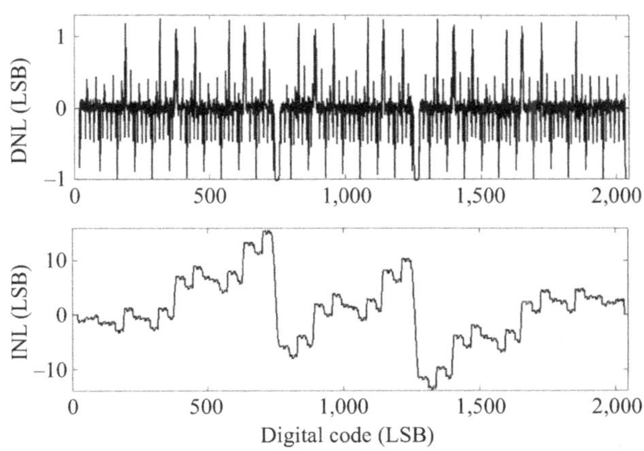

Figure 9.20 DNL and INL of nonideal 11-bit pipeline ADC Simulink model

9.6 Conclusion

In this chapter, we have presented a brief analysis of pipeline ADCs and their submodules. Basic functionality and nonidealities as well as modeling of each of these aspects have been discussed in detail. Specifically, an 11-bit pipeline ADC was considered as the main application. Nonidealities such as finite DC gain, limited bandwidth and slew rate, thermal noise and input offset of an op-amp also charge injection, clock feedthrough, switching thermal noise, nonlinear switch resistance, clock jitter, skew, as well as static and dynamic comparator offsets were described and added to behavioral model of the pipeline ADC. Further top level MATLAB/Simulink simulations were performed to demonstrate the performance loss that arises as a result of such nonidealities. As a result, this model can be used to evaluate any technique to mitigate the impact of these impairments.

Acknowledgement

This work was supported by the Scientific and Technological Research Council of Turkey Project number 115E752.

References

[1] Ahmed I. Pipelined ADC Design and Enhancement Techniques. 1st ed. New York: Springer; 2010.
[2] Chiu Y. Lecture Notes in EECT 7327-001 Data Converters. The University of Texas at Dallas; Fall, 2014.
[3] Tsang C, and Nishumura KA. Techniques to Reduce Memory Effect or Inter-symbol Interference (ISI) in Circuits Using Op-Amp Sharing. Agilent Technologies Inc.; Application Note, 2014.
[4] Tsui LHY, and Ko IT. Op-Amp Sharing Technique to Remove Memory Effect in Pipelined Circuit. Microchip Technologies Inc.; Application Note, 2015.
[5] Sahoo BD, and Razavi B. A user-friendly ADC simulator for courses on analog design. IEEE International Conference on Microelectronics Systems Education. 2009;p. 77–80.
[6] Panigada A, and Galton I. Digital background correction of harmonic distortion in pipelined ADCs. IEEE Transactions on Circuits and Systems I: Regular Papers. 2006;53:1885–1895.
[7] Malcovati P, Brigati S, Francesconi F, *et al.* Behavioral modeling of switched-capacitor sigma delta modulators. IEEE Transactions on Circuits and Systems I: Fundamental Theory and Applications. 2003;50(3):352–364.
[8] Barra S, Kouda S, Dendouga A, *et al.* Simulink behavioral modeling of a 10-bit pipelined ADC. International Journal of Automation and Computing. 2013;10(2):134–142.
[9] Palmer R. DC Parameters: Input Offset Voltage. Report No: SLOA059. Texas Instruments; 2001.

[10] Xu W, and Friedman E. Clock feedthrough in CMOS analog transmission gate switches. Analog Integrated Circuits and Signal Processing. 2005;44(3): 271–281.

[11] Malcovati P, Brigati S, and Francescon F. Behavioral modeling of switched-capacitor sigma-delta modulator. IEEE Transactions on Circuits and Systems I. 2003;50(3):352–364.

[12] Kobayashi H, and Morimura M. Aperture jitter effects in wideband ADC systems. 6th IEEE International Conference on Electronics, Circuits and Systems. 1999;p. 1705–1708.

[13] Zarkesh-Ha P, Mule T, and Meindl J. Characterization and modeling of clock skew with process variations. Proceedings of the IEEE 1999 Custom Integrated Circuits Conference. 1999.

[14] Shyu GT, and Krummenachero F. Random error effects in matched MOS capacitors and current sources. IEEE Journal of Solid-State Circuits. 1984; 19(6):948–956.

[15] Saxena S, Hess C, Karbasi H, *et al*. Variation in transistor performance and leakage in nanometer-scale technologies. IEEE Transactions on Electron Devices. 2008;55(1):131–144.

[16] Kledrowetz V, and Haze J. Analysis of Non-ideal Effects of Pipelined ADC by Using MATLAB–Simulink. Advances in Sensors, Signals and Materials. 2010, 85–88.

[17] Hanfoug S, Moulahcene F, and Bouguechal N. Contribution to the modeling and simulation of current mode pipeline ADC based on Matlab. International Journal of Hybrid Information Technology. 2015;8(3):83–96.

[18] Crippa P, Turchetti C, and Conti M. A statistical methodology for the design of high-performance CMOS current-steering digital-to-analog converter. IEEE Transactions on Computer-Aided Design of Integrated Circuits and Systems. 2002;21(4):377–394.

[19] Maloberti F, Estrada P, Malcovati P, *et al*. Behavioral modeling and simulation of data converters. 1992 IEEE International Symposium on Circuits and Systems. 1992;vol. 5, p. 2144–2147.

Chapter 10
Power systems modelling

Jindrich Svorc[1], Rupert Howes[1], Pier Cavallini[1], and Kemal Ozanoglu[2]

10.1 Introduction

The increasing deployment of electronic systems in many areas of our lives brings with it the need to provide power supplies for those applications, many of which are driven from a battery, increasingly Li-ion based, which itself may also have the requirement to be charged. For these systems, efficiency is of high importance to maximize battery runtime. High efficiency usually demands the use of a direct current DC–DC converter. This chapter starts with discussing how to construct a small-signal model of such a converter, which is inherently a large-signal time-variant system, such that it can be modelled in an electrical circuit simulator. Modelling and predicting the efficiency of the overall power system is then addressed. The behaviour of the power system is also strongly dependent on the battery characteristics, as well as the passive components, in this case, capacitors and inductors. An introduction to the modelling of all these topics is given.

10.2 Small-signal models of DC–DC converters

10.2.1 Motivation

Most of the common DC–DC converters are designed as a feedback system where the output voltage or current is compared with its reference value, and the error between actual and the reference value is used for controlling the power stage in a way that the error is minimized. The main reason for introducing a feedback system is the accuracy of such a system, which is usually better in comparison to open systems. One of the drawbacks of feedback systems is that the loop itself needs to be neatly designed and checked; otherwise, it can get unstable. If the system gets unstable, small or severe oscillations can occur, which can damage the controller or the following circuitry, in the worst case. Fortunately, the circuit theory taught us

[1]Dialog Semiconductor, Swindon, UK
[2]Dialog Semiconductor, Istanbul, Turkey

various methods on how the stability can be checked and how the feedback loop can be correctly adjusted. The first step is to adjust the DC operating point of the whole circuit. Then the small-signal model can be created. The procedure for the feedback loop adjustment is usually done in the following manner:

1. Create a simplified mathematical small-signal linear model of the loop.
2. Adjust the open-loop response in order to get correct DC gain, phase margin, and other desired input parameters from a specification.
3. Check the final open-loop response in a circuit simulator (like Spice and Spectre).

The mathematical model should be created in a way that it covers all significant contributors, which affect the final small-signal open-loop response. It means, the mathematical model usually does not need to cover all parasitic capacitances and resistances tied to all devices in the circuit. However, it may cover significant parasitic effects if needed – like serial resistance of inductors.

On the other hand, the simulator calculates the DC operating point and creates its own internal small-signal model, which takes into account all the parasitic devices covered by the models used in the simulator, for example, parasitic capacitances of MOS transistors. If any of those additional parasitic devices cause the check-in simulator to fail (due to some effects which are not predicted in the mathematical model) the loop needs to be readjusted.

One possible way how to handle these types of problems is to use an extra margin for the loop adjustment with the mathematical model. For example – if the model predicts a phase-margin of 60° but the simulator check reveals that the phase-margin lies below 50°, it is reasonable to adjust the loop for phase-margin better than 70° and expect a similar increase of phase-margin in the simulator as well. Although this method is very simple and easy, it may fail as the adjustment in the model does not aim directly for the root cause of the phase margin degradation, and it does not take into account any consequences.

A better approach is to include additional significant parasitic effects from the circuit back into the mathematical model. In that case, the additional effects will be apparent, while the loop is being tuned so the loop adjustment will be made with respect to those effects.

The previous paragraphs describe a very general procedure of adjusting a feedback system, which can be easily linearized. Unfortunately, this is not the case in switching circuits as they are non-time-continuous and not linear. However, we would like to use the well-proven and familiar methods from linear circuits for switching circuits as well. In order to do so, we need to model the switching circuit with a linearized model, and this chapter provides a brief introduction.

It should be noted here that the chapter does not cover how the switching converter should be designed, where to place poles and zeroes of the transfer function and rigorous derivation of linear models. However, these topics can be found, for example, in [1–3].

The aim of this chapter is to provide intuitive insight into how the linear model works and how it can be implemented in a circuit simulator. The example of the model-creating procedure is described with a simple voltage mode buck converter. Because it is just a model, it has its limitations and the most important of them will be discussed later in this chapter.

10.2.2 Assumptions

For the following section, it is assumed that the reader has reasonable knowledge how the common DC–DC converters work and what the terms like 'continuous conduction mode', 'discontinuous conduction mode (DCM)', 'DC operating point', 'small signal' vs. 'large signal', mean.

10.2.3 Test vehicle

As was mentioned before, the main aim of this chapter is to give an introduction on how to create a model of a switching converter, which we can use for our benefit for designing stable and well-behaved converters. Let us start with a simple buck converter with a voltage-mode controller, which will serve as a test vehicle. A buck converter is in general one of the simplest converters, and the fundamental model-creating ideas can be well described. When understood, those ideas can be used for other converters as well.

The buck converter schematic diagram is shown in Figure 10.1. It comprises input capacitor C_{IN}, power inductor L, output capacitor C_{OUT}, power stage (PWR) which usually contains some logic (DIG) and drivers (DRV) for the power switches, pulse width modulator (PWM) and an error amplifier (EA) with compensation network.

The block is powered by V_{IN}, V_{REF} is the reference voltage that sets the target for the regulated voltage V_{OUT}, and the load is represented by an external resistor R_{LOAD}.

The basic operation is as follows: EA compares output voltage V_{OUT} with reference voltage V_{REF} and based on the difference between them sets the error signal ERR. The error signal is then fed into the PWM, which creates rectangular signal DUTY with fixed frequency and duty-cycle D proportional to the error signal ERR. The DUTY signal is then transferred to the LX node in the form of a pulsating signal with the same duty-cycle D and amplitude equal to V_{IN}. For the ideal case, the DC voltage at the LX node is given by the following formula:

$$avg(V_{LX}) = DV_{IN} \tag{10.1}$$

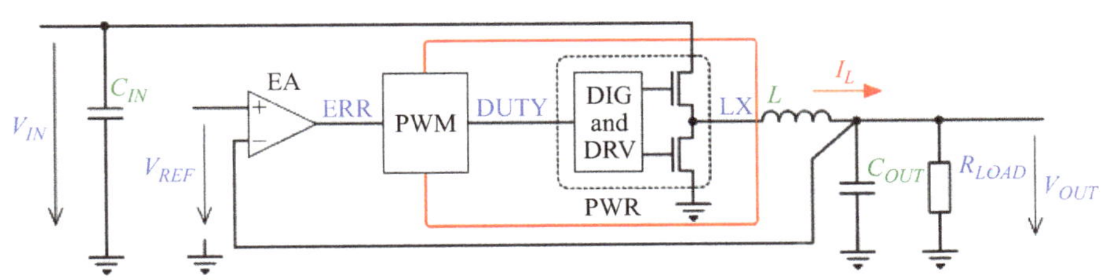

Figure 10.1 Block diagram of a buck DC–DC converter

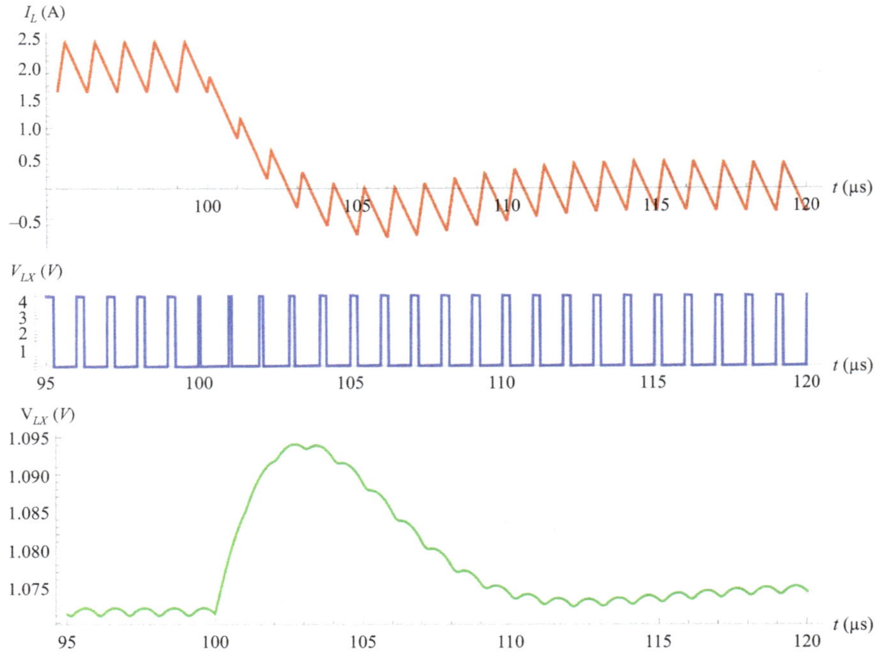

Figure 10.2 Buck converter waveforms example

The voltage at the LX node has a rectangular shape. In order to remove its switching components, an *L–C* filter is put in place, so the output voltage V_{OUT} is predominantly a DC component plus a small residual voltage ripple. The output voltage of an ideal buck DC–DC converter is then given by the following equation:

$$V_{OUT} = DV_{IN} \tag{10.2}$$

An example of the time behaviour of the inductor current I_L, the voltage at LX node V_{LX}, and the output voltage V_{OUT} in the buck converter is shown in Figure 10.2. The picture shows a load release event at time 100 µs. The load suddenly changes from 2 to 0 A, and the output voltage shoots up. The controller lowers the duty-cycle and the coil current decreases, and the output voltage goes back to its normal level.

10.2.4 Partitioning of the circuit

From the perspective of the linear model, the block diagram in Figure 10.1 can be separated into switching and non-switching parts. The non-switching part works with time-continuous signals, whereas the switching part works with switching signals. The border between switching and non-switching part is marked with a red rectangle. All the blocks, which are inside the rectangle, belong to the switching part, and blocks outside belong to the non-switching part of the system.

The PWM and the second-order filter *L–C* sit at the boundaries between continuous and non-continuous time domain. Those blocks act as a port between switching and non-switching part of the circuit.

The non-switching part of the circuit can be linearized with well-known mathematical tools from circuit theory. In addition, the non-switching part can be directly simulated in a circuit simulator, and no special modelling method is needed. It means, if we want to make a model of a buck converter, we can use the same C_{IN}, *L–C* filter, R_{LOAD}, but also the whole EA stays unchanged.

On the other hand, the switching part of the system needs special attention as it is not possible to perform standard small-signal analysis on the blocks included. In order to do so, we need to model those blocks with simple linearized models, which remove the switching and provide similar small-signal behaviour as the switching original.

In fact, there are usually no small-signal-related effects applied to the signal within the switching part of the system so in that case, it is enough to model the blocks, which pass the signal in and out of the switching part of the system. In the buck example, it is the PWM and power-stage PWR. Those blocks are our main interest and will be examined later in this chapter.

10.2.5 Model types

In order to start with a clear view of what model is being talked about, a short overview is given in the following sections.

10.2.5.1 Switching model

In general, the switching model covers the normal switching operation of the converter and might cover some additional features as well. There is no fixed distinction between simple and complex model, but for the purpose of this chapter, we can split switching models into two types with respect to the complexity level.

- The *complex switching model* is predominantly done in an advanced simulator (Spectre®, PSpice®, etc.) and evaluates the time behaviour of the final product. This model should cover all special functionalities, for example, coil current limitation, and so on. It should also cover all the parasitic effects related to all the devices. The accuracy and coverage are the most important for those models.
- The *simple switching model* can be used for initial circuit evaluation, for example, in a feasibility study. It does not cover all the functionalities and parasitic effects; however, it may include things like coil current limitation and significant parasitic effects. The number of effects, which are covered, depends on the designer's choice. Speed and simplicity are the key features. This model is a starting point for the next listed model.

10.2.5.2 Averaged model – continuous-time model

Both averaged and switching models still operate in the time domain. The important difference between the averaged model and the switching model is that the averaged model employs no switching. For the purpose of this text, we can define two levels of complexity of the averaged model.

- The *simple averaged model* is usually based on the simple switching model. The switching blocks are replaced with their averaged counterparts. All switching signals are replaced with averaged signals. The instantaneous level of the averaged signals corresponds to the instantaneous average value of the original switching signal, when averaged over one switching period. More about averaging will be shown later in this chapter. Since there is no switching, the simple averaged model brings the advantage of running native small-signal AC analysis provided with the simulator. Although the averaged model is operated in the time domain, the simulator creates internally the small-signal linear model for the *s*-domain. The simple averaged model is the model, which is of our main interest further in this chapter.
- The *complex averaged model* is a fork of the simple averaged model where the original non-switching models (e.g. EA) are replaced with the full-feature complex models, which cover all the nonideal effects. The switching blocks are still modelled as in the simple averaged model. It is apparent that this model extends the simple averaged model by employing the nonidealities from other time-continuous blocks but does not account for any switching effects.

10.2.5.3 Small-signal model – AC model

In comparison to the averaged model, the small-signal model shows the behaviour of the circuit in the *s*-domain instead of time domain. It is a model, which describes small-signal behaviour of the converter, and it is used mainly for loop adjustment, input and output impedance evaluation, and other AC characteristics. From the simulator-user perspective, there is no special model for the simulator as the averaged model serves for this purpose.

Nevertheless, it is always wise to create a mathematical model of the converter as it helps one to gain a better insight into the way in which the parameters affect the response and usually allows for very quick adjustment of the loop.

10.2.5.4 Used models

It should be noted here that unless explicitly stated, no complex model will be discussed in the rest of this chapter. Therefore, the following apply:

- switching model denotes simple switching model,
- averaged model denotes simple averaged model.

10.2.6 Basic theory for averaged – continuous-time model

10.2.6.1 Small-ripple approximation

The small-ripple approximation means that we can neglect the output voltage ripple. This is reasonable in most cases as the voltage ripple is usually much smaller than the DC level of the output voltage V_{OUT}. Comparison of V_{OUT} waveform and its DC component is shown in Figure 10.3.

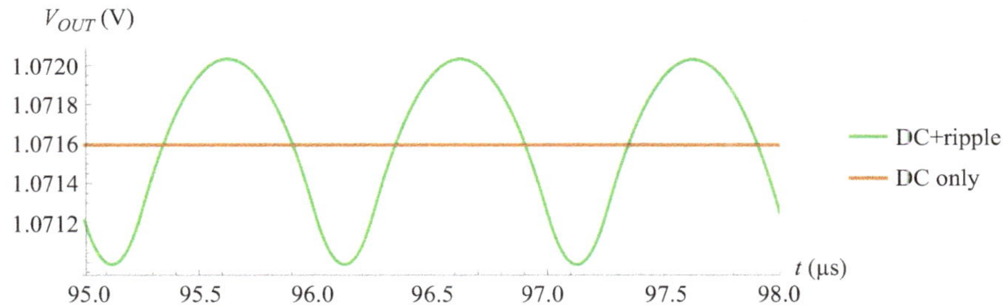

Figure 10.3 V_{OUT} waveform with the DC component

It is apparent that the peak-to-peak voltage magnitude is in millivolts, whereas the DC component is about 1,000 times larger. Hence, we can use the small-ripple approximation and neglect the ripple for our averaged model. So, for steady-state operation, it applies:

$$V_{OUT\ DC} \gg V_{OUT\ peak\text{-}to\text{-}peak} \rightarrow V_{OUT}(t) \cong V_{OUT\ DC} \tag{10.3}$$

The main reason for the small-ripple approximation is simplifying the equations for average model derivation. However, the main important consequence for our model is that we do not need to worry about ripple as it is assumed to be negligible.

10.2.6.2 Averaging

The process of averaging is very important for a basic understanding of how the averaged model works. The main goal here is to remove the switching and its artefacts by averaging a particular signal over one switching period. We will denote the averaged signal in the following style:

$$\langle A \rangle = \frac{1}{T_S} \int_0^{T_S} A(\tau) d\tau \tag{10.4}$$

where T_S is the switching period and f_S is switching frequency defined as:

$$T_S = \frac{1}{f_S} \tag{10.5}$$

The following section gives a bit of introduction into the theory. There are two state variables used for evaluation of the DC–DC converter:

- the voltage across an inductor,
- the current through a capacitor.

The reason why those variables are suitable for our evaluation is that they are changed with a step between two (or potentially more) states with constant level. It means that during each state, the inductor voltage or capacitor current is not changing, or at least they are not changing a lot, so the small-ripple approximation is still valid. The duration of each state corresponds to the duty-cycle D with respect to the switching period. An example for $V_L(t)$ is shown in Figure 10.4.

The theory says that in the steady-state operation, the average value of those state variables is equal to 0:

$$\langle V_L \rangle = \frac{1}{T_S} \int_0^{T_S} V_L(t) dt = 0 \tag{10.6}$$

$$\langle I_C \rangle = \frac{1}{T_S} \int_0^{T_S} I_C(t) dt = 0 \tag{10.7}$$

This fact is very intuitive. For example – if there is no average current through the output capacitor C_{OUT} over the switching period, the output voltage $V_{OUT}(t_0)$ at the beginning of a period is equal to the voltage $V_{OUT}(t_0 + T_S)$ at the end of the switching period – which is exactly the case in steady-state operation when the converter works in equilibrium.

A similar situation applies to the inductor voltage, so if there is zero average voltage across the inductor over one switching period the inductor current at the end of the period $I_L(t_0 + T_S)$ will be equal to the inductor current $I_L(t_0)$ at the beginning of the period. The case of the inductor voltage is shown in Figure 10.4.

The situation changes when a disturbance hits the controller and the duty-cycle changes. In that case, the average voltage across the inductor is no longer equal to 0 V and as a consequence, the inductor current at the end of the switching period $I_L(t_0 + T_S)$ is not equal to $I_L(t_0)$ at the beginning of the switching period – see Figure 10.5.

The most important observation from the previous paragraphs is that the change between the initial level at t_0 and the level at the end of the switching period $t_0 + T_S$ depends only on the change of the duty-cycle (assuming other parameters are constant) and not on the exact shape of the current within one period.

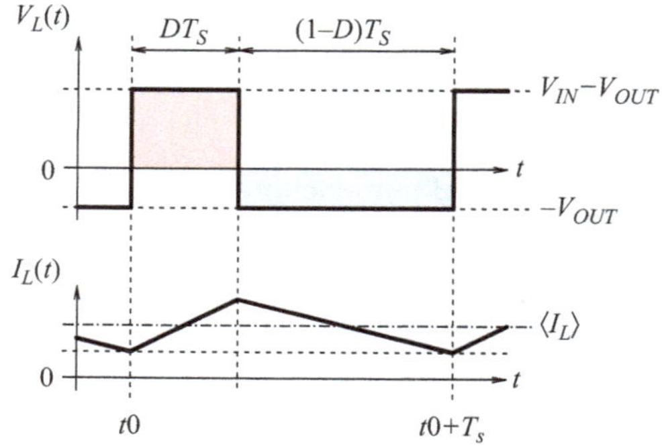

Figure 10.4 Inductor voltage and current in steady state

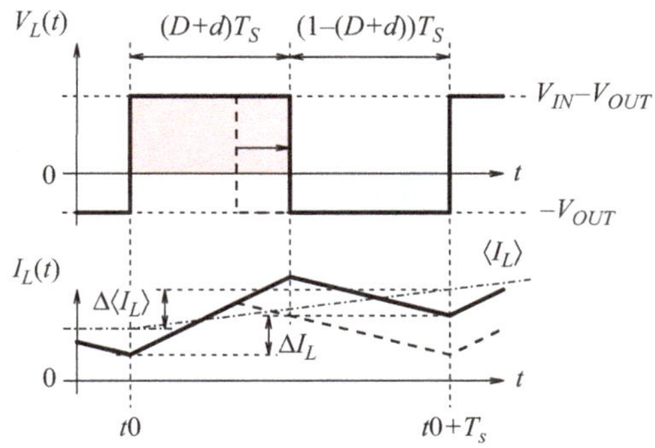

Figure 10.5 Inductor voltage and current with duty-cycle variation

In the case of the inductor, we can start with the equation for the average voltage across the inductor:

$$\langle V_L \rangle = (V_{IN} - V_{OUT})D - V_{OUT}(1 - D) \tag{10.8}$$

with some algebra:

$$\langle V_L \rangle = V_{IN}D - V_{OUT} \tag{10.9}$$

and the current at the end of the switching period is derived as:

$$I_L(T_S) = \frac{1}{L}\int_0^{DT_S} V_{IN} - V_{OUT}\,dt - \frac{1}{L}\int_{DT_S}^{T_S} V_{OUT}\,dt + I_L(0) \tag{10.10}$$

$$I_L(T_S) = \frac{1}{L}((V_{IN} - V_{OUT})DT_S - V_{OUT}(1 - D)T_S) + I_L(0) \tag{10.11}$$

$$I_L(T_S) = \frac{1}{L}(T_S(V_{IN}D - V_{OUT})) + I_L(0) \tag{10.12}$$

From (10.9) we know that $V_{IN}D - V_{OUT} = \langle V_L \rangle$ thus:

$$I_L(T_S) = \frac{T_S}{L}\langle V_L \rangle + I_L(0) \tag{10.13}$$

This equation shows that the change of the inductor current during one switching period is a function of the average voltage applied to the inductor. It implies that for the averaged model, we can replace the instantaneous signals with their averaged counterparts. This step removes the ripple from the model (which is assumed negligible).

10.2.6.3 Example of an averaged signal

Figure 10.6 shows the original triangular coil current I_L signal and the corresponding ideal averaged signal. It is apparent that the averaged signal is smooth and can be used for a small-signal analysis.

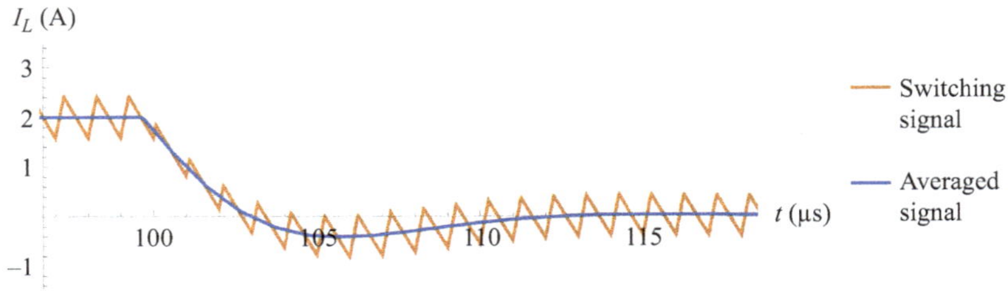

Figure 10.6 Switching and averaged coil current

10.2.7 Duty-cycle signal model

It might not be apparent how the duty-cycle signal DUTY is represented in the averaged model. Duty-cycle in the original switching model is a logic signal, which has its own logic voltage levels. However, the information, which the DUTY signal holds, is the duty-cycle, which is then transferred to the pass device and to the LX node. The duty-cycle is in the range of 0%–100% or it can be expressed as a real number from 0 to 1. The 0–1 representation is used directly in the mathematical model. In the averaged model, it is convenient to use a voltage signal with a range from 0 to 1 V. In that implementation, the voltage itself represents the duty-cycle. Moreover, it can be directly compared with the mathematical model if needed.

10.2.8 Pulse width modulator model

The PWM creates an interface from the linear domain into the switching domain. In the buck example, the PWM takes the error signal as input and creates a pulse-width-modulated logic signal with the corresponding duty-cycle. The common basic topology used for the PWM is shown in Figure 10.7.

There is a ramp generator driven with a clock signal, indicated by CLK. The ramp generator provides a saw-tooth-shaped signal RAMP that is then compared with the error signal in the PWM comparator. The output of this comparator is the DUTY signal. Typical waveforms are shown in Figure 10.8. The signals shown here are in the voltage domain, but there is no limitation to using currents instead.

The duty-cycle at the output of the PWM is derived in the following manner:

$$V_{PEDESTAL} + \frac{V_A}{T_S} D\, T_S = V_{ERROR} \tag{10.14}$$

thus

$$D = \frac{1}{V_A}(V_{ERROR} - V_{PEDESTAL}) \tag{10.15}$$

Based on the previous equation, we can create a linear model, which follows our choice of duty-cycle representation as a voltage signal in the range from 0 to 1 V (0%–100% duty-cycle). The model employs one voltage-controlled voltage source, as shown in Figure 10.9.

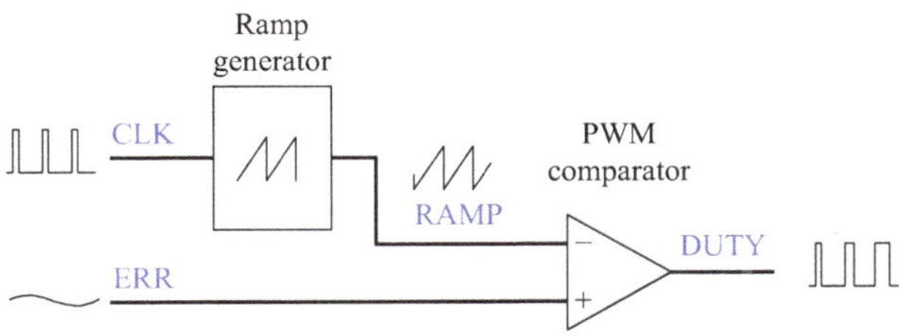

Figure 10.7 Basic PWM block diagram

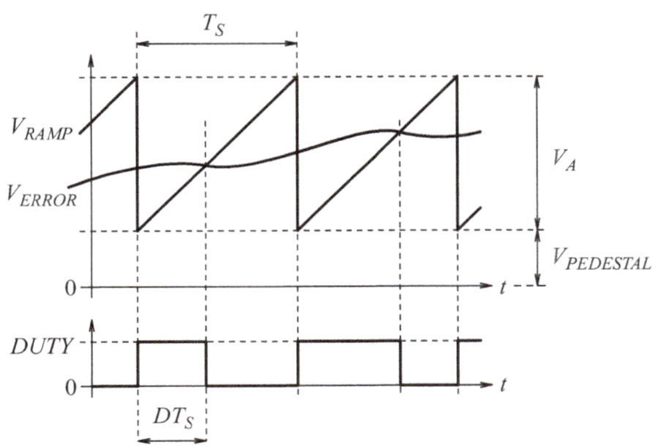

Figure 10.8 Basic PWM waveforms

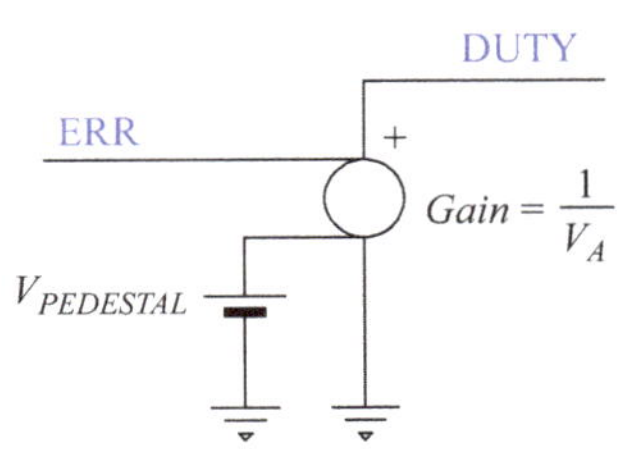

Figure 10.9 PWM linear model

The fixed-voltage $V_{PEDESTAL}$ respects the pedestal voltage from the switching model and the voltage-controlled voltage source with a gain of $1/V_A$ is the gain of the PWM.

10.2.9 Model of the power stage

The power stage, together with the *L–C* filter are the elements which create the interface between the switching and non-switching parts of the buck converter. As we stated earlier that the *L–C* filter is already a linear block so it will be kept unchanged in the averaged model. So, the block which needs to be modelled is the power stage.

The power stage in the original switching circuits simply amplifies the DUTY signal in a way that the amplitude at the LX node is in the range between the negative and positive voltage supply levels. In our example, these are 0 V and V_{IN}. The duty-cycle is kept unchanged.

It was shown before, the signals in the average model are modelled by the average of the signal over one switching period. The same is still applicable to the V_{LX} voltage. Figure 10.10 shows the V_{LX} waveform and its average signal.

Based on Figure 10.10, the following equation can be derived:

$$\langle V_{LX} \rangle = \frac{V_{IN}\, D\, T_S}{T_S} = V_{IN}\, D \tag{10.16}$$

So, the V_{LX} in the averaged model is a simple multiplication of V_{IN} by duty-cycle.

A similar exercise can be done for the input current of the power stage with the following outcome:

$$\langle I_{IN} \rangle = I_L\, D = I_{OUT}\, D \tag{10.17}$$

The equation shows that the input current I_{IN} can be modelled as a simple multiplication of inductor current I_L and duty-cycle. The duty-cycle signal is modelled with a continuous-time signal in a meaningful range from 0 to 1 V.

Therefore, by combining (10.16) and (10.17), the original power stage can be replaced with its averaged model shown in Figure 10.11.

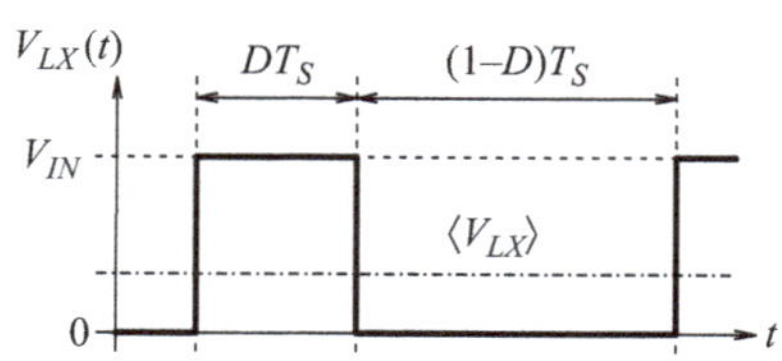

Figure 10.10 V_{LX} voltage and average voltage

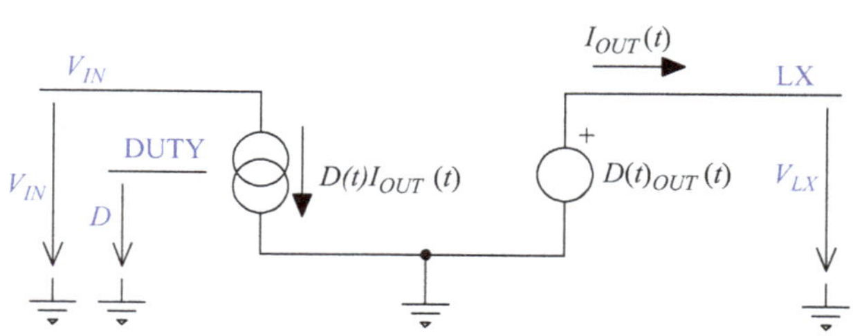

Figure 10.11 Full model of the power stage

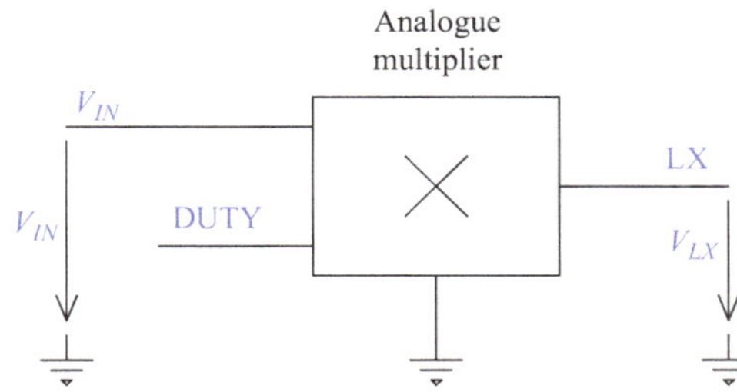

Figure 10.12 Model of the power stage with a single multiplier

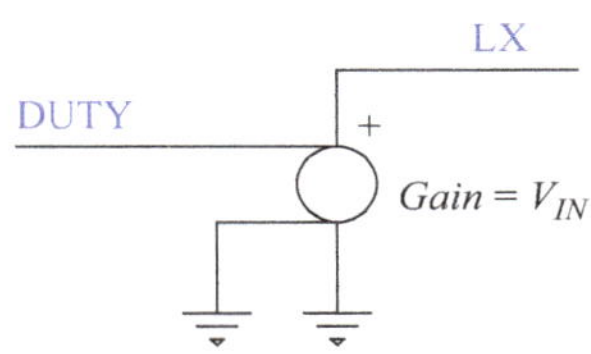

Figure 10.13 Simplified power-stage model with a voltage-controlled voltage source

The following paragraphs show an example of model simplification. If there is no need to model input current of the converter, a simplified version of the power-stage model with a single multiplier can be used. The model then changes to the version shown in Figure 10.12. This corresponds to the right part of the model in Figure 10.11.

The model in Figure 10.12 covers the change in duty-cycle as well as the change in V_{IN}. However, because the net V_{IN} is not loaded, the V_{IN} affects the V_{LX} in one direction only. In addition, an analogue multiplier might not be available in some simulators. In that case, it is possible to simplify the model further and use the one shown in Figure 10.13.

Bear in mind that the simplified model from Figure 10.13 does not cover any input-related properties like a change in V_{IN}. However, it can be used for the loop stability investigation if the input voltage is connected to a voltage source with low internal impedance and can be assumed as a fixed voltage.

10.2.10 Complete switching, linear and small-signal model

The schematic diagram of the switching model of a simple buck converter, including the ideal Error Amplifier (EA) and the compensation network is shown in Figure 10.14. The mode provides the frequency response shown in Figure 10.15.

The DC gain is set by the resistors R_1 and R_2, and there is one zero due to C_1. The C_1 provides a phase boost in order to get a good phase margin even with the

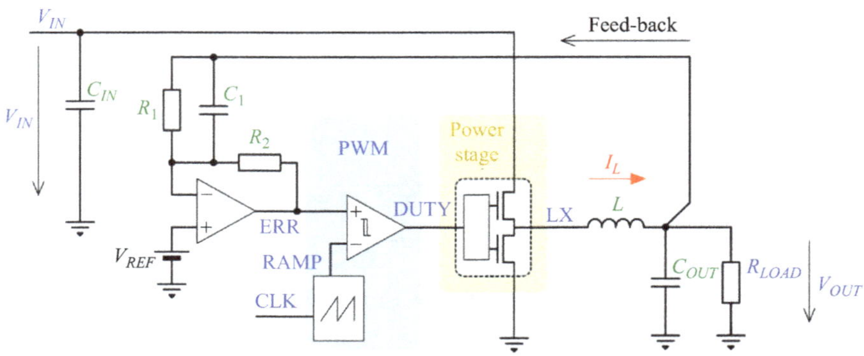

Figure 10.14 Switching model of a BUCK converter with voltage-mode control

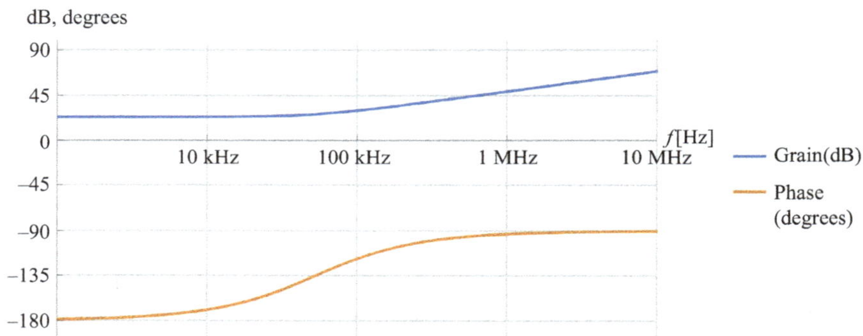

Figure 10.15 Error amplifier bode plot

L–C combination in the output filter. A rigorous theory for the compensation adjustment is out of the scope of this chapter, but it can be found in [1,2].

The switching model from Figure 10.14 is the starting point for the new averaged non-switching model. Two blocks, which need to be replaced, are marked with rounded rectangles. Those are the PWM and the power stage. If they are replaced with their non-switching models derived earlier in this chapter, we obtain the model shown in Figure 10.16. This model allows us to perform small-signal AC analysis, stability analysis, and a transient analysis with time-continuous signals. However, it does not cover parameters like input impedance, because the simplified power-stage model from Figure 10.12 is used.

The averaged model is very close to the model used for a mathematical evaluation of the loop – mathematical model. In order to create a mathematical model of the loop, the V_{IN} is set as a DC source. For the small-signal analysis, all the DC sources are replaced with their internal resistance. Voltage sources are replaced with short circuit and current sources with an open circuit. The resulting model is shown in Figure 10.17.

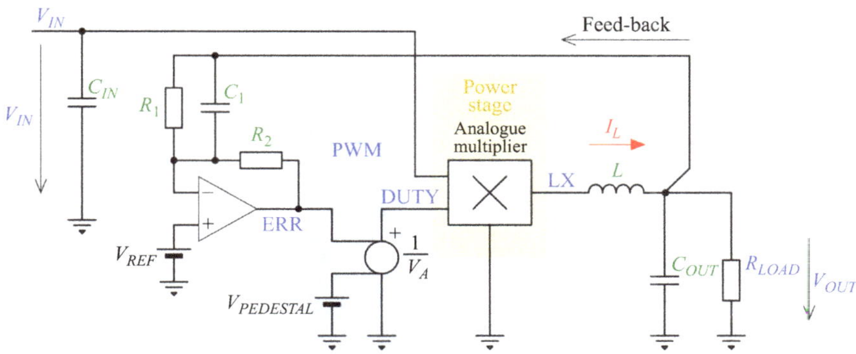

Figure 10.16 Time-continuous averaged model of the simple BUCK converter

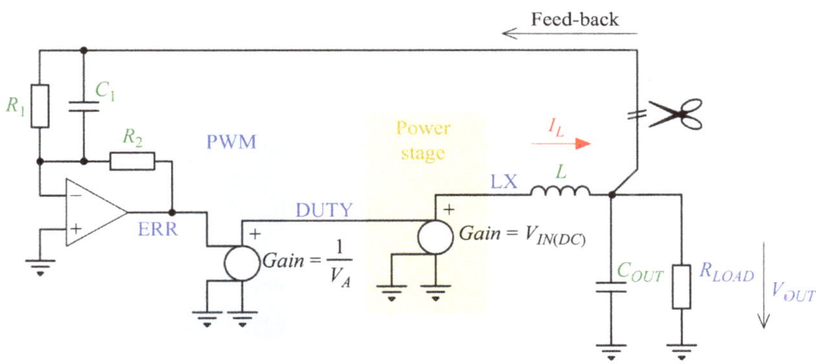

Figure 10.17 Simplified small-signal model of the BUCK converter

The gain of the PWM is simply left unchanged, but the pedestal voltage is removed because it is a DC source. The power stage is replaced with a simple gain of $V_{IN(DC)}$, which is the DC value of the input voltage V_{IN}. It replaces the analogue multiplier with the assumption that the V_{IN} is fixed, and no small signal comes from there.

The scissors mark shows a good place where the loop can be cut for the loop investigation. The impedance relation before and after the cut should stay unchanged. If the simulator offers a particular analysis for loop investigation, the impedance will be correctly handled. Otherwise, the highlighted place is a good candidate because the output of the buck can be assumed as a low impedance node, so the *R–C* network around the amplifier will not affect it.

10.2.11 The small-signal open-loop transfer function

After putting all the blocks from Figure 10.17 in the loop, we obtain the open-loop transfer function $H_{OL}(s)$. The terms in the equation are explained as follows: the first R_1, R_2, C_1 term comes from the EA compensation network. The following

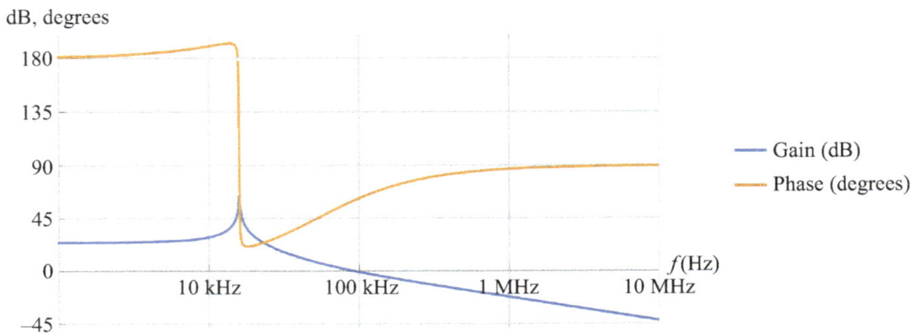

Figure 10.18 Ideal open-loop transfer function of the BUCK converter

terms are the gain of the PWM and gain of the power stage. The last term with L, R_{LOAD}, and C_{OUT} comes from the L–C filter, including the load.

There are two complex poles from the output filter damped by the R_{LOAD} resistance. In order to compensate for the double pole phase shift, a zero is introduced in the EA.

$$H_{OL}(s) = \left(-\frac{R_2}{R_1}(1 + R_1\,C_1\,s)\right)\frac{1}{V_A}V_{IN(DC)}\frac{1}{1 + (L/R_{LOAD})s + C_{OUT}L\,s^2} \tag{10.18}$$

This transfer function can be directly used for evaluating the open-loop transfer function and stability. Figure 10.18 shows the bode plot of $H_{OL}(s)$.

The open-loop transfer function shows a cross-over frequency of about 100 kHz and a phase margin of about 60°. The resonant frequency of the L–C filter is very apparent. Since there is no serial resistance of the inductor R_L, the damping of the L–C filter is just due to the R_{LOAD}.

10.2.12 Comparison of various models

10.2.12.1 Used parameters

In order to make a good overview and comparison of the various models, the following example of a BUCK is used as a case study. It is supposed to be a small buck for a battery-powered application like mobile phones. Capacitors C_{IN}, C_{OUT}, and inductor L are discrete components. All the other circuitry is supposed to be integrated into a widely used common complementary metal oxide semiconductor (CMOS) technology.

In order to make the model a bit more realistic, the serial resistance of the inductor is added. The other model improvement is that the EA is designed as an operational transconductance amplifier with finite G_M and load capacitance C_2 connected between the ERR node and ground. These changes bring the second pole of the amplifier into the picture and provide behaviour that is more realistic.

Figure 10.19 shows the simplified diagram of the improved time-continuous averaged model.

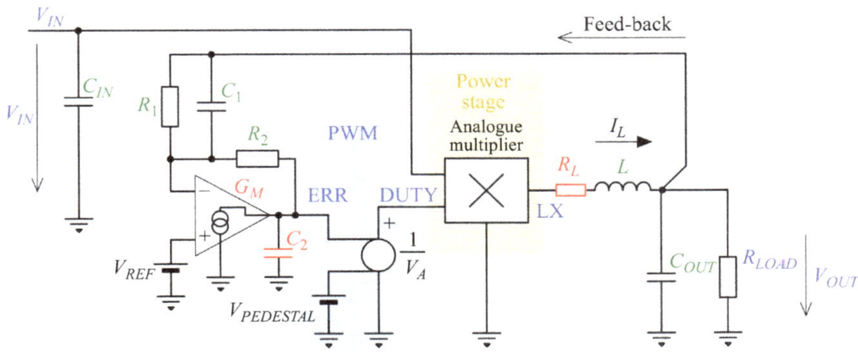

Figure 10.19 Improved averaged model of the simple BUCK converter

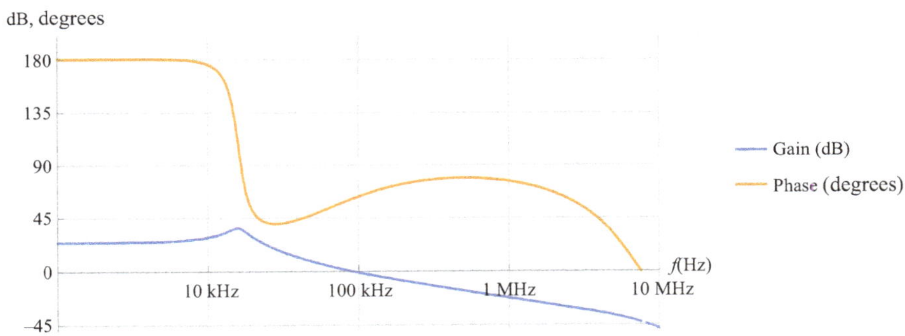

Figure 10.20 More realistic open-loop transfer function of the BUCK converter

10.2.12.2 Small-signal models

After incorporating the changes into the mathematical model, the open-loop transfer function changes. The second pole of the EA affects the phase above a certain frequency, and the resonant effect of the *L–C* filter (peak) is more damped as the R_L resistance has been introduced.

The transfer function from Figure 10.20 is obviously more realistic than the one from Figure 10.18. Nevertheless, the more elements that are added to the circuit, the greater the complexity of the transfer function; hence, a mathematical expression is no longer easy to handle with a classic manual calculation requiring instead the use of software tools, for example Wolfram, Mathematica, MATLAB, OCTAVE, or similar, capable of computing several complex equations and producing the final transfer function.

When we use the continuous-time model from Figure 10.16 and perform a small-signal analysis in the simulator, we obtain the transfer function shown in Figure 10.21. As expected, the results are identical to the mathematical model.

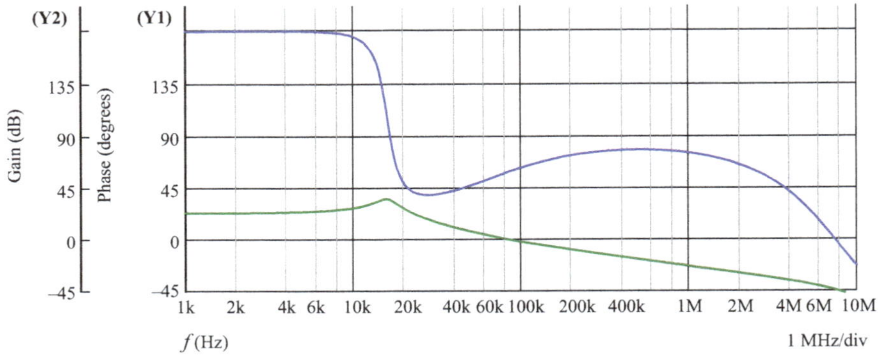

Figure 10.21 Open-loop transfer function from AC analysis in the simulator

Table 10.1 Buck model parameters

Parameter	Value	Unit
V_{IN}	4.4	V
V_{OUT}	1.1	V
f_{SW}	1	MHz
C_{OUT}	100	μF
L	1	μH
R_L	0.025	Ω
R_{LOAD}	10	Ω
V_A (ramp)	4	V
R_1	100	kΩ
R_2	1,500	kΩ
C_1	30	pF
G_M	1	mS
C_2	10	fF

Note: The same parameters – excluding the newly added – were used for the plots earlier in this chapter.

10.2.12.3 Linear and switching model

We will now compare the results of the averaged and switching models. The comparison is made with the switching model in line with Figure 10.14 and the time-continuous model from Figure 10.16. Both models use the same parameter set listed in Table 10.1. For a fair comparison, the same load conditions are applied to both models. For this example, the test pattern is a load step from 110 mA to 2.11 A within 1 μs. The condition is obtained by adding a piecewise linear current source in parallel to R_{LOAD}, as shown in Figure 10.22.

The comparison of the results of both averaged and switching models is shown in Figure 10.23.

Both models react almost identically apart from the high-frequency switching ripple in the switching model. The identical behaviour implies that the small-signal

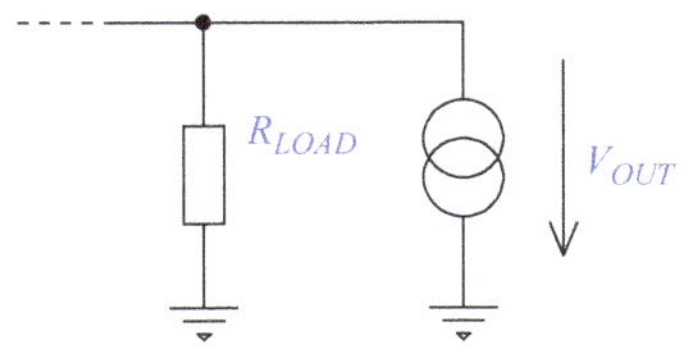

Figure 10.22 Load of the buck model for load transient simulations

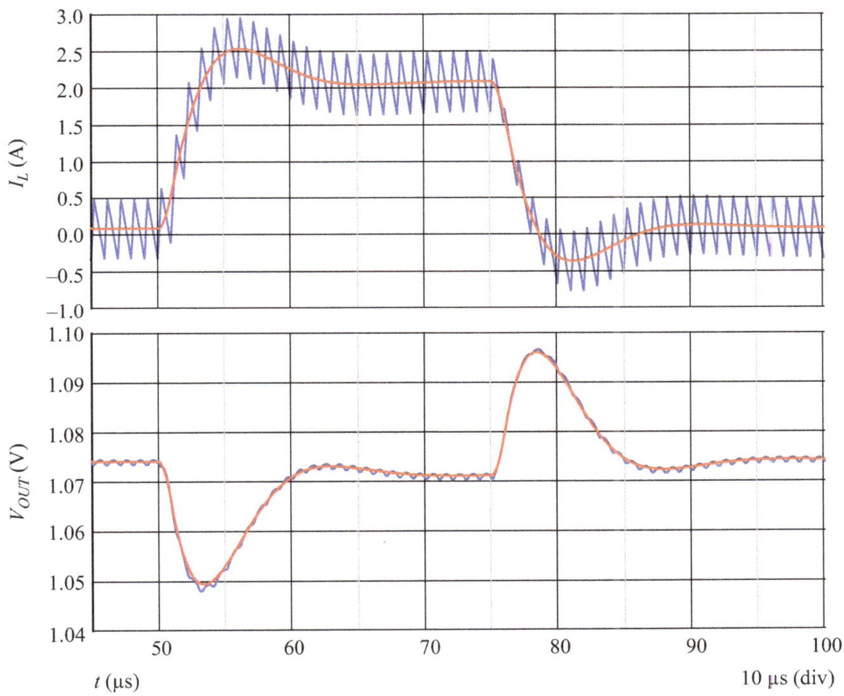

Figure 10.23 Comparison of the transient response of the switching and averaged model

linear model and open-loop transfer function correctly describe the switching system.

10.2.13 Other outputs of the small-signal model

The example of the small-signal analysis shows the open-loop transfer function of the controller loop. In general, the model can be used to extract other small-signal parameters, such as output and input impedances. Bear in mind that for input impedance simulation, the full power-stage model shown in Figure 10.11 must be used.

However, the model is just a model and does not cover all the effects of the circuit. A good designer must be aware of the limitations of the model and make sure that the model is capable of providing the expected behaviour. For instance, the simplified power-stage model illustrated in Figure 10.13 has no connection to input node V_{IN}. It means any effect from the input node is not considered. So, the model can be well used in the situation when the converter is powered from a voltage source with low internal impedance. Nevertheless, the simplified power-stage model will not provide correct results when an input filter is placed in front of the converter.

10.2.14 Switching frequency effect

Up to now, we have not investigated the impact of the switching frequency. There are many effects, which the switching nature takes into the picture, and it is beyond the scope of this chapter to try to cover them all. However, one of the most important will be briefly discussed.

10.2.14.1 Switching small-signal analysis

In order to investigate the effect of the changes in the switching frequency behaviour, we need now to introduce a new class of tool that enables the user to compare with the same model test and bench the behaviour in time and frequency domains.

One of the widely used tools in the industry is the SIMPLIS® simulator on which the analysis illustrated in the following paragraphs is based. SIMPLIS simulator is able to perform a small-signal analysis directly on the switching model. The results show the small-signal behaviour, including the effects of the switching nature.

10.2.14.2 The additional phase shift due to the sampling

Due to the way the DC–DC operates, it belongs to a sampled systems family with discrete-time. The period is given by the clock signal with fixed frequency f_S. One important effect of the sampling system is the delay. In the discussed buck example, the PWM comparator cannot make a new decision until the next period starts. If it takes more decisions within the same period, it indicates an incorrect mode of operation.

Every delay in the system is somehow translated into the phase of the response. For example, if the delay is constant, and it is equal to the switching period, the phase shift at the switching frequency is equal to 2π (360°). However, the phase shift due to the switching is more complex and does not have linear phase. The comparison between the small-signal linear model and small-signal switching model responses is shown in Figure 10.24.

The comparison of the open-loop transfer function of both models clearly shows the effect of switching frequency. A significant phase shift close to the switching frequency does not allow the designer to push the cross-over frequency of the loop too high. The switching frequency limits the possible speed of the controller. It is a rule-of-thumb that the safe cross-over frequency should be $f_S/10$ to $f_S/5$ maximum. Figure 10.25 shows a comparison of the same design with various switching frequencies.

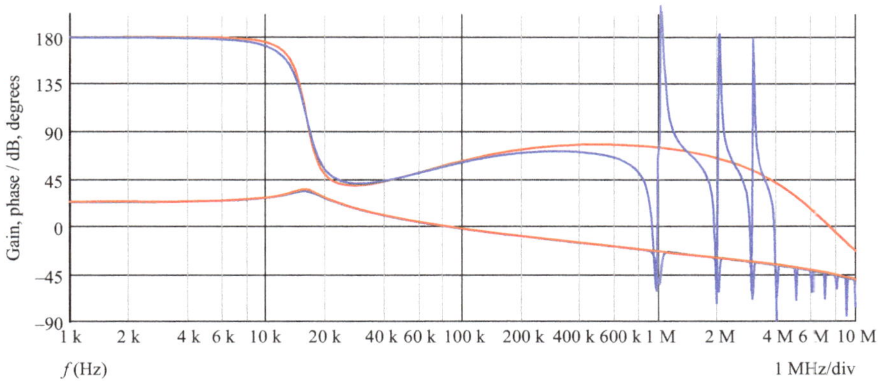

Figure 10.24 Open-loop response comparison – switching vs. averaged model; blue – switching model; red – averaged model

Lowering the frequency might have the benefit of better light-load efficiency, but from the perspective of the controller, it makes the design more difficult. The switching small-signal model predicts not just the expected lower phase-margin but also a lower DC gain. The lower DC gain should not be theoretically related to the sampling, so it is related to some other effects which are not covered by the linear model. Some of the effects which might affect the response will be shown later in the text.

10.2.14.3 The gain of the PWM

The gain of the PWM was discussed in the section about the PWM average model. The PWM linear model is derived with the assumption that the input signal (error signal) is constant over the whole period or at least that it is kept constant around the decision point where it crosses the ramp signal. If this assumption is fulfilled, the gain of the modulator is equal to $1/V_A$ as derived earlier. Unfortunately, it is often not the case in a real switching converter – especially in converters with a fast loop.

The small-signal gain of the PWM depends on the angle the input signal hits the ramp signal. A common shape of the input signal is a type of wave with amplitude and phase shift with respect to the beginning of the switching period. The shape in steady-state operation is usually the same for each period, and an offset does not affect much the shape over one period. The gain of the modulator in the discussed case is as follows:

$$G_{PWM} = \frac{\Delta D}{\Delta V_{ERROR}} \tag{10.19}$$

Figure 10.26(a) shows the ideal duty-cycle with the flat input signal, whereas (b) and (c) show the impact of different shape and phase shift of the input signal to the gain of the PWM.

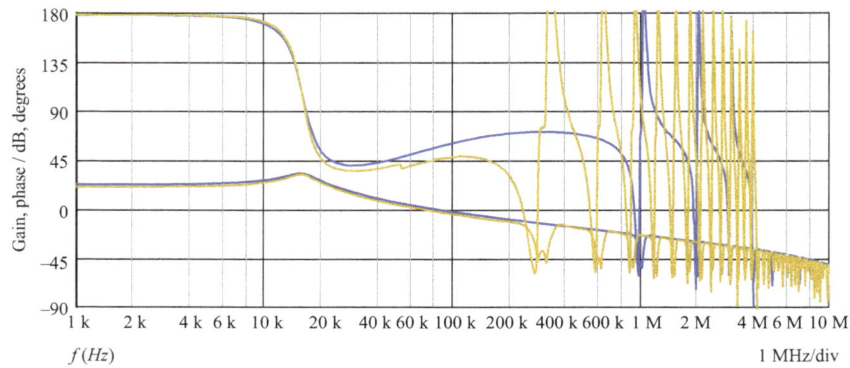

Figure 10.25 Open-loop transfer function with various switching frequencies f_S; blue – $f_S = 1$ MHz; yellow – $f_S = 300$ kHz

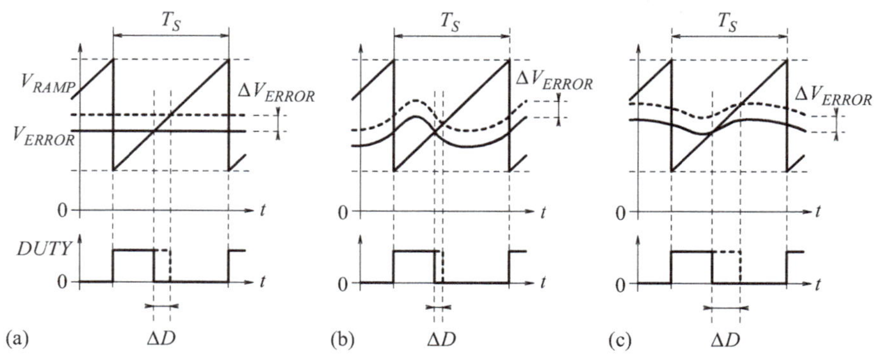

Figure 10.26 Effect of the shape of the error signals: (a) flat error signal; (b) error more vertical when reaching ramp; (c) error more parallel to the ramp

When the input signal hits the ramp signal in a more vertical manner, the gain of the modulator G_{PWM} can be significantly lower than that in the normal case. It is shown in Figure 10.26(b). Theoretically, if the input signal was vertical, the small-signal gain would be 0. Similarly, when the input signal is close to being parallel to the ramp signal, the gain is going to be very high. This case is depicted in Figure 10.26(c).

This effect can change the gain significantly. Fortunately, it is not very apparent when the loop is relatively slow with respect to the switching frequency (cross-over frequency $<f_S/10$) because the error signal is usually flatter. On the other hand, this effect cannot be easily modelled in the linear model, and it may significantly complicate the design of converters with very fast response.

10.2.15 Comparison of the averaged and switching models

The simplified mathematical model predicts the position of the important poles and zeroes accurately and allows for correct shaping of the frequency response. The averaged model of the converter connects the small-signal s-domain model with the time response, which can be compared with the switching model. The comparison is important as it proves that the linear model is reasonably close to the switching one.

There is one important rule with regards to the comparison. Both models must stay in the normal mode of operation, where none of the special features implemented in the switching model come into the picture. It means no current limit, no DCM operation and similar. The switching model needs to be in a mode of operation that the averaged model is designed for. Otherwise, incorrect results will be obtained.

A good approach for the comparison is to apply the same external parameters to both models and apply a transient event like load transient or similar. It is important that the switching model must not leave the region where we can assume close-to-linear operation – i.e. must not reach the current limit, for example.

10.2.16 Limitations of the averaged model

This section summarizes the limitations of the averaged model. A few of them have already been mentioned and a few new ones are added.

1. Phase roll-off at frequencies close to f_S: The predicted phase-margin might not be as high in the switching model, so the closer to f_S the larger margin needs to be taken. It is usually not a problem for a converter loop with cross-over frequency $<f_S/10$. This effect is clearly visible in Figure 10.25.
2. Variation of the PWM gain: This effect is more visible with fast loop. It cannot be easily modelled; however, it is possible to add a trimming constant for the gain. On the other hand, the constant should be checked over the whole duty-cycle range as the gain can significantly change with duty-cycle.
3. Large load transient applied to a switching model may lead to an abnormal mode of operation where the buck converter works with 0% and 100% duty-cycle. This mode leads to a large-signal oscillation, which is not covered by the averaged model. One of the root-causes which can cause those types of oscillation is that the buck keeps 0% or 100% duty-cycle over more than one period, which effectively lowers the switching frequency.

The following points are related to the differences between the switching and averaged model in transient simulations. Those usually do not affect the normal small-signal transfer function response.

1. The average model assumes synchronous rectification meaning that both switching elements in the power stage are controlled by the DUTY signal. In the case of the asynchronous operation (i.e. the low side switch is implemented by a simple diode with its cathode on the LX node), it may occur that certain conditions may not be taken necessarily into account in the average model

described until now. An example is the case, in which the inductor current I_L becomes negative. However, such a feature can be theoretically included in the averaged model.

2. The basic averaged model does not cover current limit features, which are essential features of almost all DC–DC converters. In the case of fast load transient, even when the current limit is reached, the models behave differently.
3. During a fast load transient, the coil current I_L might not be able to follow the error signal in the switching model. However, the averaged model can react faster, since the duty signal can be outside of its range, 0–1 V. An easy fix for this issue is to limit the signal at the output of the PWM model to the correct range.
4. In the case of very fast DC–DC converters, the reaction of the buck to a fast load transient might depend on the timing with respect to the clock signal and immediate position of the coil current. If the load transient occurs just after the PWM trips, the buck needs to wait until the next clock cycle, which makes the reaction of the switching model worse. On the contrary, if the load transient is applied before the PWM trips and the EA can elevate the error signal quickly enough, then the switching model starts reacting with coil current I_L higher than the average current, so it shows better performance than the averaged model.

In this section, an introduction into the modelling of a DC–DC converter was shown using an example of a simple buck converter. The fundamental principle of averaging and using the averaged signal was shown as well as the model for PWM and power stage. Those principles are in general applicable to other types of converters. However, it is beyond the scope of this chapter to cover all the converter types in detail.

10.3 Efficiency modelling

In most portable systems, high-efficiency switching converters are used to extend battery life [4]. High efficiency is also requested in power systems to reduce thermal loss, thus preventing the system from overheating. Thus, efficiency has become a critical design parameter in many power systems, specifically switching converters.

To design a switching converter with optimum efficiency, the conventional design approach is to generate an efficiency model of the switching converter as a primary step. The model will incorporate the key loss mechanisms as well as various operation modes of the converter (e.g. sleep mode [5], pulse-frequency modulation (PFM) mode [6], and PSM mode [7]). This model helps in understanding trade-offs between design parameters and provides fast insight into system performance.

The following text describes an efficiency model of a DC/DC converter derived from the critical loss mechanisms, followed by example output plots. The principles described can also be applied to other switching converters where similar mechanisms exist.

We shall start by introducing the concept of switching and conduction losses and progress with analysing the major efficiency loss contributors for the case example of a synchronous buck converter. Figure 10.27 gives the major mechanisms of power loss in a synchronous buck converter.

Switching losses (aka capacitive losses) dominate at low loads and occur due to switching of pass devices, pass device drivers, and the parasitic capacitances at other switching nodes [8]:

$$P_C = Cf_s V_{DD}^2 \tag{10.20}$$

where f_s is the switching frequency of the buck converter, V_{IN} is the input supply voltage, and C is the equivalent capacitance observing the switching which is a lump sum of corresponding pass device capacitors, pass device driver capacitors, layout, and printed circuit board (PCB) parasitic capacitors.

Conduction losses (aka resistive losses) dominate at high load currents, given by [1]:

$$P_R = I_L^2 r + \frac{\Delta I_L^2}{12} r \tag{10.21}$$

where r is the equivalent resistance in the load current path, the first term of the equation is the DC component of coil current, and the second term is due to coil current ripple. Conduction losses occur due to the resistance of pass devices, DC resistance (DCR) of the coil, and parasitics of PCB traces and other external components in the current path.

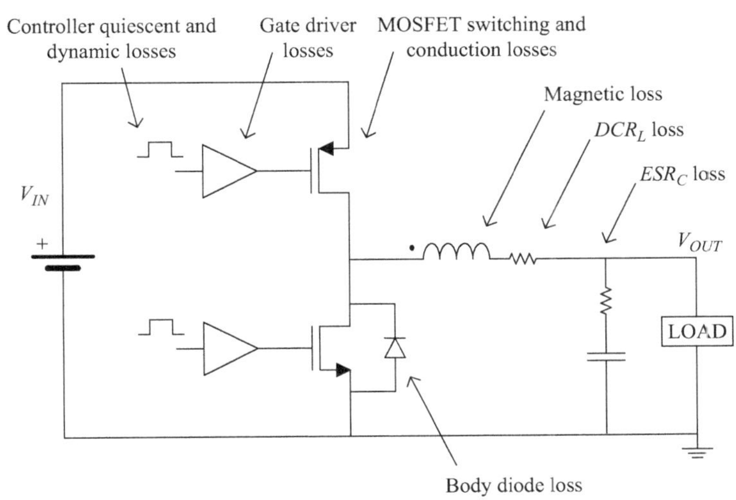

Figure 10.27 Mechanisms of power loss in a synchronous buck converter

Magnetic losses in the coil are harder to model. The traditional approach is loss estimation using a set of curves provided by the coil manufacturer. More practically, the core loss curves can be fit into a so-called Steinmetz equation:

$$P_{core} = A f_s^X B_{\max}^Y \tag{10.22}$$

where P_{core} is the power loss due to core losses, f_s is the switching frequency, and values of A, X, and Y are listed as core loss parameters [9,10]. Today, many manufacturers also provide webpage-based tools and calculators for analysis and comparison of core losses, see references [11,12]. An in-depth inductor modelling investigation is given in the following sections of this chapter.

Non-overlapping time is employed in synchronous switching converters to prevent shoot-through current among pass devices. During non-overlapping time, both pass devices are in tri-state resulting in coil current flow through the body diodes of the pass devices. The power losses due to non-overlapping time can be expressed as:

$$P_{NOL} = 2\, V_D\, t_{NOL}\, I\, f_s \tag{10.23}$$

where V_D is the body diode forward conduction voltage, I is the current flowing through the body diodes, and f_s is the switching frequency.

The equivalent series resistance (ESR) of the output capacitor also contributes to resistive losses, which can be expressed as:

$$P_{ESR,C} = R_{ESR,C}\, I_{ripple}^2 \tag{10.24}$$

Quiescent current of the analogue operational circuit blocks and control circuitry adds to the power losses with:

$$P_q = I_q\, V_{DD} \tag{10.25}$$

where I_q is the total quiescent current of the functional analogue blocks and V_{DD} is the supply voltage. This term dominates the efficiency losses at the low load current range.

Figure 10.28 gives an example of an efficiency curve of a switching converter model in PFM and PWM modes vs. load current.

In this subsection, it was shown that many design parameters exist (sizing of pass devices, switching frequency, selection of coil, etc.), which take place in the optimization of a switching converter for high efficiency. The model will help in showing the trade-offs between the design parameters and will provide an early performance estimation. The model hence enables a top-down design process methodology starting from system architecture and switching strategy to optimal transistor-level design [13].

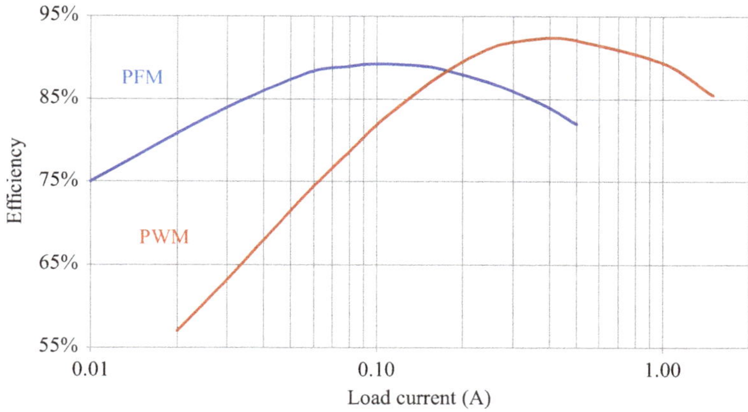

Figure 10.28 Efficiency curves of a switching converter model in PFM and PWM modes vs. load current

10.4 Battery models

The use of batteries is becoming pervasive in a whole variety of appliances. We have been familiar for some years with battery-powered portable applications, such as laptops, tablets, and phones, with more recent adoption in transportation, low power wireless sensor nodes for IoT, medical implants, energy storage to name but a few. Most of these applications use lithium-ion battery technology, which has the advantages of high energy density and low maintenance. There are several different chemistries in use, which vary in the cathode material composition, with the anode usually being constructed of graphite. These different chemistries create trade-offs between cost, performance, charge/discharge rates, and cycle times.

A battery pack consists not only of the chemical cell itself but also contains protection circuitry to prevent overcharge or deep discharge, both of which are safety issues. Battery models presented in the literature fall into one of the following categories. Electrochemical models are used to gain an insight into the physical and chemical behaviours [14]; these require substantial compute resources and detailed knowledge of the internal construction. Based on the molecular level, they solve many partial differential equations and achieve high accuracy. A next step away is the mathematical model, with perhaps some empirical content, used to predict capacity, state of charge (SOC), aging, charge, and discharge behaviour [15]. Electrical equivalent circuit models [16] using lumped parameters are of more use for the design of end applications where the charge, discharge, and SOC are of more interest. The latter will be discussed in more detail next.

In its simplest form, the battery can be modelled as an ideal voltage source in series with a resistor, as in Figure 10.29, to model the internal resistance (R_{int} model), where:

$$V_{batt} = V_{oc} - i(t) \times R_{int} \tag{10.26}$$

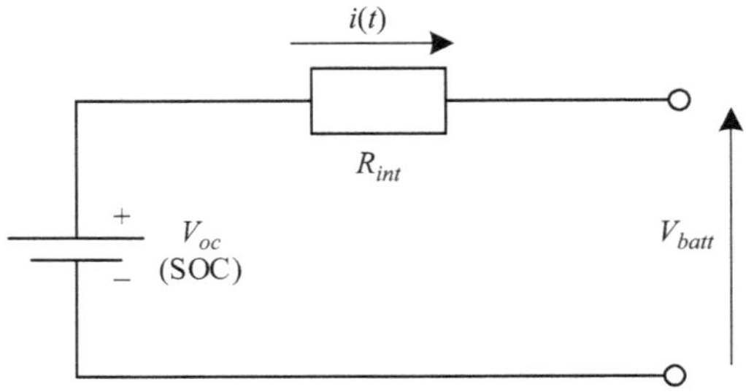

Figure 10.29 R_{int} model

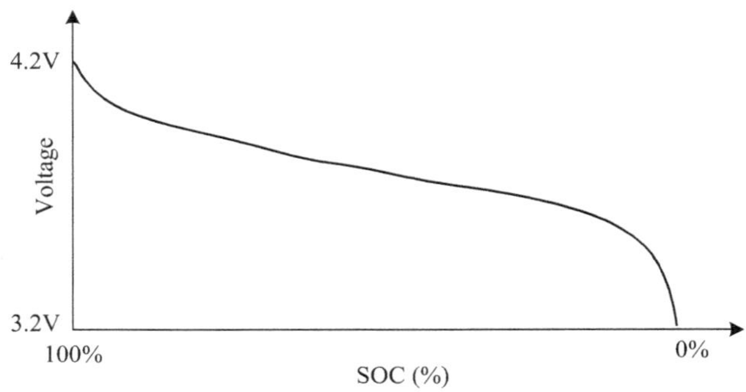

Figure 10.30 Typical SOC discharge curve

V_{oc} is the open-circuit voltage, $i(t)$ is the dynamic internal current, and R_{int} is the internal resistance comprising contact resistance, electrolyte resistance, and separator resistance [17].

V_{oc} will be a function of (i) the SOC, which is the available capacity of the battery expressed as a percentage of the full-rated capacity in ampere–hours, (ii) the state of health which is a measure of the ageing and reflects the ability to be charged and discharged compared with a new battery, and (iii) temperature. At lower temperatures, chemical activity is reduced, resulting in a perceived increase in internal resistance and reduced SOC [18]. At higher temperatures, chemical activity increases, which reduces the internal resistance and raises the SOC and V_{oc}. Some self-discharge can occur at elevated temperatures, which can diminish the V_{oc}.

Because the behaviour of the battery over temperature, charge and discharge rates will vary with its chemistry as well as its form factor, capacity, packaging, etc., the model is best fitted against measurements.

R_{int} can be determined by several methods, including the current step method [19]. V_{oc} can be determined from the discharge characteristics of the cell, which has the behaviour shown in Figure 10.30.

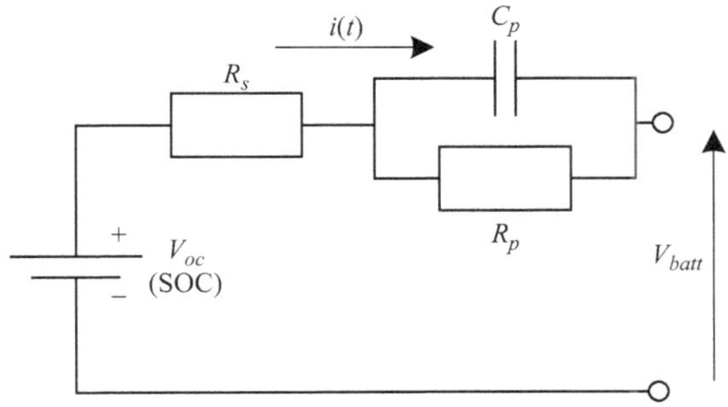

Figure 10.31 First-order RC model

The actual SOC curve is dependent on the discharge rate and temperature (T):

$$V_{oc} = V_{oc}(SOC, T, t) \tag{10.27}$$

$$SOC(t) = SOC(0) - \int_0^t \frac{i(t)}{Q} \tag{10.28}$$

where Q is the battery capacity.

The simple model does not include the dynamic effects of the battery. A next step up is the first-order RC model in Figure 10.31.

The inclusion of R_p models the charge transfer resistance and resistance of the porous electrodes. The capacitor, C_p models the diffusion capacitance.

Higher-order models [20] can be constructed with more RC branches to account for diffusion-related space charge and charge transfer effects in the porous electrode double layers, with the task then of fitting the battery characteristics to the model [21].

The second-order RC model [22–24] more accurately models the short and long time-constants of the battery's transient response.

10.5 Capacitance modelling

Capacitors are used in power systems in output filters, input load current filters, charge pumps, and bootstrap circuits. Indeed, they are of critical importance, so having a model of their real behaviour is essential to model the whole system accurately. Correct modelling of the capacitor in addition to the power delivery network is essential for determining output ripple, load transient, and input supply impedance.

Capacitors exhibit many nonidealities. An equivalent circuit model is shown in Figure 10.32.

The construction of the capacitor results in an ESR, which accounts for all the resistive losses and an ESL (equivalent series inductance). In practice, ESL is not constant but tends to decrease at higher frequencies, which can be modelled as a

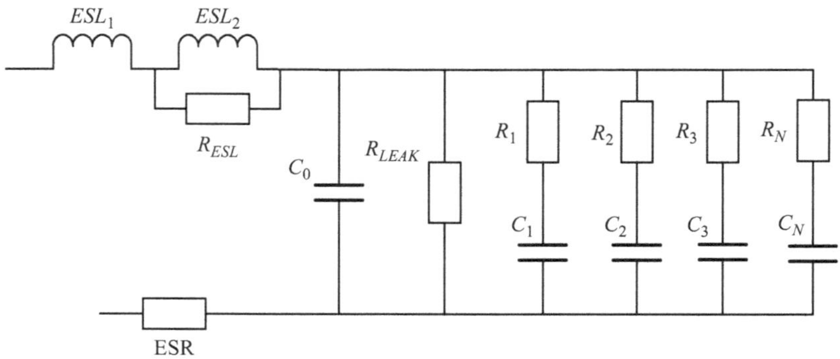

Figure 10.32 Capacitor equivalent circuit with dielectric loss

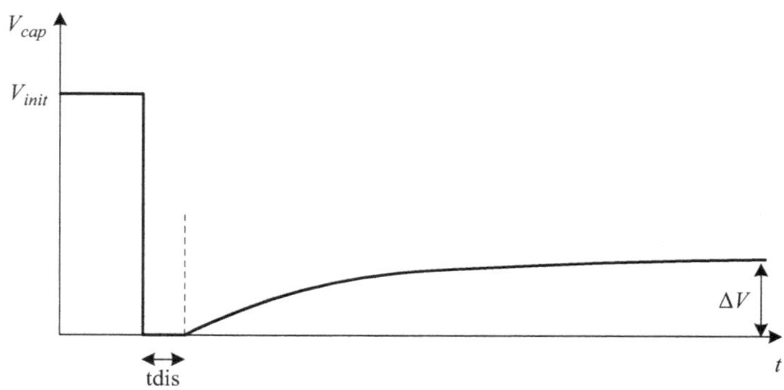

Figure 10.33 Capacitor equivalent circuit with dielectric loss

resistance, R_{ESL}, in parallel with part of the ESL. The leakage resistance R_{LEAK} results from dielectric losses inherent to the material.

C_1 and R_1 through to C_N and R_N model the dielectric absorption effect (aka soaking effect and voltage remanence). This effect is more important in sampled-data systems, rather than power systems. However, as will be seen below in some power system configurations, this may be important. When a capacitor has been charged for a long period and is then briefly discharged, the capacitor is then able to recharge itself to a small degree, due to the dipole relaxations within the dielectric material [25,26]. This reluctance of the material to give up all its charge immediately manifests itself as a type of memory. The effect is strongly dependent on the dielectric material and is more pronounced in ceramic capacitors, which are commonly used in power supply systems. The behaviour is shown in Figure 10.33.

The parallel RC branches model the different frequency-dependent dipole relaxations. This is effectively several poles of different frequencies, which can also be conveniently modelled as a fractional pole [27].

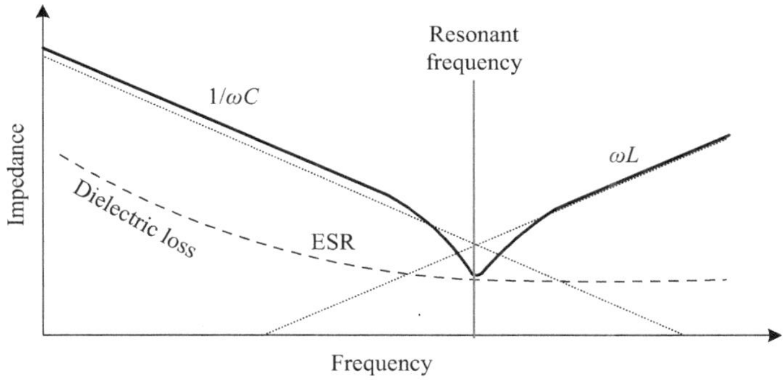

Figure 10.34 Capacitor impedance and ESR characteristic

In power systems, the capacitor is normally being driven, so the circuit can both source and sink current to and from the capacitor. In a one-sided system, such as a single-sided driven low-dropout (LDO) linear regulator, it can only source current, so if there is a request to decrease the output voltage, for example, a dynamic voltage control down command, the LDO has to rely on the load to pull down the output. In such a scenario, the dielectric absorption may well need to be considered.

Multilayer ceramic capacitors are popular for input and output filtering in switched-mode power supplies due to their high capacitance density. The Class II X7R material is popular, exhibiting high-temperature stability, but suffers from a significant DC voltage coefficient. The voltage-dependent capacitance could be modelled with a lookup table [28] or more efficiently with a non-linear (quadratic) capacitor or voltage-dependent capacitor model (varactor) available in many simulators. It is important to ensure that charge is conserved in the non-linear model to avoid simulation errors.

In a similar manner, many simulators provide models in which first and second-order temperature coefficients can be defined.

The simple model results in an impedance characteristic similar to Figure 10.34.

10.6 Modelling the inductors

Inductors are passive components for effective and efficient power management. They are typically used to store energy temporarily by building a magnetic flux generated by an AC current flowing through them, while any DC current can pass freely. The generated magnetic flux, usually indicated as ϕ, must effectively be stored in a magnetic material which represents the core of the inductor.

Inductors are manufactured in various forms and shapes following the same principle of winding a conductor in turns (various typologies include spiral, stripline, toroid, solenoid, etc.) to help confining the magnetic flux generated by the current that flows in the conductor.

In passive components, the key design challenge is to maximize energy storage while minimizing losses and volume. The magnetic properties are relative to the magnetic permeability μ (measured in Henry per meter). Permeability is linked to the magnetic vectors by well-known relationship:

$$\mu = \frac{B}{H} = \frac{\phi A_m}{NI} \tag{10.29}$$

where B and H indicate the magnetic field vectors (flux density and magnetizing force), A_m is the area through which ϕ flows, and $N \times I$ is the current times the number of turns of the wire wound in the inductor.

The measure of the performance of an inductor is its inductance (air-core inductors typically have inductances of around 1 nH/mm per length of the wire, while for core materials inductance is about 20 times or higher).

The inductance, usually indicated as L, is defined by the geometry of the magnetic core and the relative permeability μ of the core material following the classic derivation from Maxwell's laws:

$$L = \mu N^2 \frac{A_m}{l_m} \tag{10.30}$$

where l_m indicates the length of the average magnetic path. Figure 10.35 illustrates the classic representation of a magnetic core and the wire wound around it.

Cores are made of ferromagnetic materials, which can store energy temporarily and are categorized as 'hard' or 'soft' based on how easy it is to magnetize and demagnetize them.

Soft magnets make it easy to change magnetization direction; hence, they are ideal for storing and releasing energy efficiently; therefore, they are preferred for the efficient switching properties required for the passive devices. Soft materials are usually characterized by high permeability, as Figure 10.36 illustrates.

Inductive components require high μ and small volumes to store and release a large amount of energy, although the side effect of high μ magnetic core materials is that for large current handling (indicated by $N \times I$), flux saturation (B_{sat}) can be

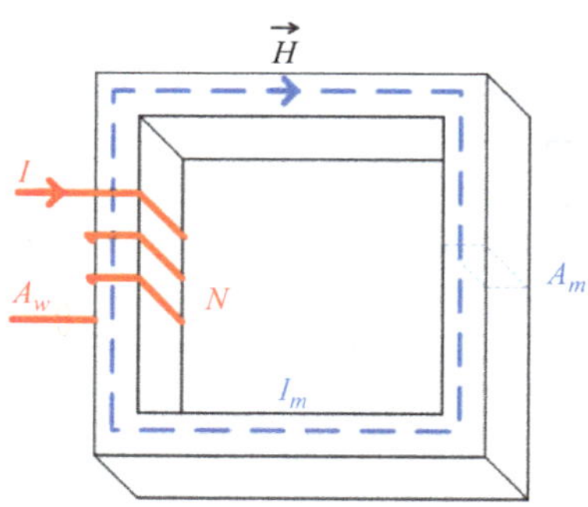

Figure 10.35 Magnetic core and geometrical parameters

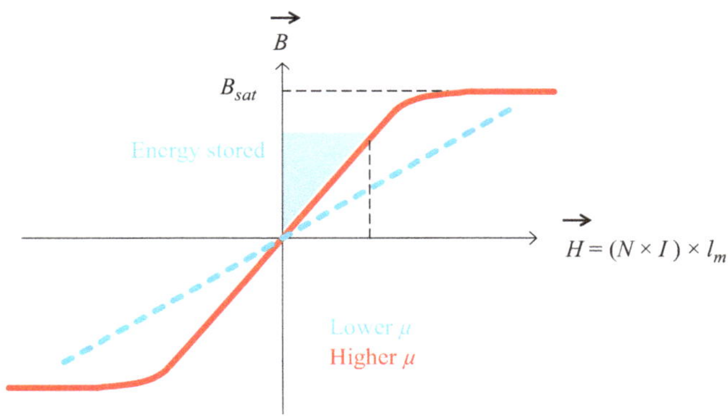

Figure 10.36 Typical B/H curve

reached earlier at which point energy can no longer be stored [31]. The transition into magnetic saturation is usually not sharp, as illustrated in the previous figure.

$$I_{sat} = \frac{NB_{sat}A_m}{L} \tag{10.31}$$

The equivalent saturation current expressed by (10.31) is a fundamental parameter in passive components and is indicated as the value at which the inductance no longer remains constant but starts to drop. Manufacturers often specify this parameter in inductor datasheets with 10%–20% margins to warn users to avoid operation in which the inductors may enter nonlinear behaviours as it will be described later in this chapter.

10.6.1 Spice modelling

To model an inductor for Spice simulations, we introduce a conventional simplified model that includes the basic components to which we can add further advanced contributions to consider losses and saturation effects.

With respect to Figure 10.37, L and DCR indicate the nominal inductance and the series DCR of the windings. R_p is the parasitic parallel resistance, which is usually neglected, as it is several orders of magnitude higher than the equivalent impedance at the terminals of the inductor model. C_p is the parasitic parallel capacitance and it depends on the package size, number of turns, coil shape, etc.

Values of the L, DCR, R_p, and C_p components can often be found in the inductor datasheet of choice.

The series DCR represents the equivalent resistance of the conductor wire in the windings and follows a simple equation:

$$DCR = \rho N \frac{l_w}{A_w} \tag{10.32}$$

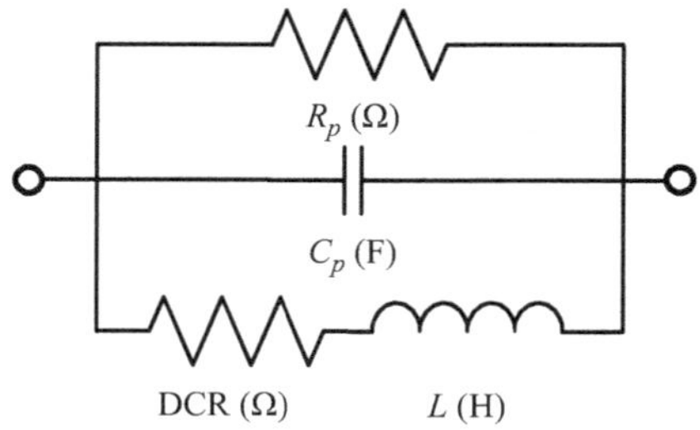

Figure 10.37 Basic Spice model of an inductor

where ρ is the conductivity of the wire material (for copper at room temperature is 1.724×10^{-6} Ω cm), N the numbers of turns, l_w the length of the turn, and A_w the area of the section on the wire.

Intuitively, larger coils built with a large N around a core material present not only a large inductance value but also higher DCR values.

DCR is an important parameter to model and except for the physical conductivity it depends on the geometry of the inductor itself; its value defines the DC conduction losses of the inductor but it also, together with the inductance value, contributes to the phase shift of the small-signal in the control loop of the converter as it can be found extensively described in the literature dedicated to the control theory of switching mode power supplies.

Let us now consider the AC behaviour of the basic model in Figure 10.37. The expression of the equivalent impedance Z_l is given by the following equation:

$$Z_l(\omega) = \frac{(j\omega L + DCR) \times (1/j\omega C_p)}{j\omega L + DCR + (1/j\omega C_p)} = \frac{(DCR + j\omega L) \times \left[(1 - \omega^2 LC_p) - j\omega DCR\, C_p\right]}{(1 - \omega^2 LC_p)^2 + (\omega DCR\, C_p)^2} \tag{10.33}$$

where the condition:

$$\omega_0^2\, LC_p = 1 \tag{10.34}$$

defines the resonance frequency at which the impedance simplifies as

$$Z_l(\omega_0) = \frac{\omega^2 L\, DCR\, C_p - j\omega DCR^2\, C_p}{(\omega DCR\, C_p)^2} = \frac{L}{DCR\, C_p} - \frac{j}{\omega C_p} \tag{10.35}$$

Figure 10.38 shows the expected behaviour of the impedance with respect to frequency as modelled by the previous equations (10.33) and (10.35). The four regions indicated are as follows:

1. Low frequency, DCR dominates but this shall be a region in which the inductor does not operate.
2. Standard inductor region of operation.

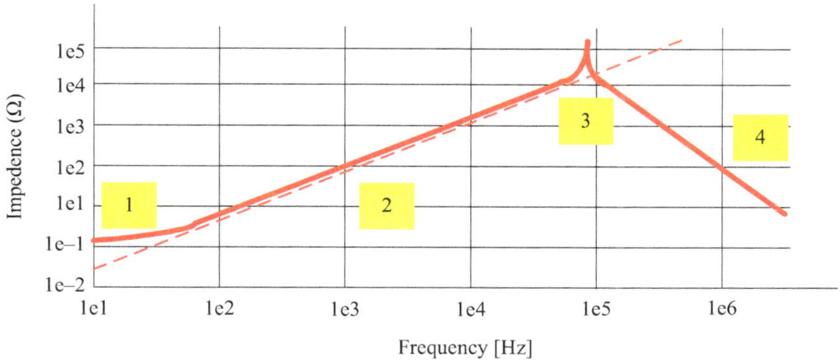

Figure 10.38 AC behaviour of a basic inductor model

3. The area close to the resonance frequency, which should be avoided.
4. Capacitive behaviour dominates, this area should be avoided.

10.6.2 Advanced modelling

As the frequency of operation increases (in the order of several kHz or MHz), spurious effects start to manifest in the form of AC losses that may be worth considering in the model of the inductor in order to give a fair estimation of the efficiency of the power conversion [30,32]. Such losses are not easy to measure and can sometimes be found as readily available data from the inductor's manufacturer. Figure 10.39 summarizes the most common AC losses and is all in various ways all dependent on the frequency of operation of the converter.

The contribution from the core losses is due to the hysteresis when the magnetic field changes direction (any amount of stored energy requires an amount of demagnetizing force necessary to demagnetize the material). In simple terms, the total energy exchanged in an inductor in a DC–DC converter, while in operation can be indicated as:

$$E = \int_0^T v(t)i(t)dt \qquad (10.36)$$

This can be transformed by applying Ampere's and Faraday's laws simultaneously:

$$E = l_m A_m \int_0^B H\, dB \qquad (10.37)$$

The integral corresponds to the shaded blue area in Figure 10.40. This energy, which is proportional to the volume of the magnetic core $l_w \times A_w$, is completely transferred to the load when the inductor demagnetizes except for the portion bounded by the hysteretic region which is lost as dissipated power proportional to the switching frequency f_{SW} of the converter:

$$P_{core} \sim \alpha B_{sat}^{\beta} f_{sw} \qquad (10.38)$$

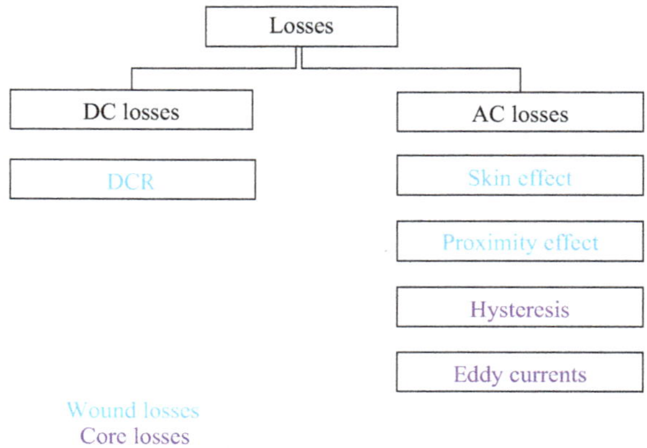

Figure 10.39 AC and DC inductor losses

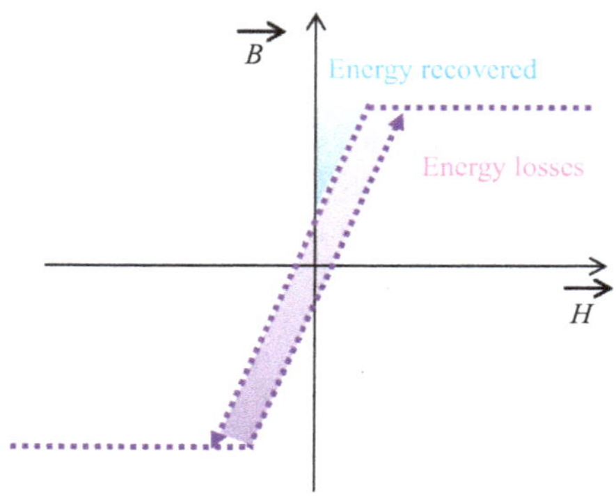

Figure 10.40 Simplified B/H diagram showing hysteretic losses

where α and β are constants. One can obtain measurement data plots of core losses in respect to various frequencies (usually applied sinusoidally) directly from the inductor suppliers.

The eddy currents are currents induced in the core material in response to the changing flux as per Lenz's law; hence, the dissipated power can be regarded as:

$$P_{eddy} = \frac{v_{core}^2}{R_{core}} \sim \frac{\gamma (B_{sat} f_{sw})^2}{\rho_{core}} \tag{10.39}$$

where v_{core}, R_{core}, and ρ_{core} are the induced voltage, resistance, and resistivity of the core material, and γ is a constant. It can be noted that also, in this case, there is a clear dependency from the switching frequency (quadratic).

The contribution of the winding losses is due to the skin effect (the density of the current increases near the surface at higher frequencies) and proximity effects

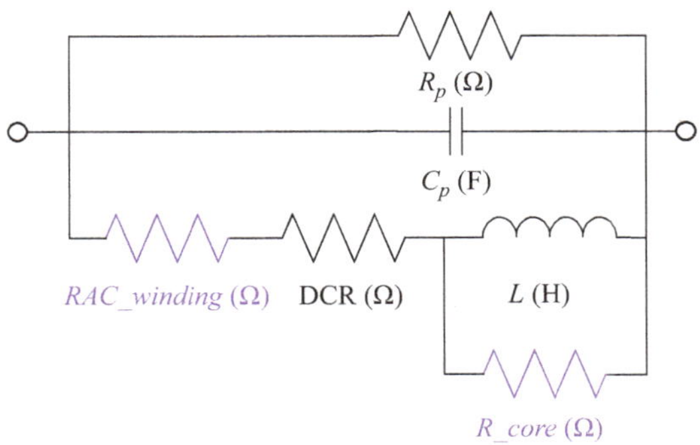

Figure 10.41 Spice model, including AC losses components

(that arise from introducing eddy currents in adjacent wires in the wound wire) both originated in the conductor wires.

The proximity effects are not simple to describe or model, but the skin effect can be represented as the reduction in the section of the conductor, which in the case of a cable of the circular section can be reduced to:

$$R_{skin} = \sim \rho \frac{l_w}{\pi(r - D_{skin})^2} \tag{10.40}$$

where r is the radius of the circular section and D_{skin} the skin depth defined as:

$$D_{skin} = \sqrt{\frac{\rho}{\pi \mu f_{sw}}} \tag{10.41}$$

To add the AC contributions to the basic model, we can group the contribution of the core losses (hysteretic and eddy losses) into an *RAC_core* component and the contribution of the winding losses (skin and proximity effects) into an *RAC_winding* component as shown in Figure 10.41.

Please note that *RAC_core* and *RAC_winding* matter only for the AC component of the current and voltage in the inductor (ripple); hence, it is often simpler in simulations to maintain the basic model and calculate the loss contribution to the conversation efficiency separately by applying (10.39)–(10.42).

10.6.3 Saturation current effects and modelling

As anticipated earlier, inductors saturate as a consequence of excessive currents. The effect is illustrated in Figure 10.42, showing how the inductance enters a roll-off region of high nonlinearity before reaching a minimum value.

The operation in weak saturation in a converter (through the example of a buck converter as given in Figure 10.43) means that the current in the coil no longer follows a linear relationship with respect to the input voltage, i.e.

$$I_{L_{PEAK}} = I_{L_{AVG}} + \frac{V}{2L} T_{ON} \tag{10.42}$$

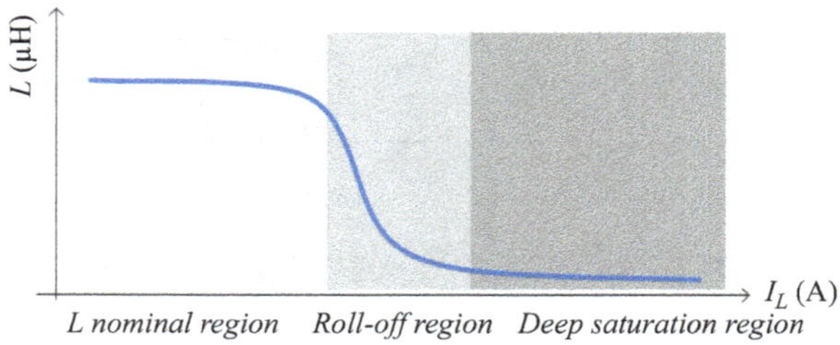

Figure 10.42 Saturation effects on an inductor

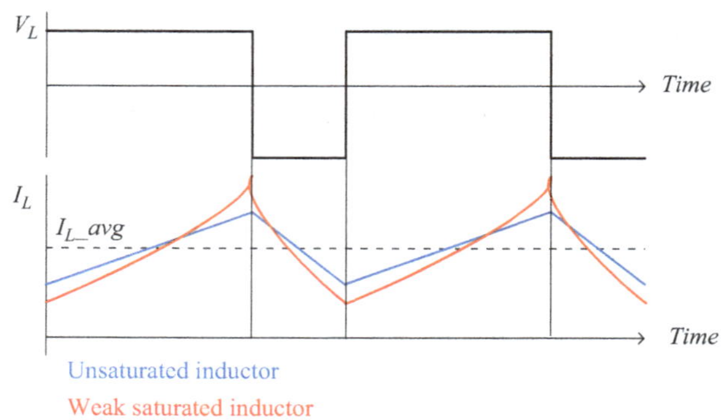

Figure 10.43 Effect of a saturated inductor on a switching converter. V_L/I_L are voltage/currents in the inductor

where the first term represents the average current in the inductor and the ripple current is represented by the second term. When saturation occurs in an inductor, the immediate inductance starts immediately to drop. The inductance drop exhibits a faster current rise at a higher current level, which leads to a higher current ripple, for which a standard current mode control often becomes impractical. Moreover, since the L drops as the current increases, the peak current may also further increase as well in response to that; this can potentially activate the over-current protection in the converter, although if the response time of such protection is not adequate, damages to the system may occur as an effect of excessive dissipation.

Attempts to model accurately the saturation conditions have been done in the recent years [29], comparing the behaviour of different core materials and related second-order effects, such as temperature, and also using advanced statistical techniques to help the user to allow to enter the zone of roll-off and therefore extend the region of operation of the currents. This is of particular interest for power-management applications, such as aerospace, in which the size and weight of the inductors can be a limitation, although in the majority of situations it is still advisable to avoid the saturation; thus, designers of power supplies may benefit from a simple but practical model of the inductance decay.

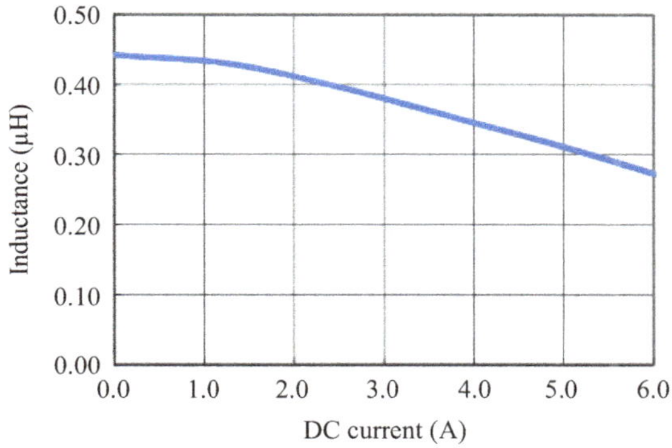

Figure 10.44 MATLAB model of an inductor using polynomial coefficients

Tools or languages for modelling, such as Spectre, MATLAB, or Verilog-A, allow defining the inductance L in the form of a polynomial vector like:

$$L(I) = L_{NOM}\left(1 + a_1 I + a_2 I^2 + a_3 I^3 + \cdots.\right) \tag{10.43}$$

Usually, a second-order polynomial approximation is a good enough representation. The extrapolation of the coefficients can be done in an empiric way by setting, for instance, L = *Lnominal-20%* at the saturation current. Figure 10.44 illustrates a model simulation based on such method.

Please note that the polynomial approximation can be extended further to model the temperature behaviour:

$$L(I) = L_{NOM}(1 + b_1(T - T_{NOM}) + b_2(T - T_{NOM})^2 + b_3(T - T_{NOM})^3 + \cdots) \tag{10.44}$$

10.7 Conclusion

An introduction to the topic of power systems modelling has been given. There are many aspects of power systems that make their modelling different from other analogue systems; however, constructing a small-signal model provides fast analysis of the behaviour, subject to certain limitations as having been discussed. The efficiency of a power system is determined by many parasitic components not just in the core converter but many times also in the external components. An insight into the nonideal behaviours and lossy mechanisms of capacitors and inductors has been given. Battery modelling is an area of active research driven by the plethora of low power wireless applications and electric vehicle adoption, and the reader is referred to the references for further reading on this topic.

References

[1] R. W. Erickson and D. Maksimovic, "Fundamentals of Power Electronics," New York. NY, Springer, 2001.

[2] R. Ridley, "An Accurate and Practical Small-Signal Model for Current-Mode Control," Bradenton, FL, USA, Ridley Engineering, 1999.

[3] R. B. Ridley, "Power Supply Design, Vol 1: Control," Bradenton, FL, USA, Ridley Engineering, 2012.

[4] G. Pistoia, "Battery Operated Devices and Systems: From Portable Electronics to Industrial Products," Amsterdam, The Netherlands, Elsevier, 2008.

[5] J. Xiao, A. V. Peterchev, J. Zhang, and S. R. Sanders, "A 4-μA quiescent-current dual-mode digitally controlled buck converter IC for cellular phone applications," *IEEE Journal of Solid-State Circuits*, 39(12), 2342–2348, 2004.

[6] B. Sahu and G. A. Rincon-Mora, "An accurate, low-voltage, CMOS switching power supply with adaptive on-time pulse-frequency modulation (PFM) control," *IEEE Transactions on Circuits and Systems I: Regular Papers*, 54(2), 312–321, 2007.

[7] H.-W. Huang, K.-H. Chen, and S.-Y. Kuo, "Dithering skip modulation, width and dead time controllers in highly efficient dc–dc converters for system-on-chip applications," *IEEE Journal of Solid-State Circuits*, 42(11), 2451–2465, 2007.

[8] S. Musunuri and P. L. Chapman, "Optimization of CMOS transistors for low power DC–DC converters," in *IEEE Power Electronics Specialists Conference*, 2005, pp. 2151–2157.

[9] M. Albach, T. Dürbaum, and A. Brockmeyer, "Calculating core losses in transformers for arbitrary magnetizing currents a comparison of different approaches," in *Proc. IEEE PESC'96*, Baveno, Italy, June 23–27, 1996, pp. 1463–1468.

[10] J. Liu, T. G. Wilson, R. C. Wong, R. Wunderlich, and F. C. Lee, "A method for inductor core loss estimation in power factor correction applications," in *APEC, Seventeenth Annual IEEE Applied Power Electronics Conference and Exposition*, vol. 1. IEEE, 2002, pp. 439–445.

[11] https://www.coilcraft.com/apps/power_tools/compare/

[12] http://www.cyntec.com/CoreLossCalculator.aspx

[13] C. Liang, "Mixed-signal verification methods for multi-power mixed-signal System-on-Chip (SoC) design," in *2013 IEEE 10th International Conference on ASIC*, 2013, pp. 1–4.

[14] V. Ramadesigan, P. Northrop, S. De, S. Santhanagopalan, R. Braatz, and V. Subramaniana, "Modeling and simulation of lithium-ion batteries from a systems," *Journal of The Electrochemical Society*, 159(3), R31–R45, 2012.

[15] L. Cai and R. White, "Mathematical modeling of a lithium ion battery with thermal effects in COMSOL Inc. Multiphysics (MP) software," *Journal of Power Sources*, 196(14), 5985–5989, 2011.

[16] N. El Ghossein, J. Salameh, N. Karami, M. El Hassan, and M. Najjar, "Survey on electrical modeling methods applied on different battery types," in *International Conference on Technological Advances in Electrical, Electronics and Computer Engineering*, 2015.

[17] M. Dubarry, N. Vuillaume, and B.Y. Liaw, "From Li-ion single cell model to battery pack simulation," in *17th IEEE International Conference on Control Applications Part of 2008 IEEE Multi-conference on Systems and Control*, San Antonio, Texas, USA, September 3–5, 2008, pp. 708–713.

[18] R. Rao, S. Vrudhula, and D. N. Rakhmatov, "Battery modeling for energy aware system design," *Computer*, 36(12), 77–87, 2003.

[19] H.-G. Schweiger, O. Obeidi, O. Komesker, *et al.*, "Comparison of several methods for determining the internal resistance of lithium ion cells," *Sensors*, 10, 5604–5625, 2010.

[20] N. Tian, Y. Wang, J. Chen, and H. Fang, "On parameter identification of an equivalent circuit model for lithium-ion batteries," in *IEEE Conference on Control Technology and Applications (CCTA)*, Kohala Coast, Hawai'i, USA, August 27–30 2017.

[21] J. Lee, J. Lee, O. Nam, *et al.*, "Modeling and real time estimation of lumped equivalent circuit model of a lithium ion battery," in *12th International Power Electronics and Motion Control Conference*, 2006, pp. 1536–1540.

[22] M. Chen and G. Rincón-Mora, "Accurate electrical battery model capable of predicting runtime and I–V performance," *IEEE Transactions on Energy Conversion*, 21(2), 504–511, 2006.

[23] L. Gao, S. Liu, and R. Dougal, "Dynamic lithium-ion battery model for system simulation," *IEEE Transactions on Components and Packaging Technologies*, 25(3), 495–505, 2002.

[24] H. Hongwen, R. Xiong, and J. Fan, "Evaluation of lithium-ion battery equivalent circuit models for state of charge estimation by an experimental approach," *Energies*, 4, 582–598, 2011.

[25] M. Kropfitsch, P. Riess, G. Knoblinger, and D. Draxelmayr, "Dielectric absorption of low-K materials: extraction, modelling and influence on SAR ADCs," in *IEEE International Symposium on Circuits and Systems*, Kos, Greece, 2006.

[26] K. Kundert, "Modeling Dielectric Absorption in Capacitors," www.designers-guide.com, July 2004.

[27] K. Kundert, "The Fracpole Suite," www.designers-guide.com, June 2008.

[28] Cadence, "A Nonlinear Capacitor Model for Use in the PSpice Environment," Cadence Application Note, December 2009.

[29] G. Di Capua, N. Femia, K. Stoyka, M. Lodi, A. Oliveri, and M. Storace, "Ferrite inductor models for switch-mode power supplies analysis and design," in *2017 14th International Conference on Synthesis, Modeling, Analysis and Simulation Methods and Applications to Circuit Design (SMACD)*. IEEE, 2017, pp. 1–4.

[30] R. Ridley, "High Frequency Magnetics Design and Modelling," APEC, 2017.

[31] D. E. Pauly, "Power supply magnetics Part I: Selecting transformer/inductor core material," *Power Conversion and Intelligent Motion-English Edition*, 22(1), 23–34, 1996.

[32] T. Eichhorn, "Estimate Inductor Losses Easily in Power Supply Designs," www.powerelectronics.com, 2005.

Chapter 11

A case study for MEMS modelling: efficient design and layout of 3D accelerometer by automated synthesis

Steffen Michael[1] and Ralf Sommer[1]

For several years, micro-electro-mechanical systems (MEMS) have been experiencing dynamic and sustained growth. This development is mainly driven by the automotive industry. One example of such a growing market is tire pressure monitoring systems (TPMS). Initiated by legal regulations in the USA and the EU (in China, laws on the mandatory use of TPMS will apply from 2020), the market has experienced its own momentum as a result of the continuous development of TPMS.

In addition, the positive market development is reinforced by the continuously decreasing costs of sensors. Costs related to sensor development are the development costs, which are reflected in the time from the customer inquiry or specification to the tape out. Although the design of MEMS has been the subject of research and development activities for years, there is still a gap compared to a typical IC design environment. This chapter provides an overview of the current state of MEMS design. Using the example of a three-dimensional (3D) acceleration sensor, new approaches for efficient design and layout are shown.

11.1 Introduction

At the beginning of the MEMS development, designs and simulations are done by the finite element method (FEM) implemented in program packages like ANSYS. For the simulation of MEMS with an electrostatic working principle, tools based on the boundary element method (BEM) like CoventorWare are more suitable with regard to the computing time. This is due to the fact that in contrast to BEM, FEM requires the meshing of the surrounding volume for electrostatics. Since the late 1990s, new MEMS-dedicated tools based on system design methods [1–4] are available on the market like MEMS+ (Coventor), MEMS Pro (SoftMEMS) and SYNPLE (IntelliSense). These tools can be used in a design flow for the sensor

[1]IMMS Institut für Mikroelektronik- und Mechatronik-Systeme gemeinnützige GmbH (IMMS GmbH), Ilmenau, Germany

system development, including MEMS and ASIC (application-specific integrated circuit), due to different hierarchical simulation level, e.g.

In addition, certain topics, like packaging aspects using mixed domain [5] and mixed-abstraction-level simulation [6], yield optimization using statistical methods [7], and even the development of new codesign/simulation platforms for heterogeneous systems [8] were the subject of investigations. These approaches reduced the number of missing interfaces between the different simulation tools. Nevertheless, some basic questions regarding MEMS design are still open. On the one hand, this concerns the selection of a suitable sensor topology with regard to the underlying technology even if a process design kit (PDK) is available. Such a PDK exists, e.g. for the simulation of capacitive inertial sensors processed in the X-FAB technology XMB10 in MEMS+. The PDK provides a library with elements where the technological caused design rules like the length/width ratio of comb fingers are integrated, but the arrangement of the sensor base elements (spring, movable mass and interdigitated comb structures) has to be done by the developer. A check for manufacturability test is not available at simulation level, so the developer must have a deep understanding of the technology with respect to processable sensor structures.

A concept for the development of an integrated design strategy, which is oriented towards the hierarchical ASIC design from the system model to the layout, is presented in [9]. A development and synthesis tool for 1D accelerometer tailored to the X-FAB inertial sensor technology assists designers with an automatic module-based determination of the sensor topology.

In this chapter, the concept respectively, the tool implemented in MATLAB® extended to a 3D accelerometer is presented.

11.2 Synthesis of MEMS designs and layouts – general aspects

At the beginning of sensor development, there is a specification on the one hand and the technology as input data on the other hand, from which one or more suitable topologies have to be derived. In the case of a 3D acceleration sensor manufactured in the X-FAB technology, there is only one basic topology which is described in detail in the following sections. By means of an automated design and layout process, a sensor is obtained from the topology (Figure 11.1).

A sensor can be characterized by its technical parameters as well as the chip area or economic efficiency. Accordingly, there are two different design scenarios – with given technical parameters, the chip area should be minimized, or with a given chip area, the technical parameters should be optimized. Relevant technical parameters in terms of an accelerometer are, e.g.,

- maximum operating acceleration (a_{max} in g),
- sensitivity (S in fF/g),
- shock survivability (A_{shock} in g),

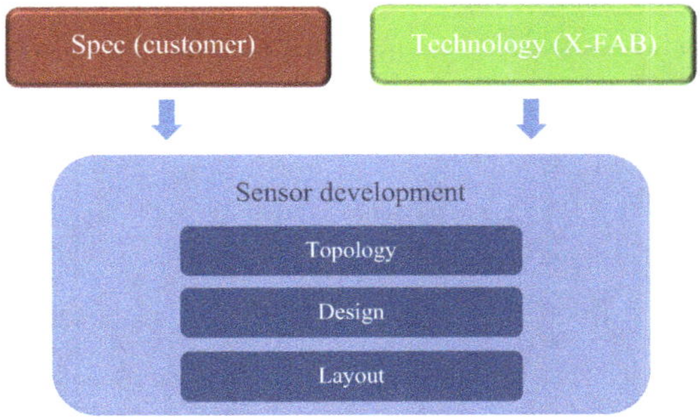

Figure 11.1 From the specification to the sensor

- resonance frequency (f in kHz),
- damping factor (ζ),
- cross-axis sensitivity – an output signal caused from accelerations on a perpendicular axis, $S_{x,cross} = S_{xy}/S_x \times 100$ in %,
- non-linearity – the maximum deviation of output voltage from a best fit straight line.

Due to the large number of input and output parameters, of which some are interdependent, an efficient sensor design by means of parameter calculations is not efficient, respectively, practically impossible in the case of given constraints due to the parameter interdependency. Hence, the use of optimization algorithms is unavoidable. Thanks to various optimization functions MATLAB is well suited for the design of an optimized sensor. An existing interface to a simulation programme, like MEMS+, enables a coupling of the optimization function with the simulation tool.

The implementation of the design tool in MATLAB offers other advantages beyond the optimization possibilities. The object-oriented programming capabilities permit an efficient high-level library design containing accelerometer base elements. Furthermore, the extensive function library of MATLAB is used for the automated generation of layout elements by Boolean functions within the synthesis tool.

11.3 Working principle and sensor structure

Capacitive accelerometers detect the displacement of the movable mass by capacitance changes resulting from a changing distance between electrode structures, one fixed to the substrate and the other one which is part of the moving sensor structure. Differential capacitance arrangements are selected with regard to nonlinearity and electronics (see Figure 11.5).

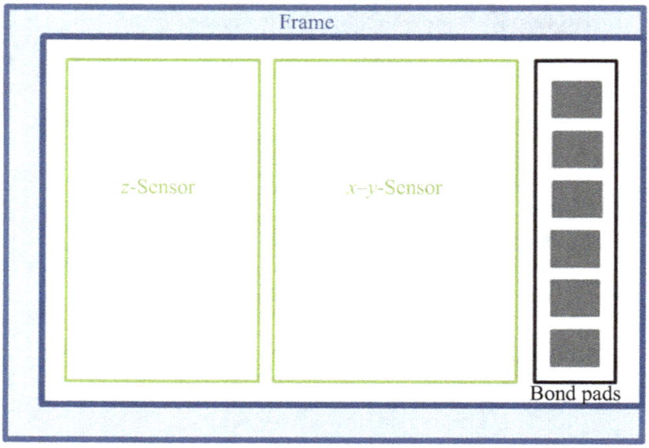

Figure 11.2 Sensor structure (principle sketch)

In principle, different basic structures are possible in the case of 3D acceleration sensors, which differ in the number of movable masses. For example, the simplest sensor structure would be a sensor with three masses, where each mass detects one degree of freedom (DoF). The most complex in terms of spring design is a sensor with a mass that detects all three DoFs. A sensor with three masses requires a large area, whereas a sensor with one mass has too much cross-sensitivity. Hence, a structure with two sensor units (a z sensor and a xy sensor) is an optimal one with respect to die size and cross-sensitivity (Figure 11.2).

11.4 Technology

The open technology platform XMB10 for inertial sensors from X-Fab is chosen as the technology for the design and layout of the accelerometer. The sensor elements are formed by anisotropic deep-reactive etching of the membrane wafer. Currently, membrane thicknesses of 15 and 30 μm are available, with regard to smaller chip areas, the technology is being developed towards thicker membrane wafers.

A feature of the silicon-on-insulator (SOI)-based technology is the electrical isolation of cavity and membrane wafers by an oxide layer. This electrical isolation affects both the capacitive evaluation of out-of-plane movements and possible sensor structures. The detection of the out-of-plane motion by a parallel plate capacitor is not possible due to the missing electrical contact of the cavity wafer. Instead, capacity changes are detected by recessed comb fingers.

Anchor structures can be realized for stopper structures, e.g. Nevertheless, anchor structures cannot be designed as stators due to the electrical isolation. Hence, sensor structures with nested movable masses which require anchored stators are not possible.

11.5 Design strategy and modelling

An essential point in the development of the design strategy is the consideration of the process-related parameter tolerances. Depending on the accelerometer technology, the thickness of the membrane layer and the underetching vary. The underetching influences the sensor characteristics by the changed width of structures like the spring as well as the changed finger gap of the comb structures in different ways. Where a narrower spring leads to a higher sensitivity, the sensitivity decreases at the same time due to a larger comb finger gap.

A two-stage approach is chosen in order to obtain a sensor, which is optimized to the process tolerances. In the first step, the optimal parameters of the comb structures are determined assuming ideal spring stiffness. In the second step, the design of the spring geometry follows. The spring width is optimized in such a way that the output parameter variations are minimized. The spring length results from the required stiffness as well as the calculated width. With respect to a minimum sensor size, the shape of the spring for the xy sensor unit is a meander.

The optimal values of the free parameters mentioned in the following sections are obtained using the MATLAB optimization functions.

11.5.1 Library approach

For the automated synthesis of the design, the differential acceleration sensor is broken down into its basic elements movable mass, capacitive comb and spring structures. With respect to an efficient layout, the higher order elementary structures implemented in the MATLAB class library already contain layout elements.

Figure 11.3 shows an example of the library element interdigital comb structure consisting of the rotor and stator fingers, the stator with metallization and the fixed-to-move transition. The fixed-to-move transition realizes the electrical coupling of the comb structure or stator to the surrounding frame – a trench isolates the silicon of the frame from that of the stator, which is electrically contacted via the metallization process.

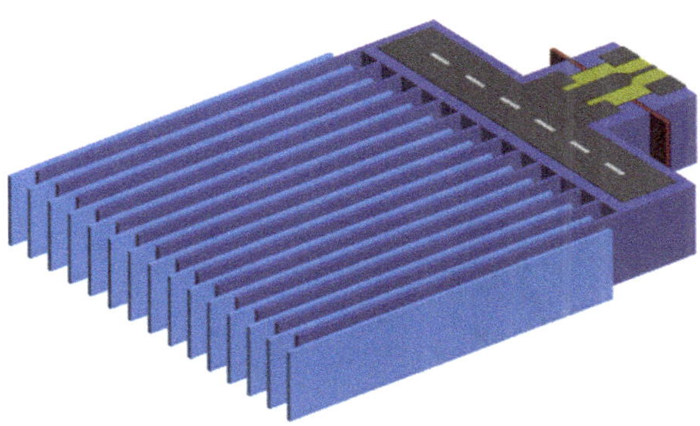

Figure 11.3 Library element comb structure (silicon blue, aluminium grey, trench red and silicon oxide yellow)

11.5.2 Modelling

Requirements for modelling and simulation are on the one hand a high accuracy, which is necessary to calculate the non-linearity of the sensor and on the other hand, a short computing time.

The demand for high accuracy leads to the conclusion that capacitive stray fields must be taken into account in the simulation, so that a simplified consideration of capacitances as plate capacitors is out of the question. A suitable method for the capacitance calculation is the conformal mapping method, where the 3D capacitive field is substituted by the superposition of several two-dimensional fields which include fringing fields [10]. The conformal mapping method is implemented in the software package MEMS+, which is used for the simulation of the electrostatic and mechanical behaviour.

11.6 MEMS design and layout example

For the exemplary demonstration of the synthesis approach, a 3D acceleration sensor is designed and laid out whose specification corresponds to the requirements of a TPMS. 3D acceleration sensors are part of the next TPMS sensor generation. While currently 1D acceleration sensors are part of a TPMS to detect the state of motion of the tire to extend battery life, the next generation of sensors will provide additional driving dynamic tire data (smart tires). The maximum acceleration is assumed to be 500 g, further specifications are a sensitivity of 0.1 fF/g and a maximum non-linearity of 1%. With regard to sensor costs, the optimization goal is a minimum area for the sensor consisting of the xy and the z acceleration units.

11.6.1 xy Acceleration unit

The xy acceleration unit is determined by a symmetric arrangement of the differential interdigitated comb structures in x as well as y to achieve a low cross-sensitivity. Meander springs at all four corners of the movable mass allow x and y displacement. The hole inside the moving mass is for the reception of the stopper structures, which prevent damage to the sensor under shock loading (Figure 11.4).

The capacitance change of a differential comb arrangement as shown in Figure 11.5 is determined by the number (N), length (l) and thickness (t) of fingers, the small (d_1) and the larger (d_2) gap between the comb fingers and the displacement of the movable comb unit Δy.

A suitable choice of parameters can be derived from the simplified analytical description (fringing fields are neglected) of the capacitance:

$$C_P(\Delta y) = \varepsilon_0 lt\left[\left(\frac{1}{d_1+\Delta y}+\frac{1}{d_2-\Delta y}\right)-\left(\frac{1}{d_1-\Delta y}+\frac{1}{d_2+\Delta y}\right)\right] \tag{11.1}$$

With a view to maximum capacitance change, the small gap d_1 should be as small as possible, which is defined by the technology (in this case 2 μm). A larger displacement Δy increases the capacitance change too, but on the other hand, results in

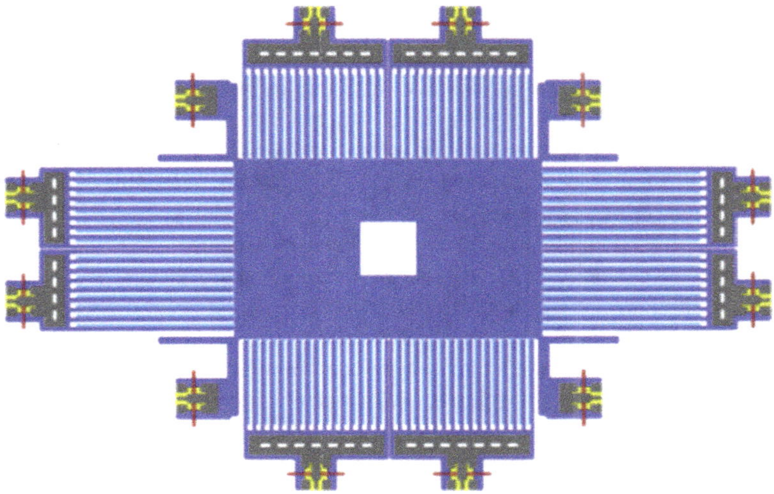

Figure 11.4 xy Acceleration unit with layout elements

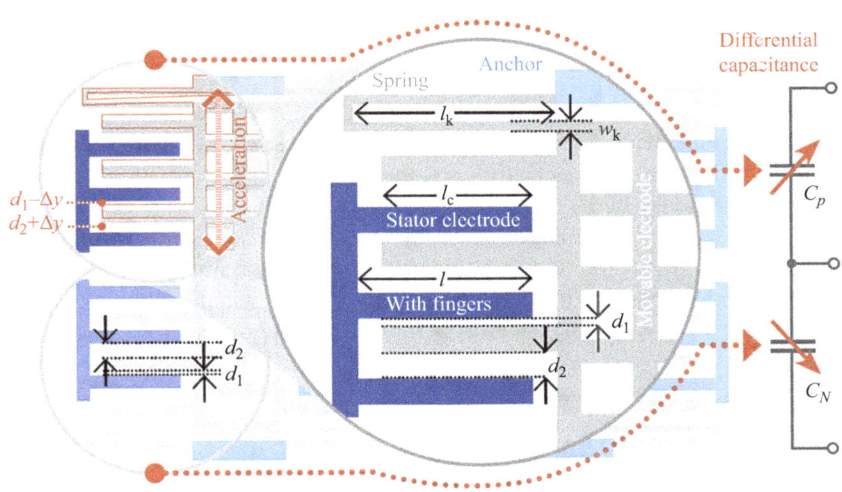

Figure 11.5 Electromechanical basic construction of a differential capacitive accelerometer

a higher non-linearity and a lower spring stiffness, respectively, modal frequency. The non-linearity is a specified parameter. Due to the given dependency between non-linearity and displacement according to Formula (11.1), the displacement is determined by the specified non-linearity.

Finally, the remaining optimizable parameters to a required capacitance value are the finger number and the length as well as the larger gap d_2 which is a trade-off between a large capacitance and a small area (a typical value is about 5 μm).

11.6.2 Accelerometer for z detection

With z or out-of-plane accelerometers, there are two possible electrode arrangements. The displacement of the movable mass can be detected by a plate capacitor. The prerequisite for this is a possible electrical contacting of the cavity or cap wafer, which is not given in the case of the used SOI-based technology. Here, recessed comb fingers as shown in Figure 11.6 allow the detection of capacitance changes caused by out-of-plane movements.

Two different operating principles exist for sensor structures with recessed fingers. The seismic mass can perform a translational or a torsional movement. In contrast to in-plane sensor structures, where the stiffness of the spring is essentially determined by its shape, the relevant parameter for the stiffness of an out-of-plane spring is its thickness which is usually defined by the technology. Accordingly, the thinner the thickness, the easier it is to design springs for translational mass movements.

In the example presented, the thickness is 30 μm. Springs for a translational movement that have a low stiffness in z and a high stiffness in x, respectively, y cannot be designed with such a thickness. Hence, a torsional sensor with symmetric differential interdigitated comb structures is chosen (Figure 11.7).

Fixed parameters are the finger gaps (2 μm) and the finger thickness (defined by technology), the ones to be optimized are the finger number and length.

11.6.3 Layout

The layout of the accelerometer includes the generation of the surrounding frame, sensor wiring and pads, as well as stopper structures. Furthermore, corners are rounded in order to avoid mechanical stresses. Mask data for the processing of the cap wafer are also created.

Like the design, the semi-automatic layout of the sensor is a two-step process. The first step is to generate the mask data for the sensor frame using the Boolean operations implemented in the higher order library. This includes the cavity and the active silicon layer with the oxide layer on top. The algorithm for the automatic

Figure 11.6 Comb structure with recessed fingers

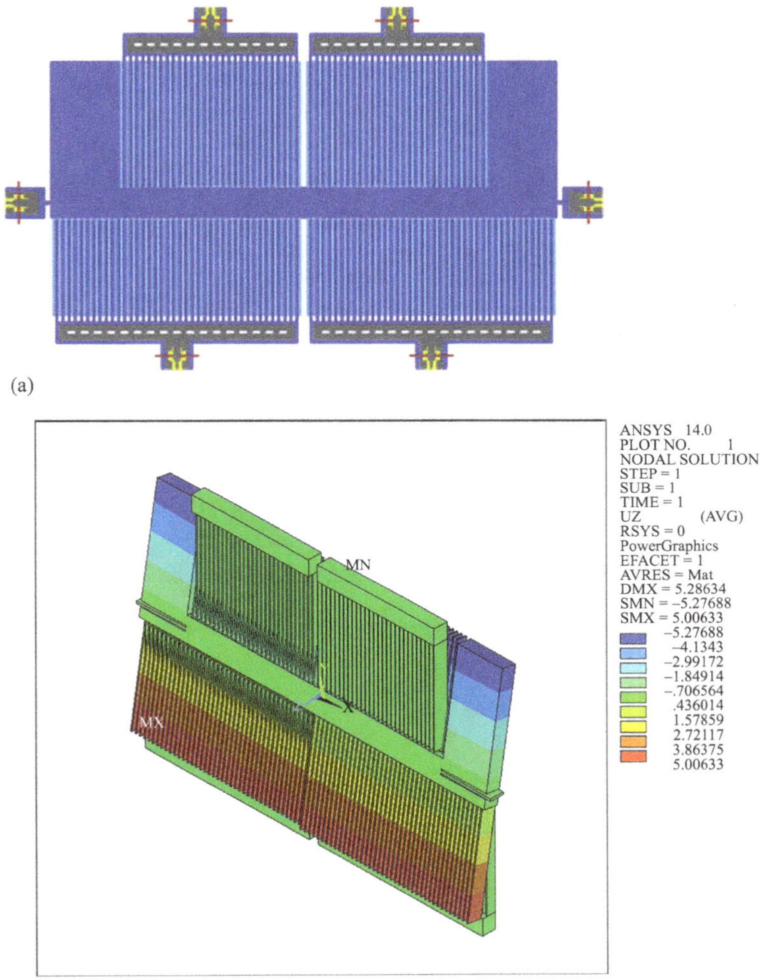

Figure 11.7 z Acceleration unit: (a) unit with layout elements and (b) static simulation

generation of the frame takes into account the design rules of the process, e.g. a minimum distance between the trenches and the cavity. The synthesized comb structures and the moving mass of the sensor already contain electrical connections. The wiring of the electrical connections (among each other and with the connection pads) is done manually; in the case of crossing cables, a layout element 'crossing' is automatically inserted (Figure 11.8).

In the final step, elements, such as bond frames and pads, are generated in Cadence Design Systems using the PDK provided by X-FAB. Figure 11.9 shows the final sensor with the dimensions 1.9×1.9 mm^2. The design rule check and the extraction of parasitic capacitances and resistances are also performed in Cadence.

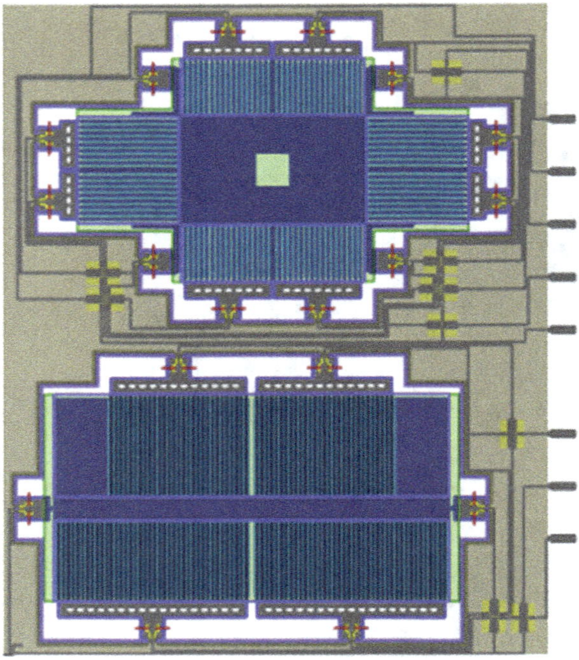

Figure 11.8 Sensor layout after phase I

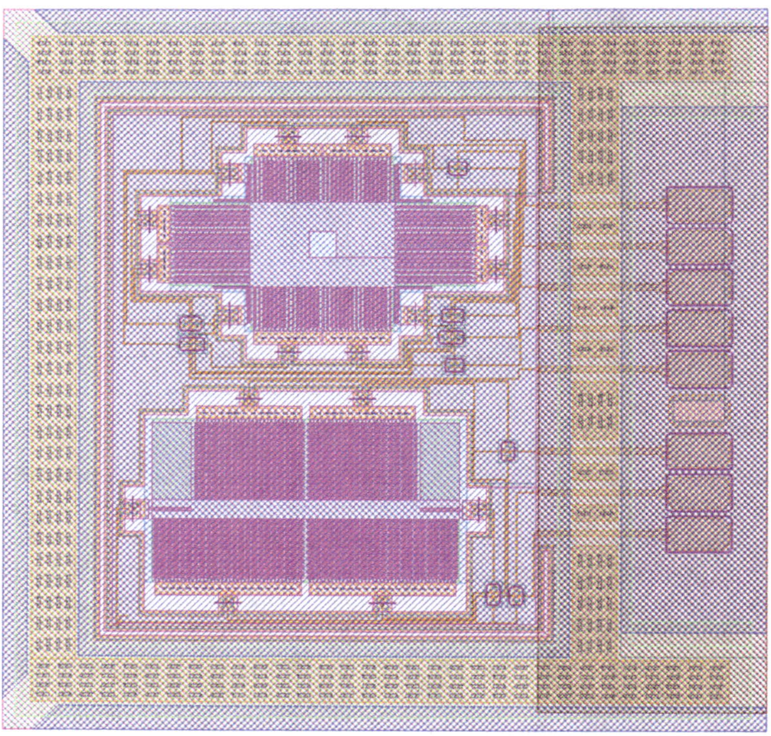

Figure 11.9 Final sensor layout

Acknowledgement

This chapter presents results of the project RoMulus that was funded by the BMBF (German Ministry of Education and Research) under the reference 16ES0362.

References

[1] G. Lorenz and R. Neul, "Network-Type Modeling of Micromachined Sensor Systems," in *Proc. Int. Conf. Modeling and Simulation of Microsystems, Semiconductors, Sensors and Actuators, MSM 98*, Santa Clara, 1998.

[2] D. Teegarden, G. Lorenz and R. Neul, "How to model and simulate micro-gyroscope systems," in *IEEE Spectrum*, 1998.

[3] G. Lorenz and M. Kamon, "A system-Model-Based Design Environment for 3D Simulation and Animation of Micro-Electro-Mechanical Systems (MEMS)," in *Proc. APCOM'07 in Conjunction With EPMESC XI*, Japan, 2007.

[4] R. Fish and S. Breit, "Bringing MEMS into the IC Design Flow," in *Proc. DAC 2100*, 2010.

[5] A. Sanginario, A. Mehdaoui, S. Zerbini, G. Schropfer and D. Demarchi, "New Design Methodology for MEMS-Electronic-Package Co-Design and Validation for Inertial Sensor Systems," in *2015 Symposium on Design, Test, Integration and Packaging of MEMS/MOEMS (DTIP)*, Montpellier, 2015.

[6] G. Schrag, T. Künzig and A. Dehé, "Enhanced Design of Microsystems by Combining Lumped and Distributed System-Level Models," in *2016 Symposium on Design, Test, Integration and Packaging of MEMS/MOEMS (DTIP)*, Budapest, 2016.

[7] M. Schwarz, J. Franz and M. Reimann, "The future is MEMS – Design Considerations of Microelectromechanical Systems at Bosch," in *22nd International Conference Mixed Design of Integrated Circuits & Systems (MIXDES)*, Torun, 2015.

[8] M. Grosso, F. Cenni, G. Gangemi, *et al.* "Enabling Smart System design with the SMAC Platform," in *2015 Symposium on Design, Test, Integration and Packaging of MEMS/MOEMS (DTIP)*, Montpellier, 2015.

[9] J. Klaus, E. Schäfer, R. Paris, *et al.*, "A contribution towards model-based design of application-specific MEMS", VLSI Journal, Volume 58, 2017, 454–462.

[10] G. Lorenz, "Netzwerksimulation mikromechanischer Systeme", Berichte aus der Mikromechanik, Shaker Verlag, Aachen, 1999.

Chapter 12

Spintronic resistive memories: sensing schemes

Mesut Atasoyu[1], Mustafa Altun[1], and Serdar Ozoguz[1]

With the end of the era of shrinking dimensions of CMOS technology, research is underway to replace this technology with emerging technology devices, such as magnetic tunnel junction (MTJ). In particular, an MTJ is the core element of embedded Spin Transfer Torque Magnetic Random Access Memory (STT-MRAM). STT-MRAM has significant advantages: high density, non-volatility, compatibility with CMOS processes, and high endurance. Recent MTJ device studies have shown that write energy can be minimized by a perpendicular magnetic anisotropy (PMA) switching [1]. In comparison to in-plane magnetic anisotropy (IMA) switching, it provides greater scalability [1]. More work is, however, required in smaller MTJ nodes for read operations.

In this chapter, we discuss the physical principles of spintronic devices and their sensing challenges and compare the different sensing schemes withing terms of the power and speed constraints. Moreover, the circuit operation and its benefits over other implementations of the proposed sensing scheme are clarified.

12.1 Background

Recent developments in the scalability of an MTJ have led to renewed interest as a promising nonvolatile memory solution due to its CMOS compatibility, and especially high endurance [2]. Information is encoded in an MTJ as a high resistance (R_{AP}) or low resistance (R_P) state based on the magnetization directions of the ferromagnets. The data reading and writing operations of STT-MRAM are explained as follows. In reading operations of STT-MRAM, tunnel magnetoresistance ratio (TMR) is used to distinguish R_{AP} and R_P states from each other. In the write operations, the switching threshold current (I_C) is a key value for the required write current of MTJ to switch its value between R_{AP} and R_P states; thus, lowering I_C is possible through scaling down of MTJ technology nodes. However, scaling down the size of the MTJ requires low current or low voltage sensing operations, and due to insufficient current or voltage sensing margin (SM), it typically results in read disturb

[1]Department of Electronics and Communication Engineering, Istanbul Technical University, Istanbul, Turkey

(RD) and incorrect results. To remedy, building MTJs with low resistance area (RA) products helps to increase the TMR such as TMR is %248 in [3]. Higher TMR values of an MTJ outcome improved SM and reduced sensing errors. Meanwhile, a resistance value of an MTJ depends mainly on the oxide thickness (t_{OX}) of the barrier layers, the value of which was taken 1 nm in our simulations.

The previous works on the sensing scheme of STT-MRAM have highlighted several sensing approaches that are structured in self-reference [4], two transistors and two MTJs cells [5], dynamic data-dependent reference [6–8], and clamped and the adequate reference [9]. These works aim to solve the vulnerabilities of a sensing scheme.

12.1.1 Physical structure of an MTJ

The physical layers of an MTJ are a free layer (FL), a pinned layer (PL), and these layers are separated by a barrier oxide layer, which is generally MgO, as presented in Figure 12.1(a). The layout of an MTJ is illustrated in Figure 12.1(b). The equivalent circuit of MTJ is shown as a variable resistor in Figure 12.2(a) and the resistance data of an MTJ are switched according to an applied bias current

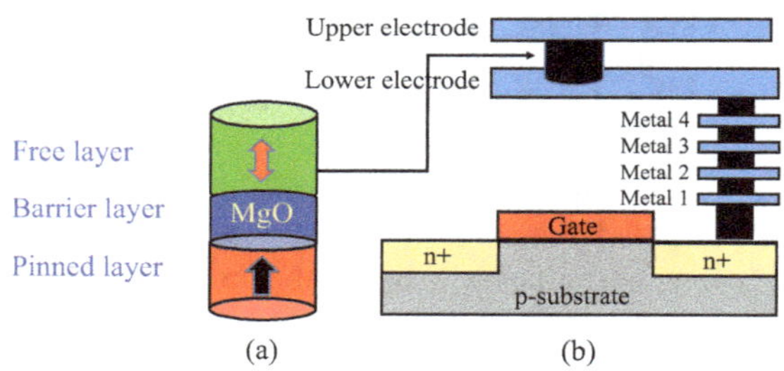

Figure 12.1 (a) The physical structure of an MTJ and (b) the layout of an MTJ with an access NMOS transistor [11]

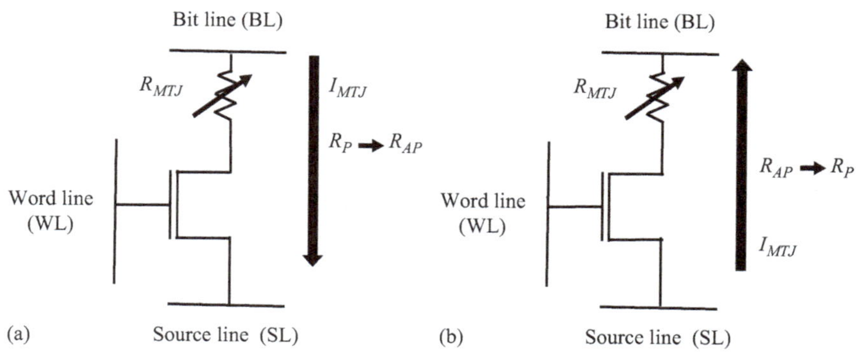

Figure 12.2 The equivalent switching behaviors of an MTJ [11]: (a) from P state to AP state and (b) from AP state to P state

direction as shown in Figure Figure 12.2(b). The conductance of an MTJ [10] is given in (12.1), and its values depend on the angle (Θ) between FL and PL.

$$G(\Theta) = \left[G_P \cos\left(\frac{\Theta}{2}\right)\right]^2 + \left[G_{AP} \sin\left(\frac{\Theta}{2}\right)\right]^2 \tag{12.1}$$

The values of average current density (J_C) change proportionally with the barrier height (E_B) values (20, 40, 60, and 80) and inverse proportionally with the pulse width (t_{pw}) values (0.2, 1, 10, and 100 ns). So these values are taken as $E_B = 60$, $t_{pw} = 10$ ns in our simulations. The values of J_C decrease with the increase in the values of E_B, whereas the switching current probability density decreases with the decrease in the values of t_{pw}, as given in Figure 12.3(a) and (b). RA dependence of an MTJ and its current–voltage relations are shown in Figure 12.4, and the hysteresis effect of an MTJ can be seen from these curves as well. Moreover, an MTJ has two different threshold currents, I_{C0+} and I_{C0-}, corresponding from R_P to R_{AP} switching driven by biasing positive currents and from R_{AP} to R_P switching driven by biasing negative currents, as illustrated in Figure 12.4. Following that, the increased area of MTJ such as 16% changes its resistance value to 500 Ω.

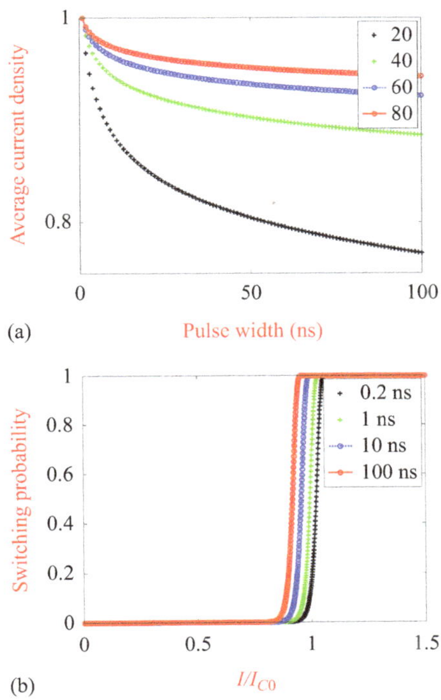

Figure 12.3 (a) The average current density of an MTJ with different pulse widths and (b) the switching probability density of an MTJ with different pulse widths [11]

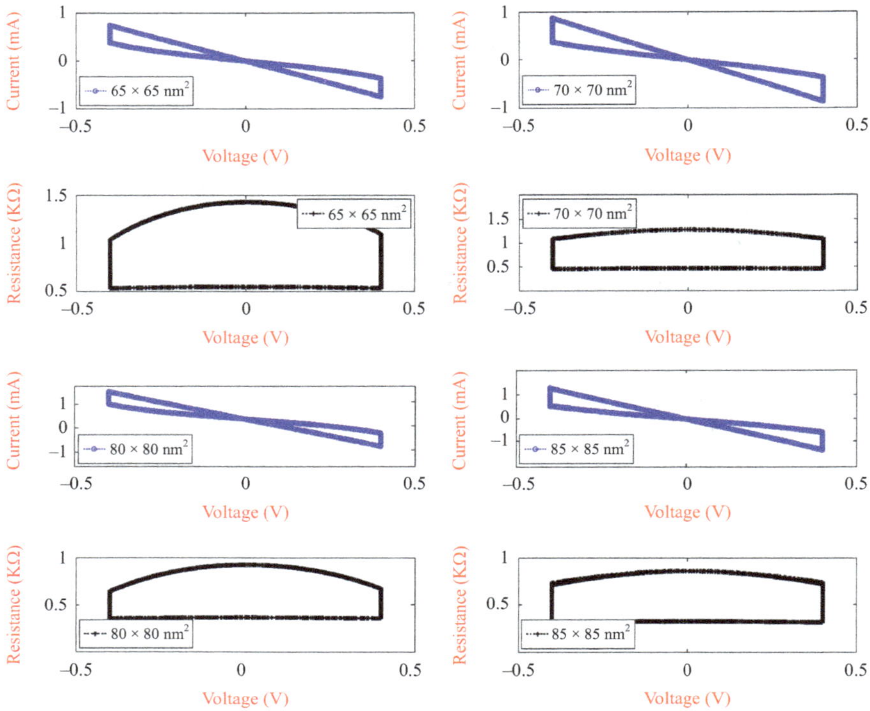

Figure 12.4 The current–voltage relations of the MTJ for its different volumes [11]

More importantly, these resistance variations are nonlinear and difficult for using these devices in a linear circuit application, such as in amplifiers [11].

12.1.2 The switching mechanism of STT-MRAM

The resistive data is switched between AP and P states by using spin-polarized current via STT. Magnetic switching dynamics of an MTJ are captured by Landau–Lifshitz–Gilbert (LLG)- and spin-polarized current-dependent STT term by LLG Slonczewski (LLGS) [12], as given in (12.1).

$$\frac{d\widehat{\mathbf{m}}}{dt} = -\gamma(\widehat{\mathbf{m}} \times \mathbf{H}_{eff}) + \alpha\left(\widehat{\mathbf{m}} \times \frac{d\widehat{\mathbf{m}}}{dt}\right) + STT \tag{12.2}$$

where $\widehat{\mathbf{m}}$ is the unit vector of magnetic moment, $\gamma = 2\mu_B\mu_0/\hbar$ is the gyromagnetic ratio, α is Gilbert's damping ratio, $\mathbf{H}_{eff}$ is the effective magnetic field, $\hbar$ is Planck's constant, μ_B is the Bohr magneton, and μ_0 is the permeability of vacuum. Different magnetic such as PMA, IMA, voltage-controlled magnetic anisotropy (VCMA), and demagnetization field, external magnetic field, and thermal field can be modeled by adding these sources to effective magnetic field term. The used simulation model, given in [10], was developed to solve the LLGS equation, considering given physical parameters in [13].

12.2 Sensing schemes of STT-MRAM

STT-MRAM cells have a resistive information. This information is converted to digital information employing a sensing scheme on STT-MRAM. In order to sense the STT-MRAM data, the current or voltage signal level applied must be lower than the switching current of the MTJ, to protect its data. A sense amplifier (SenseAmp) is dedicated to this sensing operation. A SenseAmp is categorized into a voltage SenseAmp (VSenseAmp) and a current SenseAmp (CSenseAmp). In addition, a latch in SenseAmp structure is widely used due to its high gain positive feedback mechanism. A VSenseAmp reads data slower than a CSenseAmp because the readout time of VSenseAmp is mainly controlled by a longer discharging time of bit-line and reference-line capacitances. In particular, the readout time of the VSenseAmp is faster than the CSenseAmp when the variation of threshold voltage (V_{TH}) is higher than 12 mV [14], especially, 30 mV or more in deep submicron technologies [5]. The related sensing issues originate in CMOS technology as a process, a supply voltage, and temperature variations, whereas in MTJ technology based on mainly variations of barrier oxide thickness (t_{OX}).

To emphasize, the proposed VSenseAmp follows neutralization technique on the clamped reference circuitry for more reliable read operations [15], which includes MC, MR transistors driven by VC and VR DC voltage sources, neutralization capacitances (C1 and C2), the latch is structured with M_{1-4} transistors, column selector and word-line switches, data-cell and reference-cell arrays, as shown in Figure 12.5. In our simulation framework, we used the 65 nm CMOS model and the MTJ model from [10], and some design parameters are listed as follows: power supply (V_{DD}): 1.0 V; the dimensions of MTJ: 40 (width) $\times$ 40 (length) $\times$ 1 (oxide thickness) nm^3. In addition, the process variations of MTJ and CMOS devices were stimulated: resistance variations of MTJs are modeled as a t_{OX} variation such as 2% for (3σ).

The read operations are carried out in three phases: pre-charge, evaluation, and decision; our timing strategy for these phases is adopted from [9] and operated via SAE, SAE1, and word line (WL) clock signals. The activation orders in the proposed VSenseAmp are shown in Figure 12.6(a); first, the WL is activated with 1 ns of pulse width; second, the SAE is activated with 0.5 ns of pulse width; third, the SAE1 is activated with 0.25 ns of pulse width. In the pre-charge phase, data and reference cells are activated through WL clocks, and then both bit-lines and reference-lines were pre-charged to V_{DD} and equalized via M_9 to provide the same time delays among bit-lines and reference-lines.

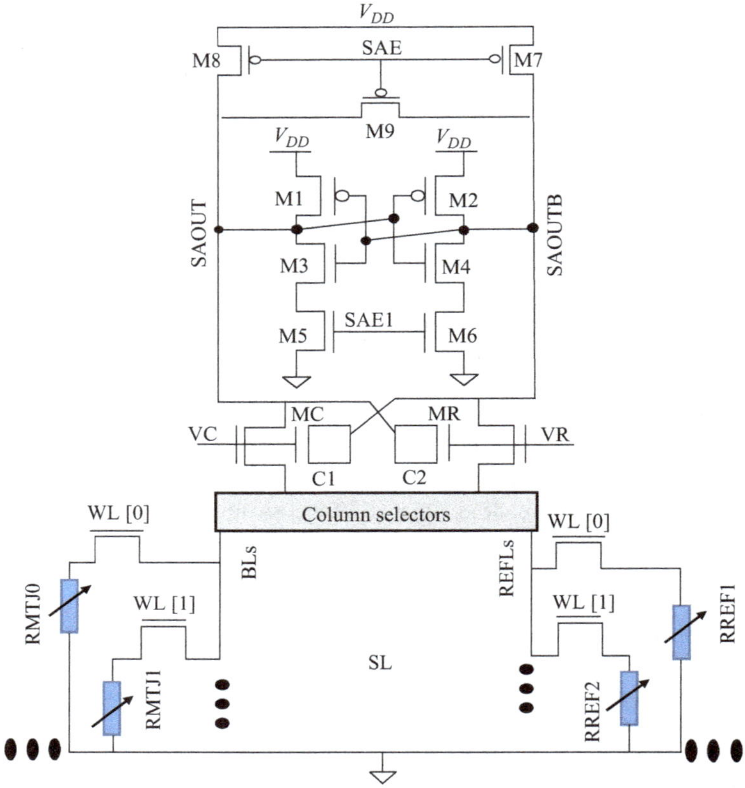

Figure 12.5 The schematic of the proposed sensing scheme [15]

In the evaluation phase, the proposed VSenseAmp is activated with SAE clock and disconnected with SAE1 clock which is for low-power operation [9], and then the output nodes of the proposed VSenseAmp (SAOUT, SAOUTB) go to supply voltage or to ground depending sensing data value, as shown in Figure 12.6(b).

The effects of CMOS and MTJ process variations for the BER performance of the proposed VSenseAmp were evaluated with Monte Carlo simulations (with 1k samples). To find optimal sensing solution is challenging because the resistance variations of R_P and R_{AP} are asymmetric as given in Figure 12.7; on the other hand, if t_{OX} values are 1 nm or less [16], such variations in resistance will be higher and deteriorate the output more, and the BER efficiency can also be degraded by other nonlinearities of the proposed sensing device. Therefore, our simulations have shown a significant decrease in BER [15] compared to [9] due to voltage variation reduction of the bit-lines and reference-lines via proposed technique. The power and speed performance comparisons of the proposed

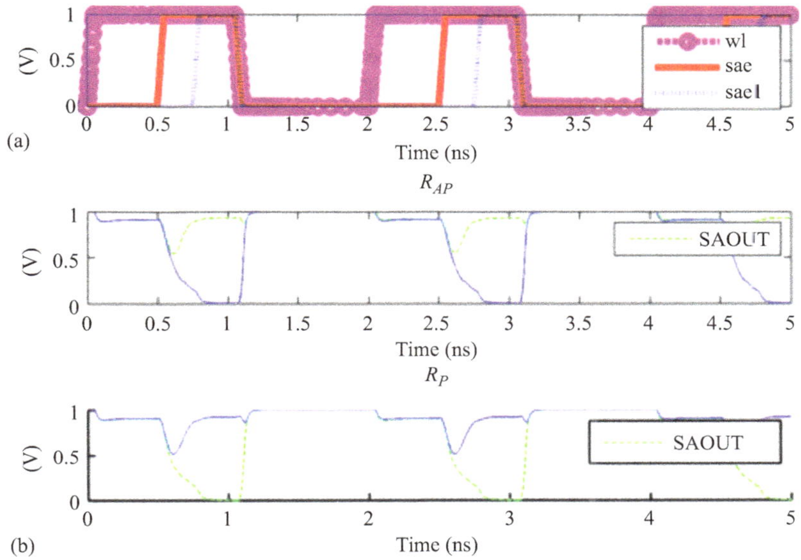

Figure 12.6 (a) The timing strategy of the proposed sensing scheme and (b) sensing results of the proposed sensing scheme for AP and P states [15]

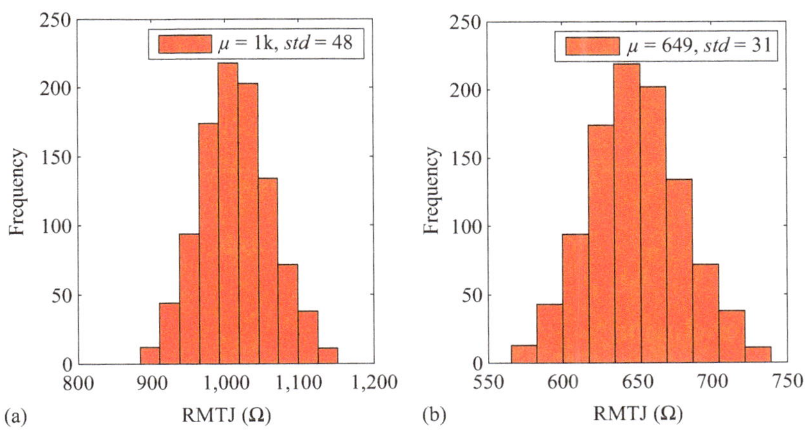

Figure 12.7 Resistance variations of the MTJ device: (a) for AP state and (b) for P state. Figure from [15,16]

sensing scheme with lower supply voltage, VDD are shown in Figures 12.8 and 12.9 for AP and P states separately. For speed and power comparisons, we took the capacitance values of bitlines and reference-lines, 50 pF in our design; as shown in Figure 12.9.

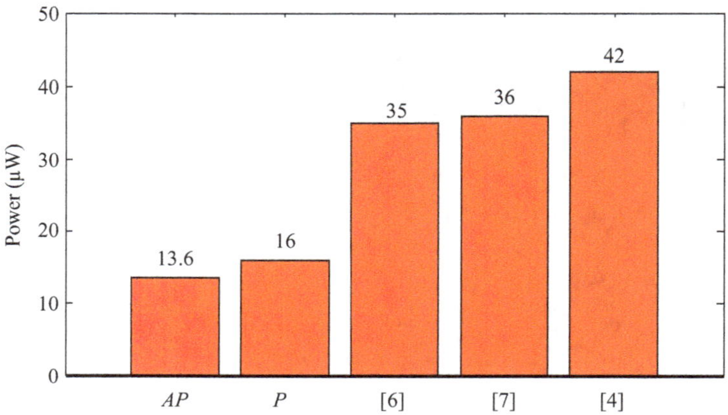

Figure 12.8 The power comparisons of the sensing schemes [15]

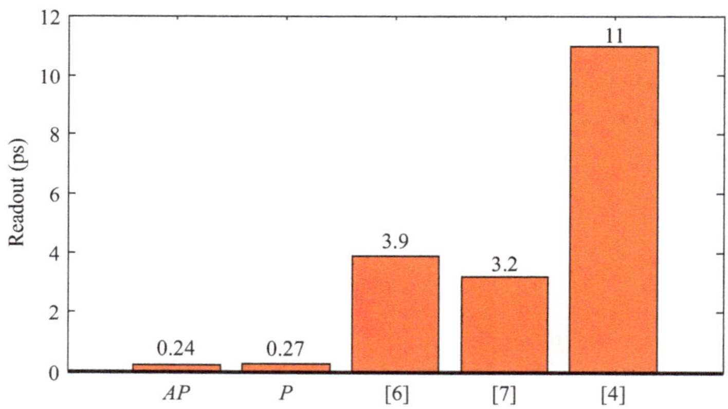

Figure 12.9 The speed comparison of the sensing schemes [15]

12.3 Conclusion

In this chapter, we compared STT-MRAM sensing schemes by briefly introducing the physical structure of the MTJ. We compared the proposed sensing schemes and other sensing schemes in terms of power and speed constraints. Especially, the power dissipation of STT-MRAM arrays significantly reduces by lowering voltage headroom as in our proposed design. Further, for ultra low power memory operations, new device technologies are needed as given in [16] which is operated with low switching current values. To enumerate in our proposed design: a read operation consumes 2.5 times less power and a sensing speed is faster 13 times as a comparison in given works. In particular, the proposed design should also be used in the sensing scheme of crossbar memory arrays.

Acknowledgments

This work is part of a project that has received funding from the European Union's H2020 research and innovation program under the Marie Sklodowska-Curie grant agreement no. 691178, and by the TUBITAK-BIDEB 2214/A. The authors would like to thank Prof. Kaushik Roy for his helpful discussions.

References

[1] Kishi T, Yoda H, Kai T, *et al.* Lower-current and fast switching of a perpendicular TMR for high speed and high density spin-transfer-torque MRAM. In: 2008 IEEE International Electron Devices Meeting; 2008. p. 1–4.

[2] Fong X, Kim Y, Venkatesan R, *et al.* Spin-Transfer Torque Memories: Devices, Circuits, and Systems. Proceedings of the IEEE. 2016;104(7): 1449–1488.

[3] Tezuka N, Oikawa S, Abe I, *et al.* Perpendicular Magnetic Tunnel Junctions With Low Resistance-Area Product: High Output Voltage and Bias Dependence of Magnetoresistance. IEEE Magnetics Letters. 2016;7:1–4.

[4] Chen Y, Li H, Wang X, *et al.* A 130 nm 1.2 V/3.3 V 16 Kb Spin-Transfer Torque Random Access Memory With Nondestructive Self-Reference Sensing Scheme. IEEE Journal of Solid-State Circuits. 2012;47(2):560–573.

[5] Ryu J, and Kwon K. A Reliable 2T2MTJ Nonvolatile Static Gain Cell STT-MRAM With Self-Referencing Sensing Circuits for Embedded Memory Application. IEEE Transactions on Magnetics. 2016;52(4):1–10.

[6] Kim C, Kwon K, Park C, *et al.* 7.4 A covalent-bonded cross-coupled current-mode sense amplifier for STT-MRAM with 1T1MTJ common source-line structure array. In: 2015 IEEE International Solid-State Circuits Conference—(ISSCC) Digest of Technical Papers; 2015. p. 1–3.

[7] Kang W, Pang T, Lv W, *et al.* Dynamic Dual-Reference Sensing Scheme for Deep Submicrometer STT-MRAM. IEEE Transactions on Circuits and Systems I: Regular Papers. 2017;64(1):122–132.

[8] Xue X, Fu Y, Zhao Y, *et al.* Dynamic Data-Dependent Reference to Improve Sense Margin and Speed of Magnetoresistive Random Access Memory. IEEE Transactions on Circuits and Systems II: Express Briefs. 2017;64(2): 186–190.

[9] Tsuchida K, Inaba T, Fujita K, *et al.* A 64 Mb MRAM with clamped-reference and adequate-reference schemes. In: 2010 IEEE International Solid-State Circuits Conference – (ISSCC); 2010. p. 258–259.

[10] Fong X, Gupta SK, Mojumder NN, *et al.* KNACK: A hybrid spin-charge mixed-mode simulator for evaluating different genres of spin-transfer torque MRAM bit-cells. In: 2011 International Conference on Simulation of Semiconductor Processes and Devices; 2011. p. 51–54.

[11] Atasoyu M, Altun M, Ozoguz S, *et al.* Spin-torque memristor based offset cancellation technique for sense amplifiers. In: 2017 14th International

Conference on Synthesis, Modeling, Analysis and Simulation Methods and Applications to Circuit Design (SMACD); 2017. p. 1–4.

[12] Slonczewski JC. Current-Driven Excitation of Magnetic Multilayers. Journal of Magnetism and Magnetic Materials. 1996;159(1):L1–L7. Available from: http://www.sciencedirect.com/science/article/pii/0304885396000625.

[13] Lin CJ, Kang SH, Wang YJ, *et al.* 45 nm low power CMOS logic compatible embedded STT MRAM utilizing a reverse-connection 1T/1MTJ cell. In: 2009 IEEE International Electron Devices Meeting (IEDM); 2009. p. 1–4.

[14] Sinha M, Hsu S, Alvandpour A, *et al.* High-performance and low-voltage sense-amplifier techniques for sub-90 nm SRAM. In: IEEE International [Systems-on-Chip] SOC Conference, 2003. Proceedings; 2003. p. 113–116.

[15] Atasoyu M, Altun M, and Ozoguz S. High Speed and Low Power Sensing Schemes for STT-MRAM with IPMTJs. CoRR. 2018;abs/1803.07037. Available from: http://arxiv.org/abs/1803.07037.

[16] Wang KL, Alzate JG, and Amiri PK. Low-Power Non-volatile Spintronic Memory: STT-RAM and Beyond. Journal of Physics D: Applied Physics. 2013;46(7):074003.

Chapter 13
Conclusion

Mustafa Berke Yelten[1] *and Günhan Dündar*[2]

Analogue integrated circuit design started with primitive transistor models that were able to capture the operation of solid-state devices. Only simple functionality could be accomplished through the circuits designed. Via aggressive scaling, however, the number of transistors that could be integrated has dramatically increased, thereby boosting the capability of circuits to achieve diverse and complex tasks. At first, more transistors meant an augmented computational cost only, since a large number of equations had to be solved concurrently to determine the circuit response. The time to find out the transistor operating points became crucial. Transistor models should be both accurate and easy to evaluate. Thus, alternative approaches have been proposed to characterize the device physics mathematically, some of which have led to the development of well-known transistor models, such as BSIM, EKV, and PSP, over the years.

As the scaling continued in its unprecedented pace, novel modelling issues started to arise. Simulations using primitive device models with few parameters were not sufficient to predict the outcomes of measurements. There were different reasons for this observation: smaller transistor sizes were triggering quantum mechanical effects, such as quantum tunnelling as well as entailing models with more complex underlying equations. Moreover, with shorter geometries, device-to-device variation of transistor parameters significantly increased. There was a need to characterize the changes induced by manufacturing steps with a separate set of parameters. Finally, devices were failing after prolonged usage due to the high vertical and lateral electric fields they undergo during regular operation. These were basically reliability issues occurring towards the end of the device lifetime. Thus, reliability phenomena needed to be described with dedicated models, as well.

Smaller sizes opened another battlefield for modelling experts. With smaller dimensions, individual capabilities of transistors were also increased. Short channel MOSFETs had smaller threshold voltages, which was leading to low power operation of analogue circuits. This feature made them appealing for biomedical applications, including cochlear implants, retinal prosthesis, and many more popular examples.

[1]Electronics and Communications Engineering Department, Istanbul Technical University, Istanbul, Turkey
[2]Electrical and Electronics Engineering Department, Boğaziçi University, Istanbul, Turkey

Besides, leakage currents of transistors become critical, especially to efficiently use the limited power stored in batteries of such devices. As a result, leakage current modelling was scrutinized and new models were proposed. What is more, short channel devices also had a dramatically high transit frequency, which supported many radio-frequency applications. A wireless boom of various products, including mobile phones, tablets, and laptops, revolutionized the means of communication across the globe. This triggered the improvement of transistor models with respect to their performance at radio and microwave frequencies. More recently, the growing capability of digital signal processing ensured that analogue signals are collected via sensors and then converted through fast analogue-digital-converters into the discrete domain. Numerous Internet of Things applications depend on cheap and efficient implementations of the sensors along with transceivers to broadcast the sensed data to the centre in order to take action. Following this trend, sensors with various sensing capabilities have been designed. These circuits have inherently an analogue nature and necessitated low noise operation for accurate sensing. Hence, accurate yet practical noise models for transistors have been developed.

Based on these emerging applications, one performance metric can be more important than the other. Thus, novel transistor types have been fabricated, which yield much better results with respect to the others in terms of a given performance metric. Each time a new transistor came out, corresponding device models have been proposed. In addition, if there is a change in the type of material, i.e., if instead of silicon other semiconductors, such as gallium arsenide or gallium nitride, have been employed, first the physical differences and new effects have been identified and then new models specific to the novel transistor are constructed.

On a parallel level, circuit designs need to be modelled as well due to their increased complexity. As mentioned before, the simulation of highly integrated circuits requires a large computational cost. In order to reduce this cost, circuit macromodels have been proposed that are built with fewer circuitry and nonlinear devices to be converted into a combination of linear components that mimic the behaviour of the original element. As the level of abstraction is not enough to reduce the complexity, behavioural models are proposed that can be just Verilog-A codes to describe the role of the circuit. If a model with a high level of abstraction is employed, then the inaccuracy of the model becomes larger. Nevertheless, the computational complexity is reduced at the same rate.

In order to both account for accuracy and computational efficiency, circuit models turned to optimization techniques. New modelling approaches have been developed, including response surface modelling, which finds the optimum model that achieves capturing the circuit operation with the least computational cost. In the same spirit, machine-learning tools have been employed to serve the same purpose. This trend also resonated with system-level thinking of microelectronic circuits. Without exactly designing the actual circuit, much of the high-level design can be accomplished by utilizing highly accurate and computationally efficient behavioural models. Nevertheless, the lack of transistor-based designs sharply increased the margins of variation for each component, thereby leading to unrealistic ranges for expected circuit performance metrics.

With all the previous events lined up, what is the outlook for modelling in analogue integrated circuits? As the prophetic Moore law approaches to its inevitable end, downscaling of transistor sizes significantly slowed down. Consequently, research in digital integrated circuits has become somewhat saturated. On the contrary, analogue integrated circuits have become popular mainly because many technology nodes have not been adequately evaluated from an analogue circuit perspective. Smaller sizes mean more integration capability and low-power operation for analogue circuits; at the same time, they result in more complexity and reliability degradations, and higher process variations. Nevertheless, such challenges just produce more research work for modelling experts. As the strengths of smaller sizes are leveraged in analogue integrated circuits, there will be more modelling applications both at the transistor and circuit levels.

A second track for analogue circuits would be the endless integration capability of sensors into microelectronics structures. One example is in micro/nanoelectronic mechanical systems as read-out integrated circuits (ROIC). Modelling the compound system of sensor-ROIC is by itself a challenging task as both parts of the compound, as well as a their interface, contain details that impact the combined performance. Furthermore, this modelling has to be performed across domains, such as physics and circuits. A second example would be extending the analogue-circuits-based sensors to flexible substrates. Modelling the circuits based on polymer and/or organic substances leads to a variety of new problems to be solved, such as heat dissipation and unevenness of substrate materials typically used in wearable electronics applications. Such details should be carefully modelled to attain the best performance. Organic electronics are still hampered by lifetime worries, whereby efficient reliability simulation is a must.

A third track is extending the validity of the models into extreme conditions, such as very high/cold temperatures, high/low humidity, and even high ionizing radiation environments. Many applications, such as space electronics circuits, radiation detectors, and certain vehicle sensors, require that circuits are operated under such conditions; nevertheless, typical ranges of the transistor models do not cover them. As a result, applications may break down earlier, operate suboptimally, or not work at all. With proper modelling, these problems can be alleviated and more applications could be conceived and realized.

In terms of the mathematical modelling methodologies, machine-learning techniques may extend to a broader application area than it has now. In neural networks, deep learning adds more hidden layers, thereby increasing the learning capability of the model. This can be leveraged for better accuracy in the design and simulation of analogue circuits. Furthermore, any approach that reduces the number of simulations will be strongly pursued as each simulation has an associated computational cost; hence, they should be conducted in an economical way. In that regard, unsupervised learning techniques offer new avenues for improving the capabilities of the models that would be employed in the design and optimization of analogue circuits. Finally, reducing the dimensionality would be a great asset for any model, which would improve the accuracy. Inevitably, as the number of transistor parameters increases unboundedly, models need to handle larger dimensionality especially in process variations analysis. The best approach to tackle this issue is resorting to techniques, which will eliminate insignificant parameters. Although there are well-known methods, such as principle component analysis, they do not take into account the nonlinear dependencies of the input to the outputs. Novel approaches are welcome and may increase the success rate of the developed models.

Modelling for integrated circuits has always been a challenging task since it is very difficult to satisfy conflicting constraints with reasonable trade-offs. Despite the complexity of coming up with models of high impact, it is a rewarding task in the sense that positive outcomes of a novel modelling approach or application can be immediately observed. Analogue integrated circuit design still plays a vital role in the future of micro- and nanoelectronics, therefore corresponding modelling studies will involve further problems to be solved, for which this book will hopefully be a helping guide to modelling specialists and analogue circuit designers.

Index

www.ingramcontent.com/pod-product-compliance
Lightning Source LLC
LaVergne TN
LVHW011848200926
848484LV00002B/416